8X

D0874141

The Care
and Breeding of
LABORATORY   ANIMALS

Four years

Three years

Two years

Five years

One year

Infant, one month old
Mother, circa fifteen years

A founding member of a monkey colony with her six daughters, ranging in age from five years to one month. The rhesus monkey (*Macaca mulatta*) may begin to reproduce in the third year and has a gestation length of six months, thus averaging one baby a year. Little is known about the duration or the close of the reproductive period in this primate.

The Care
and Breeding of
LABORATORY ANIMALS

EDMOND J. FARRIS, Editor
Executive Director and Associate Member
The Wistar Institute of Anatomy and Biology

AND

A STAFF OF FIFTEEN CONTRIBUTORS

JOHN WILEY & SONS, INC., NEW YORK
CHAPMAN & HALL, LIMITED, LONDON

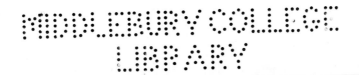
MIDDLEBURY COLLEGE
LIBRARY

$7.20. PBS 2N'50 Bul

636.4
F21

S7
77
F3

COPYRIGHT, 1950
BY
JOHN WILEY & SONS, INC.

All Rights Reserved

This book or any part thereof must not
be reproduced in any form without
the written permission of the publisher.

PRINTED IN THE UNITED STATES OF AMERICA

MIDDLEBURY COLLEGE
LIBRARY

Dedicated to the memory of
SAMUEL S. FELS, LL.D., Sc.D.
who was a true friend of science

PREFACE

This book has been prepared as an answer to the numerous inquiries from those engaged in breeding and caring for animals used in research.

The information contained in this book is based upon the practical experiences of the experts selected to write about a particular species. I am deeply appreciative of the contributors' cooperation, interest, and willingness to make available the excellent chapters.

Each chapter includes complete information which enables a research worker or technician to undertake the task of breeding and caring for laboratory animals with assurance of some success.

The selection of the species to be included in the book was based primarily upon the results of a survey of animals used by American investigators, as determined by papers presented at annual meetings of three representative scientific organizations: The American Association of Anatomists, The American Society of Zoologists, and The Federation of American Societies for Experimental Biology, which includes the following groups:

The American Physiological Society, Inc.
The American Society of Biological Chemists, Inc.
The American Society for Pharmacology and Experimental Therapeutics, Inc.
The American Society for Experimental Pathology
The American Institute of Nutrition
The American Association of Immunologists

The species employed in 1500 investigations in 1947 are listed in order of frequency of use, as follows:

Mammals	
Man	334
Rat	317
Dog	258
Rabbit	110
Mouse	108
Cat	81
Guinea pig	49

Mammals (Cont.)

Monkey	41
Cow	25
Hamster	9
Sheep	9
Pig	8
Horse	4
Others	16

Aves

Chicken	58
Other fowl	12

Amphibia	59
Reptiles	2
Fish	21
Invertebrates	118
Not stated	32

It is felt that the present selection of laboratory animals will satisfy the immediate needs of the majority of workers. Later it is hoped that the list and other information can be extended as required.

While I was serving as chairman of a symposium on "Animal Colony Maintenance," conducted under the auspices of the New York Academy of Sciences in 1945, the following thoughts were expressed:

"As investigators, we are more or less consciously substituting a laboratory animal for man—in effort to solve problems presented by man. Animal experimentation is usually justified on the grounds that the results may be carried over to man.

"In the past, this was done in qualitative terms merely. For example, a given diet proved inadequate for the growth of a mammal, and we expected that it would prove inadequate for man, too. Today, anatomical, physiological, and nutritional research has become quantitative in character. In any such investigations, a relatively uniform animal is desirable. A geneticist prefers mutations and variations, and rightfully so for his purpose. Yet it is evident that healthy, clean, vigorous stock is essential for accurate research. A standardized animal is needed, for a standardized animal is to the biologist what the pure chemical is to the chemist. Certainly to secure such an animal is next to impossible, but, by proper precautions, a nearly standardized animal or group of animals for research purposes should be the aim of all laboratories. Investigators should attempt to maintain colonies under their own control, so that the experimental animal and its breeding records are known.

"As biologists, we know that the lower mammals differ from the higher in being less able to regulate their physiological processes. They are more directly responsive to changes in environment. Respiration, pulse rate, and temperature changes are examples of this lack of regulation. It is familiar to us also that appearance of a stranger in the colony disturbs animals, as evidenced at the dairy or chicken farm, where the output of milk or eggs is lowered. The animals, though long domesticated, are readily disturbed.

"The animal is a very sensitive bit of apparatus. As the late Dr. Henry Donaldson once described the rat, 'It has more tricks than a string galvanometer, and must be treated with care and consideration.' I recall that, on one occasion, simply moving rats from one building to another prevented breeding for several months.

"The animal should be contented and happy, and while this condition is not easy to attain, it is worth striving for."

The following subjects were considered in the symposium mentioned above, and the printed report is recommended for those desirous of general information on these subjects:

Genetic purity in animal colonies, by Frederick B. Hutt, of Cornell University.

The mating of mammals, by Carl G. Hartman, of the University of Illinois.

Feeding laboratory animals, by J. K. Loosli, of Cornell University.

Infectious diseases of laboratory animals, by Herbert L. Ratcliffe, of the Zoological Society of Philadelphia and the Department of Pathology of the University of Pennsylvania.

Influence of environmental temperatures on warm-blooded animals, by Clarence A. Mills, of the Department of Experimental Medicine, University of Cincinnati.

Animal colony maintenance—financing and budgeting, viewpoint of the university, by Sidney Farber of the Harvard Medical School.

Animal colony maintenance—financing and budgeting, viewpoint of the commercial breeder, by C. N. W. Cumming and F. G. Carnochan, of Carworth Farms, Inc.

I wish to thank Drs. Carl Hartman and David Wenrich for suggestions incorporated in several of the chapters. I also wish to acknowledge my appreciation to members of my staff who so graciously aided in this endeavor: Miss Beatrice Roberts for secretarial assistance, Mrs. Claire Schultz and Mr. David Diamond for technical assistance, and Mr. Harold Baitz for photographic work.

EDMOND J. FARRIS

The Wistar Institute of Anatomy and Biology
August, 1950

CONTENTS

1· THE MONKEY

G. VAN WAGENEN *

Department of Obstetrics and Gynecology
Yale University School of Medicine
New Haven, Connecticut

INTRODUCTION

On the day a primate animal is introduced into the laboratory, changes and new responsibilities come not only to those in the animal room. The department secretary recognizes the urgency of locating the undelivered can of milk. The receiving room and the distributor hear of it, and the instructor remarks to his wife in the evening, "Remember the monkey operated upon Tuesday? She did not get her milk this morning, and each time I went into the room today she pressed against the front of the cage, watching every move I made. The young animals were whimpering, too." Thus a minor disturbance is revealed to the observant person—a far cry from the finality of the report, "A mouse dead in cage ten." The whole new attitude can be summed up, perhaps, by saying that it is the importance of the individual animal, a primate animal whose needs and wishes we can better understand, whose actions fascinate and sometimes revolt, whose intelligence and emotional life demand a more special care lest an injury to the psyche reflect on the soma. And above all, this is an animal of man's own order, whose anatomical and physiological likeness offers a most promising next step for the interpretation of that mass of information already assembled by the use, through many years, of countless smaller laboratory animals.

The Animal. The rhesus monkey, *Macaca mulatta,*† is the common laboratory primate. Its habitat is southeastern Asia, where it lives at various altitudes. The presence of this animal in the laboratory,

* With acknowledgment to the Nutrition Foundation, Inc., which, through the interest of Dr. C. G. King, made a generous grant in 1944 to Professor Arthur H. Morse and the writer for the upkeep of the Obstetrics Monkey Colony during the war years.

† Sutton, 1880; Heape, 1896; Pocock, 1932; Miller, 1933; Zuckerman and Ful-

1

probably, has been determined by the fact that it is easily caught in great numbers in native villages near the coastal towns in India. From these ports it is shipped to animal dealers in the United States. Smaller monkeys such as marmosets or green monkeys may be more desirable for some research problems because they are easier to handle and economical of space, food, and care, but basic information con-cerning the rhesus is far greater than for any other monkey. One reason for this is the well-demarcated menstrual cycle of twenty-eight days, so like that in the human, which singles it out as an admirable subject for studies in the physiology of reproduction and endocrinology. In laboratories for research in these two active fields historically important colonies have been set up for long-term study, not only of the experimental animal, but also of the normal. Even-tually, it is hoped, the life cycle will be as well known as that of the laboratory rodents.

Housing. On introducing the monkey into the laboratory it was recognized that special and spacious quarters must be arranged. The longer life span and primate characters demanded an outdoor life as well. In America, in the third decade of this century, three laboratories established monkey houses. Dr. George W. Corner's name and work are associated with The Monkey House at the University of Rochester School of Medicine and Dentistry, an enterprise which he inaugurated in the Department of Anatomy at the Johns Hopkins Medical School; the late Dr. Edgar Allen established a monkey colony at the Univer-sity of Missouri Medical School in St. Louis; Dr. Carl G. Hartman, coming to Baltimore from Texas, re-established the Corner colony within the Department of Embryology, Carnegie Institution of Wash-ington, Baltimore. All these monkey houses provided the outside paddock with an inside shelter, and their floor plans and detailed descriptions are presented in the appendix of *The anatomy of the rhesus monkey* by Carl G. Hartman and William L. Straus, Jr. (1933). Possession of this book, or at least complete perusal, is essential for anyone planning the establishment of a monkey colony; therefore these quarters, superior in many ways but entailing major building construction, or at least alteration of existing walls, will not be de-scribed here. Rather, it is the purpose at this time to describe two types of monkey housing made up of stationary or movable units which can be assembled for small or large rooms and appropriate for installation in any light, airy place suitable for such animals.

ton, 1934. Special reference should be made here to *Bibliographia primatologica* by Theodore C. Ruch (1941), citing investigative work on the macaque up to 1939.

One type of indoor caging was built in the Department of Physiology at the Yale University School of Medicine (Kennard, Ruch, and Fulton, 1946). The second type, also built at Yale and located in the Department of Obstetrics, consists of a great over-all cage within which the individual cages move freely on wheels equipped with brakes. Neither one of these setups is unique, but rather they represent two types of arrangement, many features of which can be found in other laboratories, not only in the United States but wherever monkeys have been brought into research units and the available space adapted for their comfort. In all five of the monkey colonies mentioned here animals have been bred, and they have produced young which have matured. The colony in the Department of Obstetrics at Yale was established in 1931, and five generations now live together with the usual frictions and communal interests of other primates.

The space necessary for adequate housing and the design and size of the individual cage, even the feeding and handling of monkeys, differs with the age of animal preferred and the duration of experiments. If the more readily available young monkeys, two years or under, are to be used and the period of observation is short (as in certain infectious disease research), considerable crowding with a very simple feeding program is permissible, and belts, collars, and chains may be used to facilitate the daily handling. However, if animals are to remain in the laboratory for longer periods (as in nutritional or reproductive studies) and often under less intensive observation, greater space and freedom are needed for growth and normal development. Something more than a sustaining diet should then be offered, and the hazard of a chain or non-rigid trapeze must be avoided to protect the monkey and save the greater investment of time and interest. A glance at the older monkey residents in the classic colony at the Carnegie Laboratory reveals that the well-fed adult rhesus is a more impressive animal than the familiar small zoo or circus monkey. Indeed, the animal dealers used to advertise a special kind for sale, the "giant rhesus," which is now recognized as the fully matured individual which has fortunately found adequate food and unfortunately been captured. These animals are strong and often aggressive, and bring a better price because they have to be handled separately. Sometimes they turn up at an importer's because they have become troublesome in a zoo or as pets. Thus, the six- to fifteen-year-old adult animal in full reproductive age is not known to every laboratory using monkeys. Yet, if a *self-sustaining* monkey colony is contemplated, animals of this weight and strength must be provided for, as well as their offspring. Consequently, designs for monkey quarters fall into three categories according to the nature of the investigations contemplated: (1) short-term experiments (within the academic year); (2) long-term investigations; and (3) long-term investigations which include breeding, whether as an integral part of the study, for replacement of animals, or as a means of minimizing variability and avoiding the introduction of disease.

THE SHORT-TERM EXPERIMENT

Caging. Standard monkey cages, measuring approximately 24 by 24 by 32 inches, are available (see manufacturers' list) to be used as single units or double deckers for the laboratory needing few monkeys, and are also useful for isolation or special observation with larger installations. For the monkey room these same cages can be arranged on supporting angle-iron frames so that a double series of them forms two side walls and perhaps one end of a common runway, which is closed by wire mesh overhead and provided with a door of the same material at the other end for the use of the staff. The Yale Department of Physiology cage block (figs. 1 and 2) is an example of this construction. Cages (24 by 30 by 30 inches) do not stand directly on the floor, and wire flaps completing the lower side walls may be lifted for sweeping. The cage block should be placed 3 feet from the nearest wall for protection of the wall and to allow movement for cleaning and servicing of the water fonts. Only six contacts with the floor are necessary for twelve cages, and the lowest bar of the frame is 14 inches from the floor, all of which facilitates the sweeping and mopping. The individual cage and the block described here were planned by Dr. T. C. Ruch, and he has also designed the water device which is widely used (Kennard, Ruch, and Fulton, 1946).

The short legs of each cage rest in an open rectangle formed by the angle-iron frame of the block, so that it is held firmly and is never out of line. Shallow trays of stainless steel or Monel metal, 2 inches deep, for sawdust, slide into the frame under the wire-mesh floor of each cage. Details of the cage construction are readily seen in figure 2. The cage is made of a heavy woven wire mesh, diamond in form. In the cage illustrated, the finer 1-inch diamond mesh at the back, roof, and sides prevents contact between animals, thus minimizing infection (also note distance between cages). The water fountain (fig. 18) is attached to the rear wall to keep it from obstructing the view of an observer in the runway, and with fine mesh at the back the animal cannot reach through to the cork or any movable parts. The coarse 2-inch diamond mesh on the floor allows feces and debris to fall into the refuse tray. The floor is kept fairly clean by the casual movements of the monkey walking about the cage. The door used for

FIG. 1. View down runway of cage block (Department of Physiology), consisting of two blocks of twelve unit cages each, mounted on angle-iron frames, closed at top with wire mesh and furnished with an entrance door. (Manufactured and installed by the John P. Smith Co., New Haven, Conn. This and all other photographs in this chapter by Howard Reynolds.)

catching is a sliding panel, in the lower half of which is a flap door which can swing upward. The flap door is for feeding. Each cage is equipped with a perch consisting of two galvanized rods running the depth of the cage. Like the floor, the rods do not retain urine and feces.

For sanitary considerations as well as reasons of safety and durability high mechanical standards should be required for the construction of monkey cages. By using arc welding of mesh to either angle-iron or metal-rod frames for individual walls and doors, cracks and crevices which catch dirt or allow vermin to breed are avoided. Galvanizing the walls separately after construction and before bolting to the cage frame, or, better still, galvanizing the whole cage after assembly, further reduces crevices, cut surfaces that will rust, and unfixed wires that may be worked back and forth by agile fingers.

Fig. 2. Detail of Department of Physiology unit cage before installation of watering device and perch.

Metabolism cages can be of the same size, therefore fitting at any position in the block frame. They differ from the ordinary unit cages in that the sides and back are of galvanized sheet iron or stainless steel slanted inward toward the bottom. The floor is of a fine mesh to hold back feces and food particles, and is separate in order to slide out for cleaning. The catch tray slopes to a center hole, below and around which is welded the perforated metal screw top of a fruit jar for collection of urine.

Figure 3 shows a cage used in nutritional research, equipped with food cups which lock into the cage when the door is closed.

A less expensive cage of similar size is constructed of expanded metal. Panels are made of a single piece of 9-gauge steel perforated with 1½-inch diamond-shaped openings. It is possible to cast the whole cage of this metal, which eliminates the many crevices in the fastening of the panel sides to the cage frame. Now, although this cage is advertised to look the same as the wire mesh previously used, the cut edges of the steel plate injure the hands of the climbing animal and limit the use of this metal, preferably, to the metabolism cage, where the monkey remains only a comparatively short time and is not expected to climb about.

The size of the cages described is a compromise between that best for the animal and that which will allow a two-story arrangement with the top tier not too high for catching. Such cage units are suitable for two small monkeys, a single adolescent, or a small adult for experiments limited in duration.

Feeding. For the sustaining diet one meal a day is customary, and a substantial part of this meal usually is a commercial biscuit (Purina Dog Chow, Chim Cracker) which does not deteriorate quickly. With a

Fig. 3. Cage used in nutrition research laboratory of Dr. James H. Shaw, Harvard School of Dental Medicine. Note the cups for synthetic food and for water which lock into the cage when the door is closed. (This cage was developed in the laboratory of Dr. C. A. Elvehjem, University of Wisconsin.)

supply that will last twenty-four hours and water at hand, no animal needs to remain hungry. Fruits, vegetables, or greens, according to season and economical buying, with a variation each day, are given at the same time, and they will be consumed first. Needless to say, monkeys will adapt readily if offered a pan of water only once a day. However, it is preferred that water should be available at all times.

Catching. The central space, onto which all the cages of the block face, should be kept narrow and just long enough to be covered by two or three strides of the investigator. It must be wide enough for easy removal of the catch trays and cages and for entrance and use of the feeding and cleaning trucks. If animals are to be observed each day, they soon learn a routine. At observation time when the cage door is opened, the monkey springs out and runs over the front of the cages, jumping across to the other side and pausing for breath now and then, eagerly seizing the opportunity for exercise. Observations of motor disability are made at this time. When the cage door is opened again, the animal learns to return and thus spares itself the struggle of catching. If the monkey is needed for closer examination, the observer extends his hand and with a few quick steps about the central runway keeps the animal in motion until it is exhausted. The rhesus monkey is easily winded when kept in constant motion. As it clings to the cage, panting, the observer's right hand is placed on its lower back and, while pressing the animal against the cage, first one arm and then the other is brought over its back, leaving the monkey under control in the catcher's left hand. The right hand is free then for examining the animal, picking up a syringe, or closing the entrance door.

Cleaning. Washing of the cages is of greater importance than sterilization, although the latter assumes a different status when infectious disease is present. A completely satisfactory apparatus would combine both processes, but in most laboratories, where a commercial sterilizer has been installed or there is a great metal vat into which live steam enters (as has been used many years for metal mouse cages), a deep sink is needed for soaking and scrubbing cages. Even these smaller cages are not easily handled and call for the use of dollies or overhead tracks and pulleys if lowering into sterilizers is necessary. Sterilizers measuring 6 feet long, 3½ feet wide, and 5 feet deep accommodate two cages, one on top of the other.

Daily cleaning of the catch trays, if sawdust or shavings are used, means only the dumping of contents into one garbage can and refilling with fresh material from a second can. Both cans are wheeled into the central passage of the block. If shavings are not used, the catch trays each in turn are taken to a sink and washed. Sweeping and mopping or hosing of the floor are also daily duties.

LONG-TERM INVESTIGATIONS

Caging. It is inconceivable that growing monkeys in the laboratory be denied their essential activities of swinging, jumping, and scampering about. This calls for space. Large cages have been built out of chicken wire or stronger wire on cement floors of basements and in the corners of laboratories, where monkeys have had room to move freely and have lived in health for several years. A few monkeys can be cared for in this way, but for the laboratory engaged primarily in primate research a careful housing plan is essential.

The female of the *Macaca mulatta* reaches her full growth in the fifth year of age, but the male does not reach his full growth until the sixth or seventh year. By that time the monkeys measure from 50 to 60 cm. in body length. Males in full health are very active physically, and the restriction of cage walls for an aggressive individual seems to heighten this activity. In providing for the minimum necessary physical outlet, the adolescent and adult male must have space for turning cartwheels. This looping, characteristic of the male confined to a cage, brings into play many unused muscles. If sufficient room is not provided, some males will soon show depression, sitting quietly in the part of the cage which affords the best view. For a while this may be interpreted as an adaptation, and the animal may be thought to have become more cooperative. Too often, however, this persistent posture apparently results in a pressure atrophy, bringing on a lower-limb palsy, the so-called "cage paralysis." This condition is soon at one time or another in almost every monkey laboratory and generally has another etiology. Although the repeatedly pregnant female becomes a somewhat sedentary animal, she too should have room to climb and walk.

Having dealt with the point that a larger cage is more desirable than the unit cage described in previous paragraphs for use in the short-term experiments, it is necessary to add that neither is a zoo-size or circus-wagon-size cage completely desirable or efficient. Only a limited number of monkeys can be housed together happily in any one cage for any length of time because the physically or psychologically dominant animals will attack each other; nor do the smaller monkeys thrive in such a group. Primate animals have a way of annoying each other, and such behavior always culminates in fighting with the best available weapons—in this case teeth and fingernails. A cage of its own will remove the adult bully, separate two jealous females, and keep one affectionate youngster from clinging constantly to another. So, after having been plunged into many diverse social emergencies, it is felt that with the plan of larger cages, as freely movable as hospital beds and placed inside an over-all cage as described below, a great number of adult animals can be accommodated in a restricted space. If breeding is incorporated as a part of the animal room routine, monkeys of the two sexes and of all ages, as well as animals under changing experimental procedures, can be cared for successfully.

The second type of indoor caging to be described here in some detail can be adapted to any sizable room, preferably with a floor having a central drain but most emphatically a hard-surfaced floor which can withstand constant mopping. The over-all "great" cage should be in the center of the room, leaving a 3-foot encircling passageway between cage and walls for movement of tables, cages, and trucks,

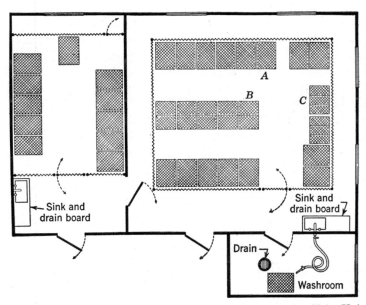

Fig. 4. Floor plan of the Obstetric Monkey Colony at the Yale University School of Medicine. The breeding animals live in the large room, where males occupy the four corner cages (mating cages) and the females live in pairs in between. *A* indicates a basic cage. The divided cages in center (*B*) are separation cages for pregnant animals, and the cages of three sections to the right (*C*) are for young animals. All cages are on wheels and move easily into washroom. A quartz-mercury vapor lamp is mounted in center of ceiling. (This and all other drawings in this chapter by Armin Hemberger.)

providing access to windows, and giving room for permanent equipment, such as weighing scales and sink. (See fig. 4 for a floor plan.)

Figures 5 and 6 are views into the over-all cage in the large and small rooms shown in the floor plan. For the great cage a 9-gauge chain-link wire fencing in a 2-inch mesh which has been hot dip galvanized after weaving was used. Uprights are of hollow galvanized-metal piping $2\frac{1}{4}$ inches in diameter. The doors hanging from the uprights are of the same wire mesh, and the frames of a lighter $1\frac{5}{8}$-inch piping. The doors are 6 feet 6 inches high to accommodate the long-handled catching net. If any pipes, especially steam ones, traverse the room and come within the cage, they should be covered so that the escaped monkey cannot use them. The size of the room determines the size of the outside cage, which in turn can accommodate only a certain number of individual cages of the type to be described.

Fig. 5. View down outside aisle of large room (fig. 4) with center row of separation cages shown to left. The pregnant monkey has hand on water font and is about to drink. Cages on the right are in front of two large windows. This room faces south and west, with big windows on both exposures giving excellent light, sun, and air, but making photography difficult. (Movable cages and overall cage manufactured and installed by the John P. Smith Co., New Haven, Conn.)

Figure 7 shows the cage used for adult monkeys. It furnishes a living space 4 feet by 2 feet 6 inches by 3 feet and has been found adequate for the largest male monkey or for two friendly females. The 1½-inch angle-iron frame, 5 feet 6 inches high, 2 feet 6 inches wide, and 3 feet deep, is mounted on rubber-tired swivel wheels, each one of which (two suffice) is supplied with brakes. The sides and top panels of the cage are of 1-inch diamond woven-wire mesh; the floor of 1½-inch diamond mesh.

The upper and larger of the two doors which close with snap locks is used in catching, the lower for placing food pans within the cage or allowing attachment of a transfer or weighing box onto the cage.

Slots are cut parallel to the floor, in the middle of the two side walls of the cage, to allow a hard pine board to be inserted as shown. After assembly of frame panels, doors, and the metal parts for holding the wheels, the whole cage is hot dip galvanized. It may be of interest that the particular cage shown in figure 7 was redipped after having been in constant use for 10 years.

Fig. 6. Small monkey room (floor plan in fig. 4), showing some bad features; i.e. cages placed directly against wall are difficult to see into, and walls become soiled, needing frequent washing. Also, the exposed pipe shown at top center, which enters the great cage, gives refuge to the escaped monkey.

Accessory to the cage is the Ruch water device, the catch tray under the wide-mesh floor, the board used as a bench, and the urine pan. The shallow catch tray is of 22-gauge Monel metal 1 inch deep and pressed by two radiating creases for the guidance of urine into a 1-inch hole cut in the tray immediately above a urine-collecting vessel, which slips into a metal run attached in front to the cage frame. For collection of urine an 8-inch aluminum pan 4 inches deep with a ½-inch flange slips easily into place. Notice the toggle on the righthand side of the cage frame to prevent pushing out of the catch tray from within. For cleaning, the tray slips out over the lower bar of the frame without disturbing the urine-collecting pan. The board used as a bench is held in place by two ½-inch iron pins outside the cage. The pins may be bent in at right angles to the top ring so that they catch into the mesh of the cage. This is a precaution necessary only for certain

FIG. 7. The typical cage used for adult monkeys, one male or two females and on days of mating a male and female. Using the same frame, the cage has two modifications. For the separation cage it is divided into two parts by lowering a metal partition through a slit in the top panel along runways down the center of the side panels, as shown to the left in figure 5. Or, the cage may be divided by two removable wire-mesh partitions into three compartments for use by young animals, as in figure 10. In the latter modification, the frame is turned so that the greater surface of the side becomes the front, which then has three doors. (Cages made both by John P. Smith Co., New Haven, Conn., and by Norwich Wire Works, Inc., Norwich, N. Y.)

animals. Sitting on the board, facing the center of the room, is the favorite position of the monkeys. At this height these intensely alert animals have a better view of activities within the room, and they can meet visitors on the same eye level. The board seems to be more comfortable than the wire floor not only for their sensitive feet and hands but also for the better-protected callosities, and they sleep on the board at night.

Feeding.* In feeding the monkey, the main problem is that of furnishing a source of sufficient protein to an animal which refuses to eat meat. Although they are natural vegetarians, monkeys enjoy drinking milk and eating eggs (lacto-, ovo-vegetarians). Eggs may be raw or cooked. They like all vegetable proteins and cereals. Potatoes, rice, and macaroni have been convenient sources of carbohydrate and the principal ingredient of the main meal. To all else, monkeys prefer greens and fruits—bananas are as ice cream to the human child. With the exception of meat and butter the monkey will eat everything the human being does; its preferences are similar, and its enjoyment and need for variety as great. The food consumption of the adult monkey is similar to that of the adult human being, and the male differential the same. The appetite of the young adolescent monkey can compete with the appetite of a boy coming into the kitchen from the baseball lot. It was surprising in our laboratory to find that the plateau in growth which seemed to characterize the pubertal male was eliminated by a third meal per diem. The infinite variation possible when feeding this animal makes convenient the planning of a weekly menu which will be related, of course, to costs and seasonal supply. If monkeys are to grow, develop, and live out their natural span of life, it is not possible to concentrate their feeding into one meal a day. Two meals a day are customary. The morning meal is postponed until the cage trays are cleaned. But before cleaning the trays it is routine in our laboratory to pass quickly around the room, offering to each individual monkey a loaf of sugar which has been dipped into cod-liver oil or has received several drops of concentrated vitamin D. Hands reach eagerly for this. The breakfast which follows the cleaning consists of bread and milk with a half-orange for each monkey. Thus the essential vitamins not always sufficient in natural foods are supplied.

---

* Before setting up a monkey regimen the consultation of two references is suggested. Hartman and Straus (1933) tell something of the food and feeding plan in the Carnegie Colony, and Ratcliffe (1940) describes a food mixture containing horse meat which has been used successfully at the Philadelphia Zoo. This semi-dry cake, which disguises meat so that it is accepted by adult monkeys, also is soft enough to appeal to nurslings. The mixture is made daily or every second day and is accompanied by fresh vegetables. Suggestions are given concerning the schedule and supplementary feeding for pregnant animals.

The afternoon meal is built around the main carbohydrate; potatoes (raw or cooked) are accompanied by an egg; when rice is cooked, beans or peas are added. Casein and liver powder are protein sources which can be added to rice, but many animals will not eat that meal unless they have been introduced to it when hungry youngsters. Fruit or greens and vegetables are added to the main dish. Sprouted grains are much relished and are superb sources of accessory food factors but are time consuming to prepare. In short, the food and menu planning requirements are much like that for any vegetarian group, but since the monkey has not learned to be unwasteful or fastidious and is so well equipped with keen eyes and skillful hands to separate the unwanted parts from the whole, he can be offered the less choice outer lettuce leaves, shop-returned bread, overlarge carrots, and the like, which come at lower prices. However, washing of vegetables, lettuce particularly, cannot be neglected, and this is the argument in favor of using a standard grain mash. A standard mash would be a complete food entailing no special preparation or shopping, and would be already familiar to the animal moving from one laboratory to another, as well as eliminating the hazard of contaminated or bad food. But if such a food is ever achieved, it is hoped that the monkey will really relish it, because the approach of the animal's friend bearing food brings out all those attractive responses of warm acceptance, knowing enjoyment, and fraternal feeling toward man that justify the lower primate's presence in the laboratory.

When natural foods are to be prepared and some cooking is planned, a kitchen must be improvised in the laboratory. A convenient sink is usually at hand, and substantial gas plates can be installed near by. Figure 8 shows a simple two-burner gas plate with space between for shifting pots placed under a chemical hood. Note the low level of the burners at a height to minimize the effort of lifting the ten-gallon cast aluminum cooking pots. And a food truck (fig. 9) is needed to convey these heavy containers from sink to stove to animal room, where each pan of food is served from the truck as it is pushed from cage to cage. A refrigerator is necessary for milk, and a cool room for vegetable storage. Costs of foods, in this case admittedly liberal, and equipment and time for preparation are usually underestimated.

Catching. For catching within the cage the short-handled net (fig. 19B) in the right hand of the catcher is inserted while the door is held firmly by the left arm and open only far enough to allow free movement. The hoop and bag are brought down over the animal's head and body, and as it is withdrawn the bag is twisted and lowered to the floor. The monkey sits on the floor within the bag, and when his head is held from behind with the left hand the net can be opened with the right hand, grasping first one monkey arm and then the other behind the animal's back.

The escaped monkey certainly is more difficult to catch in the open room or great cage. Two persons equipped with long-handled nets

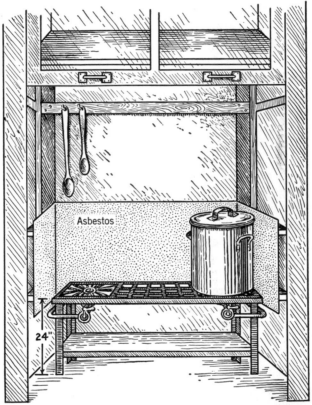

Asbestos

24"

Fɪɢ. 8. Low gas burners (24 inches high) installed in chemical hood in laboratory.

(fig. 19A) may be necessary, but the principle of winding the monkey remains the same. There are fewer escaped animals among old residents.

Cleaning. Daily cleaning of the cages consists of washing catch trays, boards, and urine pans. The trays and pans from one cage at a time are washed at the sink. The boards are collected from all cages, leaned against the wall, and hosed, first one side and then the other, and left to dry. An alternate set of dry boards replaces them. During the time boards are away the animals will take advantage of the space for various activities. Daily cleaning also includes sweeping and mopping the floor.

The washing of cages mounted on wheels is a simple process. They are pushed into the nearby washroom two or three at a time, thoroughly wetted down, allowed to soak for thirty minutes, and vigorously hosed with steaming water under pressure. All cages are checked off once a month, but some cages need washing every two weeks, and in

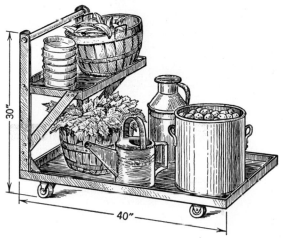

FIG. 9. The food truck used in the Obstetric Monkey Colony, Yale University.

moist warm weather those of young monkeys require weekly washing. Water bottles and fonts are cleaned weekly.

LONG-TERM INVESTIGATION WITH BREEDING

Caging. Two modifications of the basic cage used for adult monkeys are found convenient when animals of grossly different ages and conditions are to be housed. Dividing the cage into halves makes possible the isolation of the pregnant animal near term, and dividing it into thirds makes room for three young animals instead of two adults.

In the first modification (fig. 5, cage on left) division was accomplished by sliding a metal partition down between two half-panels on the cage top into metal runways inside and bisecting each side panel. Slots for the benches are placed on both sides of the partition (benches parallel the partition), and both ends of the cage are furnished with doors and water fonts. These cages have to stand out from the wall,

and in the Obstetric Colony have been placed in a row down the center
of the room, giving access to both sides of the cage. The removal of
the partition returns this cage for the use of two congenial animals
at any time.

In the second modification (fig. 10) the cage frame is used sideways.
Two removable wire-mesh partitions divide the cage into three long,

Fig. 10. Cage used for young monkeys. Here the doors for catching are on the
side of the original frame. There are no feeding doors, and the bench passes
through the partitions. The long narrow shape of the individual compartment
is fortunate, allowing the young animals to swing from side to side, hang from
the top, and further the variety of their constant motion.

narrow compartments especially attractive to the actively climbing
younger monkeys. The partitions should be of 1½-inch mesh, which
makes a diamond large enough for arms to reach through. The bench
passes through the partitions, and each compartment is furnished with
one long door (no feeding doors are necessary).

A definite arrangement of the animals and their cages in a breeding
colony is arrived at in the interest of peace and quiet. Very important is the placing of the males where they cannot see each other to

challenge every movement, as in figure 4, the larger room, where males occupy the four corner cages. The females of breeding age live in pairs in the cages between, except when near the end of pregnancy or nursing an infant, when they then are placed in the separation cages. The infant animals, when removed from the mothers to be fed by formula or after weaning, should be conveyed beyond the hearing of the colony. Both infant and mother will fret only a day or two. When returned to the animal room it is well to place the young monkey a bit away from the mothers. Better discipline is exercised if young animals are placed near a male's cage, but not so near that the less gentle parent can bite a baby's finger when he becomes exasperated.

Catching. When animals have been raised in the laboratory, gloves and nets can be discarded—catching becomes a varied procedure. Young animals spring out into one's arms, seeking to be petted, and if held firmly by one wrist they will walk upright to the examining table. They should, of course, be kept under control at all times. Some animals are allowed a swift excursion around the room if they can be relied upon to return promptly. Older animals stand up on the board, turn their backs, and are lifted out. Imported monkeys must be caught by nets, but finally even they learn after a year or so to head into or back into the net—a little patience with them, allowing them to choose the manner of catching, is rewarding.

Mating. The term "mating" is used here in the transitive sense, and by it is meant simply the placing of the female monkey in the cage of the male at the optimal time for fertilization. In the monkey there is no overt behavioral manifestation (estrus) of the approach of ovulation; the female may accept the male at any time during the menstrual cycle and during any season of the year. The menstrual cycle and the time of ovulation have been widely studied in this animal (Heape, 1896; Corner, 1923a, 1923b; Collings, 1926; Allen, 1926, 1927; Hartman, 1930, 1932; Zuckermann, 1930; van Wagenen, 1945a, 1945b, Farris, 1946).

Of primary importance is the day-to-day record of the menstrual history of each female. "Rounds" should be made at a given time every week day for the purpose of observing the perineum of each animal (the board on which she sits may give the clue), and a record of bleeding or not bleeding should be made. When this is done by the same person at approximately the same time daily, a remarkable cooperation on the part of the monkeys will be noticed. If there is a gentled monkey in the group, she can be turned about without lifting from the cage and the record made; then at the next cage the animal can be encouraged to pass from one side of the cage to the other by simply extending the hand on the outside of the cage nearest the animal. The monkeys soon learn that they are not to be caught or fed at this time, so they remain quiet and observant, and before long most of them learn to "present" the desired view. Do not pass the males by too quickly, if they wish to take part. Pertinent to this routine, a further suggestion may be added that when distinguished guests arrive someone other than the person who makes these records should accompany the guests through the colony in order to insure the more attractive view of each monkey.

With the menstrual records at hand, a weekly schedule can be made, allocating females to the male cages on the eleventh or twelfth day of their cycles. Of course, the first day of menstruation is day one of the cycle. As an administrative expediency monkeys are mated at noon each day and removed at noon the next day or second day, and this also places the animals together at a time when there is no immediate competition for food. When selected females were used, it was found that one-third of the pregnancies followed a single mating (van Wagenen, 1945a). Hartman (1928a) has described rectal palpation of the uterus and ovaries as an important measure to give definite information on the physiological condition of the animal. Although it requires some skill to detect ovulation, much is learned, and the animal with infantile uterus can readily be eliminated. Repeated palpation is necessary because uterine size changes surprisingly. To circumvent other factors which limit successful impregnation, multiparous monkeys are mated with the fathers of their last babies, and nulliparas with a preferred male or one not openly antagonistic. Unfavorable times for conception should be avoided; young animals should not be mated during puberty; parous animals not before three months after a full-term pregnancy even though menstruation has been resumed. And, too, in the summer months, it is believed, this monkey may have frequent anovulatory cycles. By mating arbitrarily, a convenient laboratory schedule has resulted when females are allowed to rest over the summer months and the matings begun in September. This arranges the pregnancies in the winter months, when they can be more carefully followed, and the birth of the infants in the early spring with a concentration at that time on their care and feeding problems.

Pregnancy. Rectal palpation is necessary to determine early pregnancy because this monkey almost always menstruates once after conception (implantation bleeding) (Hartman, 1930). The bleeding is most often delayed for a few days; that is, the conception cycle is longer than the menstrual cycle characteristic at the time for that particular monkey, and the implantation bleeding also is longer. Together, these two signs enable one to diagnose pregnancy around the twenty-third day, but palpation of the uterus is needed to confirm it. The Aschheim-Zondek test has not been adapted for use in the monkey (Hamlett, 1937).

Pregnancy in the *Macaca mulatta* has a duration of around six sexual cycles, about 164 days (Hartman, 1932). Infants which have survived have been born as early as the 147th day and as late as 180 days. A good check on whether the pregnancy is progressing normally is an x-ray (lateral) about the 70th day, when outlines of the skull and the axial skeleton are visible. Parturition in this monkey has been studied and well described (Hartman, 1928a, 1928b). The pregnant animal should be isolated about the fifth month, or earlier, if she has been caged with a younger or more active animal.

The Infant. Don't count your monkeys before they emerge; and even then, better wait to plan for their use until after the neonatal toll. The chapter that could be written between these two headings, "Pregnancy" and "The Infant," would be a long one. Just the same obstetrical complications are found in the management of the pregnant monkey that are met in the wards and delivery rooms of a hospital; and just as many infants after a few days of independent life come to autopsy, many of them presenting no apparent pathology. In our colony, full-term infants who have survived weighed from 300 to 670 grams at birth and measured in body length 17.4 to 22.5 cm.

Statistics deny that more human babies are born at night, but the long laboratory night from 5 P.M. to 8 A.M. does see the birth of more monkey babies. With good luck the baby is found clinging to the ventral surface of the mother, and it may have already been cleaned up with hair nicely parted (it grows this way; the mother does not *really* do it), or on the other hand the mother may be looking for the placenta, the cord may still be attached, and the soiled baby blinking fluid from its eyes while gasping its first breaths through mucus. Whether or not to take the baby from its mother at this time depends upon the competency of the person who catches and on the known reactions of the mother. It is well to avoid too great a struggle until the process of parturition is completed. A second person should be ready with a small turkish towel to lift the clinging baby from the mother's body and to take it immediately out of the animal room. At this point all the monkeys present are protesting. An infant monkey will cling to the warm human hand and struggle when released, but turkish towelling also satisfies many of the reflexes of the little hands and lips expecting to grasp the mother's furry body. The unfailing use of a towel when handling will facilitate smooth transference of infant from mother to basket; and, during examination or tying of the cord, the infant will lie quietly when placed on its back if both feet and hands are buried in the towel.

In order to follow the progress of the infant it is removed from the mother soon after birth, weighed, and thoroughly examined for defects. If it remains with the mother it should be weighed, inspected, and given water every second day (or daily). On day three at the time of second weighing it will be below birth weight, but on day five it should show a gain over the previous weight. Lactation in the colony rhesus is not always satisfactory. Just how much the periodic removal of the young, or the exciting rather than placid group life in crowded quarters, is responsible for failure of lactation cannot be estimated. Anyway, the isolation of the nursing mother to a quiet place is suggested, but careful check on the baby's weight is still necessary to avoid the dehydration and collapse of an apparently active and supposedly nursing infant. Even under the best of conditions all monkeys do not lactate, and one must watch for the impatient mother, who, when her milk fails to come or is too little in amount and the baby importunate, rejects her infant, for she may inflict irreparable injury.

Not all mothers wish to care for their babies. The occasional mother will apparently resent the clinging, squalling, nuzzling infant and roughly cast it off. One monkey was seen to take her newborn baby by its legs, hold it at arm's length, pull out the umbilical cord, which was still attached to the undelivered placenta, and drop the infant to the floor. A careful watch for these times when the instincts of motherhood fail results in the salvage of a satisfactory number of these infants.

When feeding and raising an infant away from its mother becomes necessary, if newborn or under two weeks, it is wrapped in a turkish towel and placed in a basket (fig. 11) with the lid closed. The rough towel and the closed basket mimic the texture, warmth, and restriction of the mother's arms. The infant will stir and call (squeak) about every two hours; and, when lifted to the feeding position with hands on the edge of the basket, its mouth will seek impetuously in every direction for the mother's nipple. The feeding position is most important because unless both feet and hands are firmly engaged in the rough surface of the turkish towel, as on the mother's hairy body, the sucking and swallowing reflexes may become confused. Rhesus monkey's milk has been analyzed (van Wagenen, Himwich, and Catchpole, 1941), and a convenient modification of cow's milk is given on page 31. The Brecht feeder designed for use by the premature human infant holds a suitable amount of milk, is easily sterilized, and is

FIG. 11. Two-week-old infant placed in correct position for feeding. Hands and feet are engaged in the turkish towel and to steady himself he grasps the edge of the basket. At this age, the full-term baby is able to keep contact with the nipple until satisfied, then slips to the bottom of the basket to sleep. During the first week someone must hold the feeder to the lips because the baby's head turns to one side or the other. It is preferable (not from the baby's point of view) to move the feeder rather than to hold the head and accustom the baby to the warmth of the human hand.

furnished with a rubber bulb opposite the nipple end of the bottle, which is under the control of the person feeding, so that milk may be expressed to start the sucking and swallowing reflexes in the newborn. The nipple, of course, is too large to fit the tiny monkey baby mouth, but the baby soon learns to purse the lips around the central perforation and suck efficiently. It is necessary for several days, however, for the nipple to be held to the baby's lips, for the incomplete myelination of the infant's nervous system does not give him control, as yet, of directive head movement; and his head position changes first to one side and then the other, and up and down. But happy is the day when the feeder can be mounted so that all the infant needs is to be placed in position, where he fills his little stomach until he is noticeably rounded and then slips into the basket to sleep. Monkey babies sleep on their stomachs with head to one side, legs sprawled in all directions, if in a warm room, or feet and limbs folded under, when cool. If too

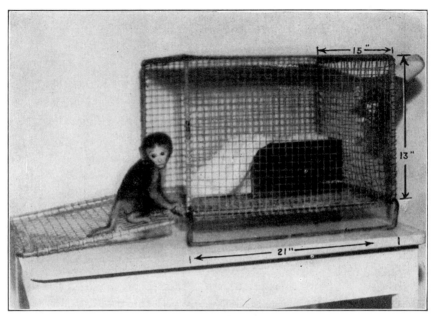

FIG. 12. An ordinary rabbit cage is furnished with a metal mouse box for a bed, a metal holder for the nursing bottle, and a cafeteria tray with newspaper, which slips under the mesh floor. Here the monkey infant lives from three weeks to three months while learning to eat solid foods and to walk and climb. It is no use putting the youngster monkey into a tall cage—for he goes straight to the top and does not know how to come down.

warm the infant will turn on its back and expose its ventral, less hairy surface.

The infant remains in the basket two weeks until the grasp reflex of the hands disappears. Then it is moved to a small cage, a rabbit cage as shown in figure 12. It now can walk. It is able to lift its body from the floor surface because the adductors of the extremities have acquired some tone (Hines, 1942). Even before this time, the infant is able to climb upward by reaching and grasping (transient reflexes also seen in the human) ; but it is necessary to limit the size of the cage to his standing and reaching height, because although he is able at two weeks to release the hand grasp he is unable to retract his legs independently or to turn around and progress head downward. When the infant is left with its mother, she takes care of this emergency by reaching up and removing the baby whose upward progression on the cage has been prevented by the top.

Although the infant monkey has passed from a basket to a cage, he still needs some protection, particularly when asleep, so body temperature must be maintained by some shelter. The ordinary metal mouse box (preferably rustless metal because of the urine) has been useful for the bed. A towel of the type which he has been using should be draped over the side of the box to the floor of the cage so that the infant can use it in climbing into the bed (fig. 12). The larger feeding bottle (the usual 8-oz. bottle) is placed outside the cage, being held by a simple metal band with the nipple of the milk bottle pushed through the wire wall into the cage, at a height allowing the animal to assume the same position while feeding which was customary in the basket. His hands then are engaged in the wire mesh as they were on the towelling. The bottle should be placed so that the animal cannot reach it from the box, in order that the products of urination and defecation, which accompany the smooth-muscle stimulation of nursing, will not soil the bed. After a few weeks when the baby monkey does not actually need the warmth of the box he will sleep outside, always on the towelling. When a few weeks old his activity obviously increases; he will spring up and down in place; he will use his hands to pick up a desired morsel of food; and if the box is removed at this time, he can arrange himself comfortably on the towel for sleeping. He is now able to take care of himself and needs more space for activity. Therefore, at the end of three months, the young monkey graduates to the nursery cage, a standard monkey cage which has been adapted for the occupancy of three animals, where he remains until the end of the first year. All monkeys under a year should live alone in order to subvert the clinging reaction which is characteristic of the young monkey before it is completely rejected by its mother in favor of her next baby. This clinging reaction, undoubtedly initiated by the grasp reflex in the newborn, is unrelated to it physiologically— rather it is an expression of infantile emotional dependence. At the end of the year young animals can be placed three together in an adult cage without much fear of "hugging," for they have learned to live separately.

Puberty. In the laboratory-raised *Macaca mulatta* the first menstruation occurs, on the average, around the second birthday; but, since the record of earliest menarche is 1 year 5 months and 4 days, all young females should be placed on the daily observation list about that time.

Males mature later, and the onset of the growth and development of the testes, which coincides with the adolescent spurt in body weight

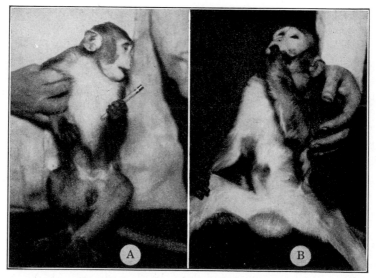

Fig. 13. A. A year-and-a-half-old monkey at the time of first menstruation. Note the two small swellings in pubic region, the first indication of sex skin development. B. The same monkey two cycles later.

and length, is seen about the third birthday or in the early fourth year. There is a great increase in physical activity at this time and also in food intake. An extra meal at noon has been found to eliminate a plateau in the growth curve at first thought to be characteristic of adolescence.

The early changes in the sex skin of the female rhesus monkey may alarm the person who is not expecting the transformation. The initial changes may be overlooked, as is the first menstruation of most monkeys. Sex skin development begins at two points in the ventral skin on either side of the midline in the pubic region. Figure 13A shows these geminate swellings, which are a definite sign that the first sexual cycle has been entered upon and that the bleeding which will mark its end will begin within two weeks. Watch carefully for it; it may be scanty. What usually happens is that someone hurries in from the animal room to report a great blister on one of the animals (fig. 13B). Since the monkey is then brought under inspection, the associated bleeding is reported, and this bleeding may be erroneously regarded as the first (Zuckerman, van Wagenen, and Gardner, 1938). Actually,

FIG. 14. Unusual extent of developing sex skin in a normal monkey. The hair has been cut to expose the skin folds. Sex skin corrugations under hair are not always so large or so extensive; however, the examination of most old females shows that the entire skin of back has been through this stage of development.

this adolescent animal may be in its third to sixth cycle. All skin is under the influence of estrogen, and the edematous thickening is an early phase through which the sex skin passes. The thin ventral skin and underlying tissues take on the translucent turgid appearance of a blister, but the thicker skin of back and tail, although thrown into tense folds during development, is not so distensible or translucent.

From the two pubic centers the swellings spread during the estrogenic phase of each succeeding cycle, subsiding during the premenstrual phase. Spreading fanwise, the swellings coalesce over the pubis and extend back to the anus in two central ridges which include the vulval lips. The perineum is influenced first, and the swelling and folding eventually pass over and down the tail; above, they extend over the thighs, down the dorsal surface of the legs; and, ascending further, they spread over the lower back and around the sides to the edge of the groin. In some instances the swelling and corrugations cover the back completely and pass down the dorsal aspect of the arm to end in a cuff at the wrist. From the back they may pass over the top of the head to the brows, forming at times a heavy ridge over the eyes. While the corrugations on the back are not often so definitely seen as in figure 14, the wrist and brow swellings are easily observed. All sex skin passes through the swollen edematous stage before the rich vascular ingrowth, spreading from the original two pubic centers, gives the brilliant red color recognized as the mature phase of the sex skin. Examination of a twelve-year-old animal will most likely reveal this red coloration extending over all the areas just described, although not as concentrated in color as the exposed mature flat area over the pubis and perineum. The intensity of this red color is a good index to the reproductive state of the monkey—the sex skin of a hypogonadal monkey is pale.

GENERAL CONSIDERATIONS

Inbreeding. The improbability of having litter-mate controls for most monkeys (only the marmoset has litters of two, and twinning in the rhesus occurs no more often than in the human) emphasizes the importance of trying to have a minimum of biological variability in a monkey colony. No breed, variety, or strain has yet been developed for *Macaca mulatta*, and, since brother-sister matings for sixteen generations are necessary to approach homozygosity, this fact is readily explained. Inbreeding is usually accomplished in small mammals by a method of rigid selection from a relatively large initial start. By this means, unfavorable combinations are eliminated and only the best selected and continued. This process of selection from a large colony is impractical when applied to monkeys. However, some reduction of variability can be accomplished by breeding within a colony when no new blood is introduced. Although it does not result in rapid inbreeding, if a limited number of sires is used the heterozygous variation is reduced approximately according to the formula $1/8N$, N being the number of males used and the number of females "unlimited" (Wright, 1931). Hutt (1945) explains that if in a colony of thirty females ten males are used, the existing heterozygosity will be reduced about 1.25 per cent, and if in the next generation three males are used, the remaining heterozygosity will be reduced about 4.25 per cent. The latter sexual ratio is somewhat low for the mating schedule of animals with a twenty-eight day sexual cycle if arbitrary seasonal mating (e.g., mating all animals in the fall months so that young are born in the spring) is planned, in which case the lowest convenient number of males is five. However, if mating throughout the year is permitted, then three of these fearsome creatures to twenty females is a sufficient number.

Dating the birth of monkeys overrides some of the difficulties of using commercial animals, since many biological differences are due to time or the aging factor. This is best accomplished in a controlled colony.

Constant selection should be maintained in the monkey colony as in the barnyard; late breeders, poor breeders, or socially difficult animals should be discarded, as well as the runts and sickly animals. Discarding usually means the allocation to experimental use. The adaptation of the monkey to laboratory life and constant improvement in its diet and care by the breeder have produced with each generation longer and heavier animals at any given age. In the fourth and fifth generations of laboratory-bred animals the magnitude of this increment is less.

Hair. Healthy hair is soft and glistening and lies flat against the head and body; one certain sign of a sick monkey is the fluffing up of the hair about the animal's face. The old monkey, too, can be recognized by the graying and coarser hair on head and face.

During the spring of the year the monkey sheds its hair and gains a new coat, not in the smooth manner of the cat but irregularly. One animal may look merely shaggy with old long hair underlain by short new hair; another monkey may lose the old coat in alopecia-like patches, bare areas of skin appearing usually on upper arm, thigh, and lower back; while one or two out of fifty animals may present a naked appearance, in which case the red sex skin can be seen to be widely distributed up the back and around the face. Added to this, the skin's blue pigmentation, in irregular splotches, makes the animal a colorful, if somewhat startling, sight. The latter two groups of animals are placed out of sight in small zoos. They are often thought to be sick animals, and in one instance, when inquiry was made about an animal whose cage had been turned away from the easy view of visitors, the answer was, "Oh, I don't know what's the matter with him. He has been pulling his hair out." And so the slender, long fingers of the monkey *do* find work in shedding season (early in Florida and late spring in Connecticut), plucking out the single dry hair or a small handful from its own pelage or a cage mate's.

The young monkey born in the spring does not obviously lose its hair the following spring, but in the third spring at the time of puberty the hair thins more evenly over the body than in adult life. The hair is shorter in the young monkey.

Dr. Shourie of Coonooe, who has done extensive nutritional research with rhesus monkeys in South India, says that his monkeys, beginning at maturity, always shed their hair in the spring of the year. This shedding then is not a phenomenon of laboratory life, as was feared at first, but apparently it is accentuated in the monkey who has lived indoors for a generation or so. The relation of humidity and temperature to this condition is not known.

Handling. *All Monkeys Bite* is the legend posted repeatedly above cages in one tourist attraction in Florida. And a number of laboratory workers bear scars to prove that some monkeys *have* bitten. The bite of the young monkey or adult female is similar to that of the human, a crushing bite. But the fully grown male has great saber-like canine teeth which can inflict a lacerating wound, and, realizing his superior quickness and great strength, he may turn vicious even after years of comparative docility. The youngster monkey, when this new use for his teeth occurs to him or happens reflexly, can be disciplined by a tap on the nose. There is the same difference as with children between the well-controlled group of young monkeys and those that are allowed to literally "get out of hand." When monkeys are to be freed for a run, give them the release generously, but when they are to be caught, be determined and well equipped. A monkey which repeatedly bites and snaps should be eliminated.

A twelve- to sixteen-kilo male monkey which has escaped from his cage into the outer cage or room is aggressive and dangerous. If there is no possibility of his escaping further, as through a window, one should at once retreat and try to plan some way to outwit him. An open cage, containing a banana lure and with a string-pull on the door, can be pushed into the room. After a while the monkey may enter this cage. If night comes, he is almost sure to return to his own cage and will be found sitting there with the cage door still open in the morning. If for some reason it is necessary immediately to catch such an animal free in a paddock or escaped into a room, two men wearing heavy gauntlet gloves and equipped with long- and short-handled nets for action at any range can make short work of the job by keeping the monkey in motion until winded.

Far more frequently, however, do monkeys bite each other. If fights between cage mates are noticed, it is obvious the animals should be examined for open wounds, the wounds cleaned and treated, and the animals separated. Never cage adult males together, for there could be a fight to the death.

Infant Feeding. Monkey's milk tends toward that of the human in composition and differs from cow's milk in having a lower percentage of protein and ash and a higher percentage of milk sugar. A modification of cow's milk for infant monkey (*M. mulatta*) feeding follows:

> 4 ounces of cow's milk
> 2 ounces of water
> 4.1 grams of sucrose (1 level teaspoon)

Bottle and sterilize. The same care and technique for sterilization of nipples and bottles customary for human infant feeding are necessary to avoid thrush, diarrhea, and other infant infections.

The infant weight should be followed every day, and the amount of milk given recorded. Figure 15 shows a convenient chart to be constructed on any calibrated paper. The quantity of milk given at each feeding and the weight response are shown for an infant which did very well with artificial feeding. There is, of course, an initial weight loss in the nursed infant, while there is no loss here.

Tuberculosis. Many monkeys arriving in shipments from India are suffering from tuberculosis and, because of the exigencies of the trip, may be in an actively infectious state. Until recently the disease has been uniformly present in all primate colonies (Schroeder, 1938a, 1938b) but, with precaution, it can be eliminated as it has in several laboratories and zoos. If any monkey coughs, loses weight without apparent cause, or becomes apathetic, or if the hair about its face becomes rumpled, it should be isolated, and above all no recently purchased monkey should be added to the animal room until after an isolation and observation period.

The tuberculin test (Schroeder, 1938b; Kennard, Schroeder, Trask, and Paul, 1939) consists in the subcutaneous injection of *old tuberculin* into the loose tissue of the eyelid, where the erythema and edema of a positive reaction may readily be seen. The old tuberculin is diluted 100-fold in sterile physiological saline, and 0.1 cc. is used as the inoculum, which thus contained 1.0 mgm. of old tuberculin. Positive reactions (fig. 16B) consist in local redness and swelling. These vary in degree; in severe reaction the eyelids might become closed and necrosis and ulceration of the skin might appear, as the reaction subsides. Usually the reaction is visible in 24 hours, increases for 2 or 3 days, and then subsides gradually. Occasionally positive reactions are not appar-

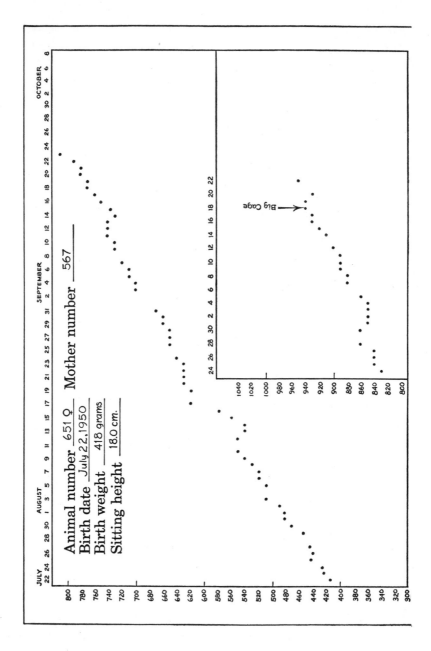

Animal number 651 ♀   Mother number 567
Birth date July 22, 1950
Birth weight 418 grams
Sitting height 18.0 cm.

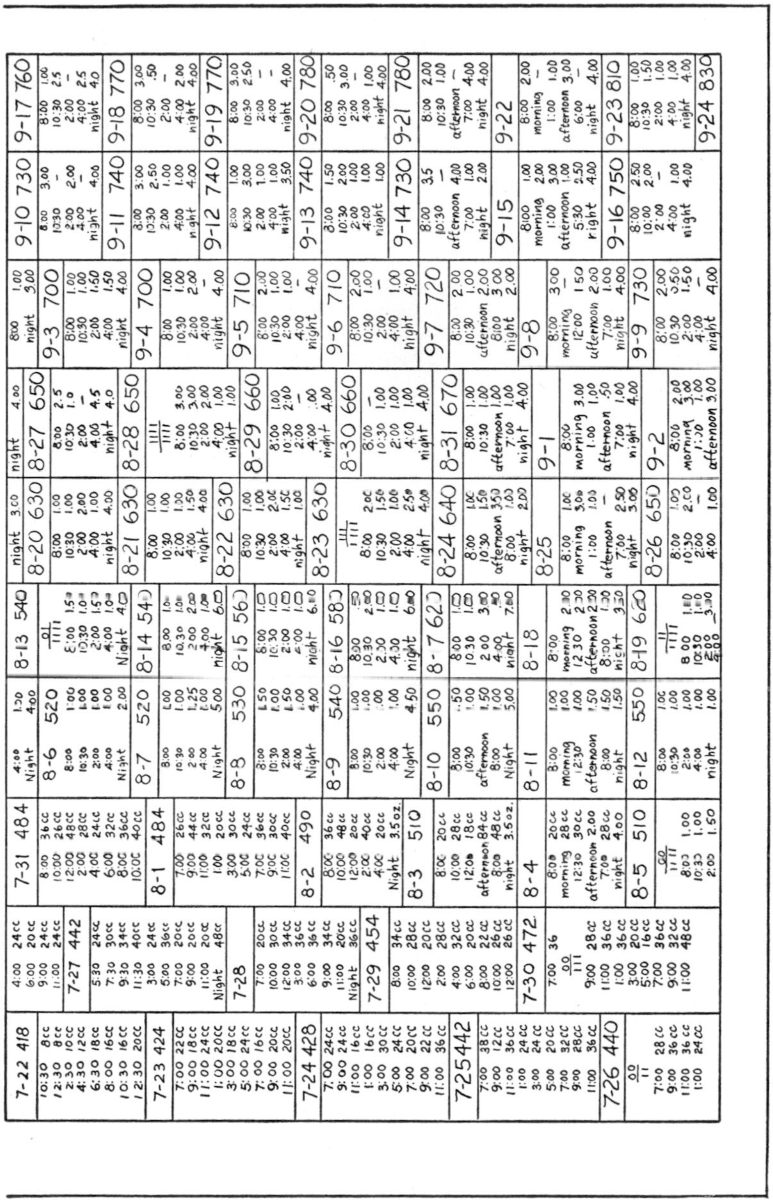

Fig. 15. Infant feeding chart. Each day begins with date and body weight (July 22; 418 grams). Growth is charted when infant is weighed before 8 A.M. feeding. Daily record of the amount of milk is important to insure maintenance and increase. On the first day this infant accepted more milk than is usual. Water is also given. Records in cubic centimeters indicate use of the Brecht feeder; ounces, the ordinary nursing bottle. Note that on July 26 (7–26), at 440 grams, the first teeth appeared, the formula indicating the presence of two lower incisors.

ent until the third or fourth day. If no reaction is seen by the fourth day, the test is considered negative.

This test is used as a preliminary weeding-out measure of new arrivals but is not accepted as final by the Philadelphia Zoo, where it is customary to study each individual monkey by taking the temperature every 4 hours over a period of several days. In some laboratories the old-tuberculin test is repeated at intervals.

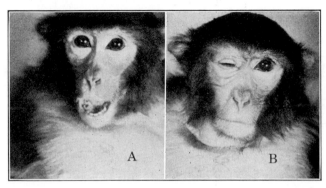

FIG. 16. Test for active tuberculosis. A. Adult female, control picture, November 18. The animal was injected subcutaneously in right upper eyelid November 20 with old tuberculin. B. Same animal November 23 with positive reaction.

Cage Paralysis. The first evidence of this disabling condition can be seen when a monkey begins to "favor" one foot. He steps on it reluctantly and finally not at all. The same process may begin in the other hind limb, and contractures will follow. In two laboratories it is believed that early treatment with vitamin B_1 or dried yeast in generous quantity will stay this degenerative process which is likely to appear in recently imported animals, in rapidly growing animals, and in large animals too restricted in space for exercise.

Treatment. The modern antibiotic and chemotherapeutic agents can be used on the same indications as in the human. The old favorite gentian violet has not been given up for surface wounds and mouth lesions because it can be painted on quickly, reveals clearly which parts have been covered, and has the added advantage of showing up the next day to remind one to follow up the treatment.

A reference (Kennard, Ruch, and Fulton, 1946) is suggested for a discussion of surgical technique for the monkey, but two practical points could be stressed here. First, the skin should be closed by subcutaneous sutures, since exposed stitches are often picked out; second, a carrying table is suggested for transport of the anesthetized animal from cage to operating table because heavy animals have been known to expire when carried ever so carefully in the arms.

In caring for the sick monkey, an important point to keep in mind is the increased ratio of surface to body mass in this smaller animal. Injection of fluids parenterally often is followed by a striking response. A tray in the laboratory should be set up for intravenous, intraperitoneal, and subcutaneous infusion of both saline and glucose. An infant which persistently refuses food may accept it after a clysis.

Loaf sugar and the common square after-dinner mint are excellent carriers for vitamins or medicine to be orally administered.

IMPORTATION OF MONKEYS

There is a seasonal importation of rhesus corresponding nicely with the academic year, opening late in October and closing the end of May. During one season (1948–1949) the foremost dealer reports handling 12,500 rhesus monkeys. In April and May pregnant animals and mothers with babies are most frequently seen in the shipments.

Most imported monkeys are from 1 to 3 years old, weighing 1500 to 4000 grams. Males and females are represented; the adolescent and adult females may be recognized by the swollen or red sex skin in the pubic region. Adult males are infrequent.

1950 (May) prices were as follows: rhesus monkeys (up to 8 lb.) $25.00 each; 8 to 10 lb. $30.00 each; over 10 lb. $40.00.

Recently a hundred monkeys were brought by air from India in 5 days, and it is possible that this will be the preferred method of importation.

Importers of Monkeys

Trefflich's Bird & Animal Company, Inc.
228 Fulton Street
New York 7, N. Y.

Meems Brothers & Ward
P.O. Box B
Oceanside, N. Y.

Louis Ruhe, Inc.
853 Broadway
New York 3, N. Y.

Chase Wild Animal Farm
Lawson Road
Egypt, Mass.

Wilbur-Ellis Company
334 California Street
San Francisco 4, Calif.

Southern California Bird & Pet Exchange
P.O. Box 7
Bell, Calif.

Warren Buck
420 Garden Avenue
Camden, N. J.

Shipping.   For shipping monkeys, any well-made wooden box of appropriate size can be modified.   Animals between 1500 and 4000

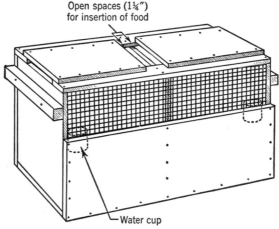

Open spaces (1¼″)
for insertion of food

Water cup

FIG. 17. Shipping box. Any appropriate strong wooden box may be adapted in the above manner. Young monkeys can be shipped two or three in a compartment, and adults should be separated. See text for description of box construction.

grams may be shipped in pairs, unless known to be antagonistic. Young animals cling together, causing eye lesions, and the old animals resent sharing small spaces and food. Before inserting wooden partitions for separation, several holes should be bored at eye level so that the monkeys can see each other. One modification of such a box is shown in figure 17. The upper third of one side is covered with ½-inch wire mesh, the edges of which are secured by the top and sides of the box. Cups for water are nailed in place as shown so that they can be filled from the outside through the wire. The center top board is in sections, the removal of one section allowing release of only one animal. These sections overlap side sections and are 1¼ inches short toward the center of the box to allow for insertion of food. Metal handles or strips of wood should be placed on both ends for ease of handling. If the journey is under 36 hours, cups can be eliminated and food (apples, uncooked potatoes, carrots, bananas, etc.) placed within the box. For longer journeys food must accompany the animals, the food parcel being nailed or fastened to the box. "Please water" should be painted in black on the box outside each cup. Monkeys have been shipped all over the United States and to Europe, both by train and by air; but in the latter case food for 2 extra days should be included even on short flights.

To release the monkey from the box remove one section of the top board and bring the opening to the cage door. A wire probe passed through the mesh of the box induces the monkey to transfer quickly.

EQUIPMENT AND SUPPLIES

Caging. Manufacturers who offer standard monkey cages or who are prepared to design and manufacture equipment to fulfill individual specifications are:

> George H. Wahmann Manufacturing Company
> 1123 E. Baltimore Street
> Baltimore 2, Md.

> Norwich Wire Works, Inc.
> Norwich, N. Y.

> A. S. Aloe Company
> 1831 Olive Street
> St. Louis 3, Mo.

> Bussey Pen Products Company
> 5151 W. 65th Street
> Chicago, Ill.

Feeding Pans. For cages, these pans may be free or fixed, and for paddocks Hartman has used an 8-pound iron dish which is so heavy that the ordinary monkey does not lift it, and which can be hosed out during the daily cleaning. Free pans for cages are usually of aluminum, deep cake size, and should be shaped to stack; round pans are less liable to be twisted than square ones. Cast aluminum, although more expensive, resists denting and cracking. The work of collecting these pans after the animals have lost interest in the food and before they become interested in who can make the greatest noise is preferred in our laboratory to any method devised for fixing the pans in place and releasing them for cleaning. There is available, however, an excellent fixed feeding dish used by nutrition laboratories and adapted

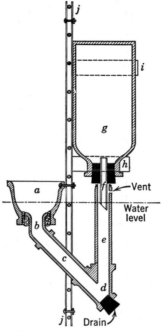

Fig. 18. Water fountain. "The chief details of construction are apparent from the diagram. The apparatus is made entirely of 'black metal' plumber's fittings and galvanized after assembly to prevent rusting. The actual drinking cup (a) is a 2 x 1 inch reducing couple from which both threads have been lathed and into which is brazed a ½ inch, 45 degree elbow (b); a ½ x 3 inch 'nipple' (c) passes through the wire of the cage and is connected by a 45 degree 'Y-branch' (d) to a ½ x 5½ inch 'nipple' (e) which receives the delivery tube (f) of the water bottle. Brass tubing $\frac{5}{16}$ inches in diameter, if beveled at the end, serves for the delivery tube; a smaller size does not allow the water to flow freely.

"The third opening of the Y-branch (d) serves as a drain, so that the device can be drained, cleaned, and refilled without opening the cage. The use of 45 rather than 90 degree fittings and the large internal diameter at the critical bend (b) makes it easy to pass a test-tube brush into the cup to clean it or to free it from obstruction by debris. The device is fixed to the cages with wire mesh walls by bolting the rim of the cup to two heavy galvanized iron plates on either side of the mesh. To these plates are also bolted the metal straps (h) and (i) that support the water bottle." (From Kennard, Ruch, and Fulton, 1946.)

for use in a particular cage (fig. 3). It is a 1-pound butter crock encircled firmly by a metal band so shaped that it can be hung on the threshold of the open door and locked into place when the door is closed. A pair of these contains a daily supply of water and food for one animal. There are two drawbacks to the fixed food container. The first is the question of how long it will remain fixed if an active adult animal begins to work with it, as they all do with an object which extends into their cage, and the second is that young animals are wont to perch on any object available and soon foul the contents. These feeding dishes work very well, however, in the position in which they hang as originally used.

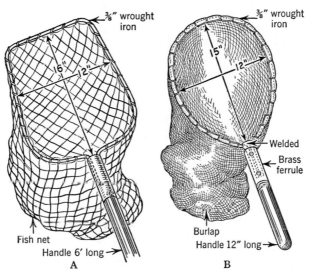

FIG. 19. Catching nets. A. Long-handled net for catching in room, heavy cord, hand-knotted in 1½-inch mesh, 30 inches deep, and lashed to frame. B. Short-handled net for catching within cage, washed burlap potato bag caught with a simple loop-stitch to frame.

Water Supply. The Ruch watering device (Kennard, Ruch, and Fulton, 1946) is the best to date and is shown in figure 18 with excerpts from the original description accompanying it. For some reason the flat gallon wine bottle, so convenient for placing against the side of a cage, has disappeared from old bottle markets; so, when estimating the length of the "c" arm of the y-branch, it is necessary to know the size of the bottle which will be in use and likely to be replaceable.

For the well-drained floor or paddocks, small tubs of water can be used, and, for warm weather, the fine spray suggested by Hartman falls into the tub and adds enjoyment (Hartman and Straus, 1933).

Catching Nets. A net similar to that in figure 19A can be ordered from W. A. Augur, 35 Fulton Street, New York 7, N. Y. The second net can be fashioned by someone in the laboratory. When the iron hoop is secured to the top of the handle, a washed burlap potato sack is caught with a simple loopstitch to the hoop, using a woven cord and sailor's needle.

Scales. The ordinary scale used for weighing human babies but made for the export trade and therefore calibrated in grams is an inexpensive instrument. If monkeys are placed in canvas bags (laundry type and size), the scales can be used without alteration. If a transfer box is used, it must be made of light metal and then can replace the metal trough designed to receive the baby.

Mercury-vapor Lamps. A central ceiling lamp is useful in cutting down the air-borne infections as well as furnishing the antirachitic action. For preventive therapy a floor lamp used outside the over-all cage is convenient. Rickets and tuberculosis are diseases frequently found in the rhesus. Watch for inflamed eyes to indicate too much radiation.

Tattoo Machine. Ink (black or red) as well as the machine can be purchased from Owen Jensen, 120 W. 83rd Street, Los Angeles 3, Calif.

For short-term experiments the head may be shaved just over the eyes and the animal's number placed there, but after a time the hair grows in. For long-term work the upper chest region just below the clavicles, where the hair is sparse, is preferred. The infant is tattooed inside the thigh with the last two digits and when two years old receives its full number on the chest.

Electric Clipper. A heavy-duty electric clipper for removing hair before tattooing the serial number or in preparation for sterilizing the skin before operation can be ordered from John Oster Manufacturing Company, Racine, Wis. This clipper works on both alternating and direct current. Local barber supply houses may carry similar instruments.

Hair Remover. The following is a formula for a powder which can be kept in a storage bottle. To use, add enough water to make a moderately thin paste.

Barium sulphide	35 grams
Flour or cornstarch	35 grams
Talcum powder	35 grams
Powdered Castile soap	35 grams

When worked into the hair with a narrow spatula and allowed to dry, this paste will accomplish depilation.

Feeder. The Brecht feeder (for infant feeding) can be purchased from Becton, Dickinson and Company, Rutherford, N. J.

REFERENCES

Allen, E. 1926. The time of ovulation in the menstrual cycle of the monkey, *M. rhesus. Proc. Soc. Exp. Biol. Med.*, vol. 23, pp. 381–383.

——. 1927. The menstrual cycle of the monkey, *Macacus rhesus.* Observations on normal animals, the effects of removal of the ovaries and the effects of injections of ovarian and placental extracts into the spayed animals. *Contrib. Embryol., Carnegie Inst. Wash.*, Vol. 19, pp. 471–475.

Asdell, S. A. 1946. *Patterns of mammalian reproduction.* Comstock Publishing Co., Ithaca, N. Y.

Collings, M. R. 1926. A study of the cutaneous reddening and swelling about the genitalia of the monkey, *M. rhesus. Anat. Rec.*, vol. 33, pp. 271–278.

Corner, G. W. 1923a. The relation between menstruation and ovulation in the monkey. Its possible significance in man. *Jour. Amer. Med. Assoc.*, vol. 89, pp. 1838–1840.

——. 1923b. Ovulation and menstruation in *Macacus rhesus. Contrib. Embryol., Carnegie Inst. Wash.*, vol. 15, pp. 73–101.

Farris, E. J. 1946. The time of ovulation in the monkey. *Anat. Rec.*, vol. 95, pp. 337–346.

Hamlett, G. W. D. 1937. Positive Friedman tests in the pregnant rhesus monkey, *Macaca mulatta. Amer. Jour. Physiol.*, vol. 118, pp. 664–668.

Hartman, C. G. 1928a. Description of parturition in the monkey, *Pithecus (M.) rhesus*, together with data on the gestation period and other phenomena incident to pregnancy and labor. *Bull. Johns Hopkins Hosp.*, vol. 43, pp. 33–51.

——. 1928b. A readily detectable sign of ovulation in the monkey. *Science,* n.s., vol. 68, pp. 452–453.

——. 1928c. The period of gestation in the monkey, *M. rhesus,* first description of parturition in monkeys, size and behavior of the young. *Jour. Mammal.*, vol. 9, pp. 181–194.

——. 1929. Uterine bleeding as an early sign of pregnancy in the monkey (*M. rhesus*), together with observations on the fertile period of the menstrual cycle. *Bull. Johns Hopkins Hosp.*, vol. 44, pp. 155–164.

——. 1930. The corpus luteum and the menstrual cycle together with the correlation between menstruation and implantation. *Amer. Jour. Obstet. Gynec.*, vol. 19, pp. 511–519.

——. 1932. Studies in the reproduction of the monkey *Macacus (Pithecus) rhesus,* with special reference to menstruation and pregnancy. *Contrib. Embryol., Carnegie Inst. Wash.*, vol. 23, pp. 1–173.

Hartman, C. G., and William L. Straus, Jr., ed. 1933. *The anatomy of the rhesus monkey (Macaca mulatta).* Williams and Wilkins, Baltimore.

Heape, W. 1896. The menstruation and ovulation of *Macacus rhesus. Proc. Roy. Soc.*, vol. 60, pp. 202–205.

Hines, Marion. 1942. The development and regression of reflexes, postures, and progression in the young macaque. *Contrib. Embryol., Carnegie Inst. Wash.*, vol. 30, pp. 153–209.

Hutt, F. B. 1945. Genetic purity in animal colonies. In animal colony maintenance, ed. by Roy Waldo Miner. *Annals N. Y. Acad. Sciences,* vol. 46, pp. 5–21.

Kennard, M. A., Theodore C. Ruch, and John F. Fulton. 1946. The housing, care, and surgical handling of laboratory primates. *Yale Jour. Biol. Med.*, vol. 18, pp. 443–471.

Kennard, M. A., C. R. Schroeder, James D. Trask, and John R. Paul. 1939. A cutaneous test for tuberculosis in primates. *Science,* vol. 89, pp. 441–443.

Miller, G. S., Jr. 1933. The groups and names of Macaques. In *The anatomy of the rhesus monkey,* ed. by C. G. Hartman and William L. Straus, Jr. Pp. 1–9. Williams and Wilkins, Baltimore.

Pocock, R. I. 1932. The rhesus macaques (*Macaca mulatta*). *Jour. Bombay Nat. Hist. Soc.,* vol. 35, pp. 530–551.

Ratcliffe, H. L. 1940. Diets for a zoological garden: some results during a test period of five years. *Zoologica,* vol. 25, part 4, pp. 463–472.

Ruch, Theodore C. 1941. *Bibliographia primatologica, a classified bibliography of primates other than man.* Charles C Thomas, Baltimore.

Schroeder, C. R. 1938a. Acquired tuberculosis in the primate in laboratories and zoological collections. *Amer. Jour. Pub. Health,* vol. 28, pp. 469–475.

——. 1938b. A diagnostic test for the recognition of tuberculosis in primates; a preliminary report. *Zoologica,* vol. 23, pp. 397–400.

Short, D. J., and A. S. Parkes. 1949. Feeding and breeding of laboratory animals. X. A compound diet for monkeys. *Jour. Hyg.,* vol. 47, pp. 209–212.

Sutton, J. Bland. 1880. Menstruation in monkeys. *Brit. Gynaec. Jour.,* vol. 2, pp. 285–292.

van Wagenen, G. 1945a. Mating in relation to pregnancy in the monkey, *Yale Jour. Biol. Med.,* vol. 17, pp. 745–760.

——. 1945b. Optimal mating time for pregnancy in the monkey. *Endocrinology,* vol. 37, pp. 307–312.

van Wagenen, G., H. E. Himwich, and H. R. Catchpole. 1941. Composition of the milk of the monkey (*M. mulatta*). *Proc. Soc. Exp. Biol. Med.,* vol. 48, pp. 133–134.

Wright, S. 1931. Evolution in Mendelian populations. *Genetics,* vol. 16, pp. 97–159.

Zuckerman, S. 1930. The menstrual cycle in the primates. I. General nature and homology. *Proc. Zool. Soc. London,* no. XLV, pp. 691–754.

Zuckerman, S., and John F. Fulton. 1934. *The nomenclature of primates commonly used in laboratory work.* Privately printed: Yale University School of Medicine.

Zuckerman, S., G. van Wagenen, and R. H. Gardner. 1938. The sexual skin of the rhesus monkey. *Proc. Zool. Soc. London,* vol. 108, pp. 385–401.

ACKNOWLEDGMENTS

The responsibility and work with the Obstetrics Monkey Colony have been shared over the eighteen years since its establishment with Joseph Negri, Frank Caruso, Ruth Vogel, and Natalie MacCarthy; their individual observations, with my own, form the basis of this chapter.

I wish to thank Dr. Carl G. Hartman for reading the manuscript and Dr. Leonell C. Strong for checking the paragraphs on inbreeding.

The photographs by Mr. Howard Reynolds and the drawings by Mr. Armin Hemberger are gratefully acknowledged.

2· THE RAT
AS AN EXPERIMENTAL ANIMAL
With Practical Notes on the Breeding and Rearing
of the Albino Rat and the Gray Norway Rat

EDMOND J. FARRIS

The Wistar Institute of Anatomy and Biology
Philadelphia, Pennsylvania

INTRODUCTION

The albino rat (*Mus norvegicus albinus*) has become the most commonly used laboratory mammal. Albinos are gentle, clean, easily handled, readily trained, and inexpensive to maintain. They live approximately 3 years and begin breeding at about 3 months of age. The litters are large and can be secured at any season. The rat is highly resistant to the usual infections and is particularly satisfactory for surgical experiments. The rat has been used in practically all fields of investigation.

In this chapter, brief practical information is given, based on the extensive experience of The Wistar Institute in the breeding and care of albino and other rats for research purposes. For more detailed information, the reader is referred to a comprehensive monograph (Farris and Griffith, 1949). This contains chapters by twenty-nine contributors on the breeding of the rat, general methods of handling, gross anatomy, experimental methods, embryology, dietary requirements, teeth, digestive system, metabolism, central nervous system, techniques for investigation of psychological phenomena, circulatory system, use in biologic assay of hormones, dosage of drugs, hematology, radiology, surgery, histologic methods, osseous system, eye, protozoan parasites, metazoan parasites, and spontaneous diseases. This publication may be referred to when the rat or other small mammals are employed for experimental work.

COLONY BUILDING, CAGES, AND EQUIPMENT

Colony Building. The Wistar Institute constructed, in 1921, a colony building which has proven very satisfactory for raising rats (fig. 1). The building contains thirteen well-lighted individual rooms: a record and a surgery room, nine rooms in which the animals are kept, a kitchen, and a cage-cleaning room (fig. 2). Each room is maintained

FIG. 1. Animal colony building of The Wistar Institute. An isolated building constructed of brick, concrete, steel, and glass.

in cold weather at a temperature of about 72°, the range being 65 to 75° F. Rooms for newborn animals should be maintained at 78° F., with a relative humidity of 50 per cent, to prevent ringtail. Excessive heat in summer is compensated for by rapid exchange of air in the rooms, which aids also in decreasing odors, moisture, and heat generated by the rats themselves. Air conditioning has solved the problems of temperature, humidity, and ventilation in some other colonies. The entire building is screened and protected against entrance of wild rats, cockroaches, and any other animals which may carry infections.

The cleaning room is equipped with a tank for washing and sterilizing cages (cage-washing machines and automatic bottle washers are available commercially). A large steam sterilizer is located in an auxiliary building. All materials such as bedding, shavings, and stored cages are sterilized before being taken into the colony.

The kitchen is equipped with a 50-gallon steam cooker, an auxiliary steam drying apparatus, a power meat grinder, and other kitchen equipment.

The colony office serves as a record room where the complete history of each rat may be found. A small room adjoining the office is used for surgery.

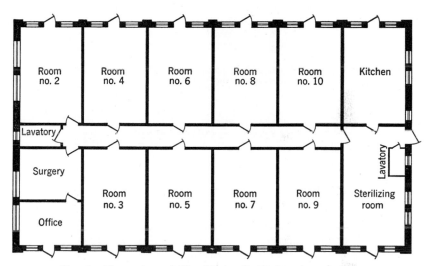

Fig. 2. Floor plan of The Wistar Institute colony building.

Cages. The Wistar Institute adopted two forms of cages to satisfy its needs, one for the breeding of large numbers of albino rats and the other for long-term genetic studies.

For long-term experiments, the cage shown in figure 3 is recommended. Cage supports may be placed on casters satisfactorily. Each cage is 10 inches high, 10 inches deep, and about 57 inches long. The end of the cage is all Monel metal, 10 inches by 10 inches, with a 1-inch rim turned on all four edges. The wooden top, back, and front of the cage are screwed to the metal ends. The floor of the cage, which is also Monel metal, is detachable from the upper portion for cleaning. The entire cage is metal lined when housing wild gray rats.

The dormer cage illustrated in figure 4 is most satisfactory for breeding and housing not more than ten adult rats. The cage is constructed of oak ½ inch thick, 35 inches long, 12½ inches high, 17 inches deep, outside measurements. It is divided by a middle partition in which is located an opening 3 inches in diameter, to permit communication between the compartments. The floor of each compartment consists of a movable galvanized-wire cloth screen, 15¾ inches square (no. 16 wire, ⅜-inch mesh). This floor is supported along the front and back by Monel metal angles. Beneath the removable floor

FIG. 3. Type of cage recommended for long-term experiments.

is a Monel metal tray, 15¾ by 15¾ by ½ inches deep, which serves to catch the dirt falling through the cage floor. This cage may be constructed and used without the middle partition.

The wooden cage has been selected in preference to one of metal, for the making is simple and the upkeep less since it does not rust. The cage is odorless, contrary to the usual belief, in any well-kept colony.

Other laboratories have found that all-metal cages are satisfactory for rats. They are useful for breeding, weaning, and housing of experimental rats, and are placed in racks or on movable stands mounted on casters. When such stands (fig. 5) are used, it is customary to place the cages, which have no bottoms, on a wire floor which is mounted above the cage pan. All excreta and food falling through the mesh are collected in the pan below, which is so placed that the rat cannot reach through the mesh or touch any of the material col-

FIG. 4. Dormer type of cage.

lected in it. The catch pan is approximately 2 inches deep, so that it may be used with nesting material as a suitable floor for cages when they are used for breeding purposes.

By a slight modification of the usual construction, paper may be substituted for the cage pans. Rolls of 40-pound waxed Kraft paper or a similar paper are mounted on the end stand, one roll of paper at each cage level. The paper is passed under the wire floor of the entire row of cage stands. Cleaning is accomplished by withdrawing the soiled paper from beneath the cages at the opposite end of the roll

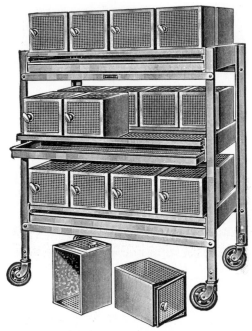

Fig. 5. Movable stand for metal cages.

Fig. 6. Rolls of paper for rapid cleaning of cages.

stand. This automatically pulls in clean paper. In a test room in The Wistar Institute this paper system has proven efficient with the dormer cages, as illustrated in figure 6.

Stands may be constructed to hold six rather than four individual cages between uprights, and in many cases the height may be increased to accommodate six or eight tiers of cages. Cages close to the floor must be protected from drafts.

Cages of various sizes should be provided to take care of the different needs of the investigators. For keeping one or two animals in a cage, the type shown in figure 5 is preferred, the rats being readily visible and easily accessible. For breeding and housing individual litters a cage longer and larger is advocated (fig. 4). When fully assembled, one portable stand will hold twenty-four cages in a space approximately 2 by 3 feet.

METABOLISM CAGE. A metabolism cage used in many laboratories is shown in figure 7. This is designed for collection of urine and droppings separately.

CAGES PERMITTING EXERCISE. There are two types of cages for measuring running activity in the rat, as follows: (1) The ver-

FIG. 7. Metabolism cage.

tically revolving drum; (2) the horizontally rotating turntable. An excellent and inexpensive turntable (fig. 8) may be mounted in the dormer cage. This type of exerciser permits a variety of movements and is easier on the feet of the rat. The turntable (Farris and Engvall, 1939) consists of a Monel metal disk, 12 inches in diameter, which can readily be removed from the cage for cleaning. From the brass hub a wire runs directly to a cyclometer (Veeder-Root Co., Hartford, Conn., rotary ratchet counter no. 6) which registers the number of revolutions caused by the rats' running.

Revolving Drum. A satisfactory cage with a revolving drum for measuring quantitatively the variations in running activity of the rat is shown in figure 9. This cage permits exercise, which is essential to the health of the rat. It permits the animal to live in the small cage section and exercise on the drum. The number of revolutions of the drum produced by the running of the animal is totaled on an automatic counter, as illustrated.

Other exercising drums of larger diameter have wooden nesting boxes, as illustrated in figure 10.

FIG. 8. Turntable equipped with cyclometer, for recording, in revolutions, running activity of rat.

FIG. 9. Exercising cage, with activity recorder.

FIG. 10. Exercising drums, showing wooden nesting boxes.

It has been observed that rats run farther in large wheels; therefore, in comparing results, it is important that exercisers of the same size be used for all animals in an experiment.

Tables. Every large colony should be supplied with three substantial tables. One should hold the scales for weighing rats. If the animals are very tame, it is feasible to use an automatic scale. (The Toledo scale with platter scoop or pan is recommended.) If the rats are not docile, it is more expedient to place an individual in a small wire-mesh cage and to use a counterpoise weight equal to the weight of the mesh cage. An accurate weighing of the rat can then be obtained by using any form of scales having a capacity of at least 1000 grams. When cages are equipped with movable stands, they may be brought to the scales and the rats weighed. If the colony contains only large cages, arranged in tiers and having stationary supports (fig. 3), the table containing the scales should be mounted on casters. A second movable table is necessary when cages are of the type shown in figure 3. The caretaker uses this table for containers with food and water supplied daily to the rats.

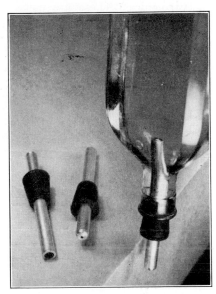

FIG. 11. Water bottle, showing details of brass mouthpiece.

While not a necessity, a table with a drawer is a convenient accessory in any laboratory animal room. It may hold the various small instruments that are used frequently and serve also for taking notes and recording data.

Drinking Devices. The device shown in figure 11 is satisfactory. Standard pint or quart bottles serve as water reservoirs. Each has a rubber stopper into which is inserted brass tubing $\frac{3}{8}$ inch outside diameter with $\frac{1}{32}$-inch wall and 4 inches in length. This is tapped $\frac{1}{4}$ inch deep at one end with a $\frac{5}{16}$-inch tap, 24 threads to the inch. A plug made from $\frac{5}{16}$-inch brass rod is screwed into this tapped end and cut off flush with the end of the tube. A $\frac{3}{32}$-inch hole is drilled in this closed plug and finished with a special lathe tool to form a smooth licking surface. The tube is inserted into the rubber stopper 1¼ inches; the remainder of it is within the bottle to serve as a support, should the bottle be jarred. The tip can be very easily replaced, lasting as a rule about 2 years even with wild gray rats, which constantly gnaw it.

Other drinking devices such as nursery bottles, hair-tonic bottles, test tubes, and glass containers of all sorts have been observed in many colonies. These are usually supplied with tips of bent glass polished only on the licking end and inserted into rubber stoppers attached to the water reservoir.

Fig. 12. Non-scattering food cup.

Fig. 13. Earthen feeder, glazed surface.

Fig. 14. Schematic marking for pedigree records.

Feeding Devices. The feeding device shown in figure 4 has proven very satisfactory for dry ration (pressed biscuit). A ledge on the lower part of the cage catches smaller particles of scattered food. The animals will eat the food caught on this ledge. For cooked diet a small pan about 4 inches in diameter and 3 inches deep is very satisfactory. In quantitative feeding it is important to have a device which prevents food from being scattered, yet allows the rat to secure food without great difficulty. Figure 12 shows a satisfactory non-scattering food cup. It is made of Monel metal and has a perforated disk which prevents the animal from scattering and digging into the food. An earthenware feeder of the type shown in figure 13 is used in many other animal colonies and may serve for a water cup as well. Cold-cream and mason jars of various sorts are also used as food cups.

Bedding. Two chief types of bedding are used in The Wistar Institute colonies. Shavings, pine or poplar, are used in cages with solid bottoms; in dormer cages, with ⅜-inch screen mesh bottoms, shredded bond paper is preferred. These materials are absorbent and deodorant, and make excellent nests. Straw, hay, and wood wool or excelsior also may be used for bedding, but wood wool is not advised because it gives off a fine dust which is irritating to the rats.

CARE OF THE COLONY

Washing Cages. The cage-cleaning technique used in the animal colony consists of:

1. Scrubbing cage with soap and water.
2. Dipping in hot water.
3. Dipping in 2 per cent cresol solution.
4. Rinsing with cold water.
5. Setting aside to dry.

The ordinary commercial dish-washing machine provided with an oversize pump may be used for washing cages. This machine sprays the cage from top to bottom first with very hot soap solution and then with scalding water. This procedure leaves the cage and its parts so hot that they dry almost immediately after being withdrawn from the machine.

Sanitation. The Aeroil Torch (no. 00), giving a temperature of about 2000° F., may be used effectively as a sanitary agent. The lighted torch is moved over the surface of the walls and floor until the room has been covered completely. This method is rapid and inexpensive. It is as effective as the high-steam-pressure method that is used in some colonies. W. F. Wells (1940) has developed a lamp which destroys bacteria by letting down a curtain of ultraviolet rays in the room. It is inexpensive, can be substituted for an ordinary electric light, and has proven useful in preventing air-borne infection.

Permanent Records. FILING. Many methods have been described for filing pedigree records. King's method (1918c) has proven very satisfactory. The scheme as outlined is as follows: A letter (A or B) places the individual in its proper series. The serial letter is preceded by a number which signifies the generation to which the animal belongs. An index number (2, 3, or 4) following the serial letter shows in which of the mother's litters the animal was born; if no index number is used, the rat was a member of its mother's first litter. The subscript after the serial letter is a number which serves to distinguish each particular rat from any other rat belonging to the same generation and litter group. To indicate the sex of the individual, its number is enclosed by the sex symbol. In King's schematic illustration (fig. 14), the symbol denotes a female rat belonging in the seventh generation of the strain. She was a member of the second litter cast by her mother, and her individual number in the series of rats belonging to the second litters of the seventh generation was 20.

MARKING. Several satisfactory methods for marking and identifying rats have been developed. One method consists of cutting a small V-shaped wedge into the border of the ear. Two, three, or four wedges may be cut in each ear to give a large number of marking combinations.

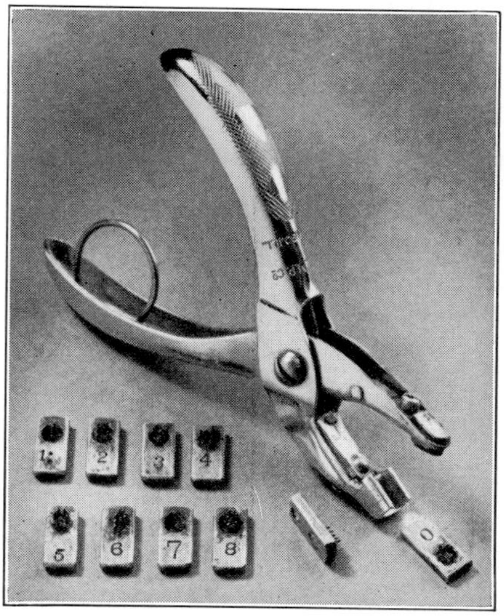

FIG. 15. Punch for tattooing numbers into rats' ears.

A new satisfactory punch was developed in The Wistar Institute for tattooing individual numbers in the rats' ears by means of India ink. An ordinary 6-inch ticket punch was used as a holder (fig. 15). The die on the end of the punch was filed off and replaced with a piece of brass channel $\frac{3}{8}$ inch long by $\frac{3}{8}$ inch wide by $\frac{3}{16}$ inch high. The number blocks were slid into place through the open end. A rubber pad of equal size was cemented to the opposite side. This comprised the holder. Individual digits from 0 to 9 were made by mounting steel pins in the form of the desired number on a piece of medium-soft rubber about $\frac{1}{8}$ inch thick. A small block of brass, approximately $\frac{1}{4}$ inch wide by $\frac{3}{4}$ inch long by $\frac{3}{16}$ inch high, with a hole drilled through it large enough to accommodate the steel-pin number, was placed over the rubber form. The hole containing the number was then filled with hot solder, and the rubber foundation drawn away from the pins. The pin number block was then filed to fit snugly into the channel holder. Two digits to each ear allow numbering up to 9999.

Another method is to paint various parts of the rat with a basic carbolfuchsin stain. It is necessary, however, to re-mark the animal about once every six weeks. Another simple method is to perforate the ear with a needle or punch in combinations of one or more dots and then paint the ear promptly with carbolfuchsin or India ink, as described above.

Cage 409

Female __1A$_{52}^{5}$__ ⚭ Born __1-5-50__ Parents __4C✕3C__ _____

Diet

Male	Date	Wt.	Wt. 3 wks. later	Date litter born	Born dead	♂	♀	Wt. litter	Fed	Date weaned	♂	♀	Wt. weaning	Litter no.
1A$_{45}^{2}$ ⚭	5/8	178	251	6/9		6	4	62	✓	6/29	6	4	313	2A' 271-278

FIG. 16. Card No. 1: Female breeder card for breeding-room file (actual size: 4 by 6 inches).

Female no. __1A$_{52}^{5}$__   Cage no. __409__

Born __1-5-50__   First mating __5-8-50__

FIG. 17. Card No. 2: Female breeder (cage card) (actual size: 2 by 4 inches).

Male no. __1A$_{45}^{2}$__   Cage no. __426__

Born __1-5-50__   First mating __5-8-50__

FIG. 18. Card No. 3: Male breeder (cage card) (actual size: 2 by 4 inches).

Still another method of marking experimental animals is as follows: the generation is represented by toe amputation, beginning from right to left; the litter number is marked on the right ear; the individual number is marked on the left ear.

RECORDING DATA. The system used for making permanent the data pertinent to the maintenance of the colony has been refined during years of experience to the present well-functioning one. The method consists of keeping on a printed card all essential data concerning an individual rat. The male card (fig. 18) is kept on the cage in

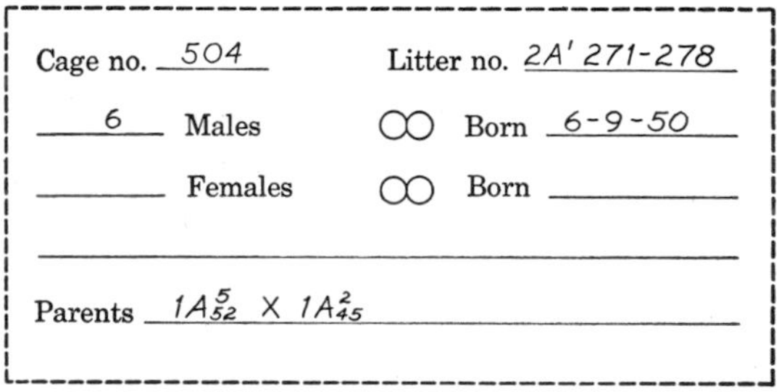

Cage no. _504_   Litter no. _2A' 271-278_

_____6_____ Males   ∞ Born _6-9-50_

_____ Females   ∞ Born _____

Parents _1A$_{52}^{5}$ X 1A$_{45}^{2}$_

FIG. 19. Card No. 4: Litter cage card (actual size: 2 by 4 inches).

which the animal is housed. It contains all the information needed on the male; it is the only individual record on him and is usually discarded when he dies. There is a corresponding female card (fig. 17) which functions in the same way, but the history of the female is kept on an additional breeder card (fig. 16) which is handled in the following manner:

1. The breeder card is made out for the female on the date when she is first mated. It is filed by number in a file provided for that purpose.

2. The mated females are examined biweekly for the placental sign indicating pregnancy. In the event of their being removed to the maternity room the card is brought up to date and refiled (by number) in a file for cards of pregnant females.

3. When the female gives birth, the history on her litter is recorded and the card is placed in a third file, which contains only the cards for nursing mothers. This time it is simply placed at the back of the file and is moved to the front in the three weeks allotted for nursing.

4. The breeder card continues to carry the record of the cumulative data on each female in this manner throughout her life. When

she is killed or dies, the card is placed permanently in a fourth file. This file is built up chronologically according to the date of death.

Although the system outlined above would suffice in yielding any necessary information, it has been found that in a large colony a daybook is an important adjunct for easy reference. The data concerning the animals are entered in the daybook in three ways. In one section the females in the colony are listed by number. Their dates of mating and the dates on which they give birth are recorded. In a

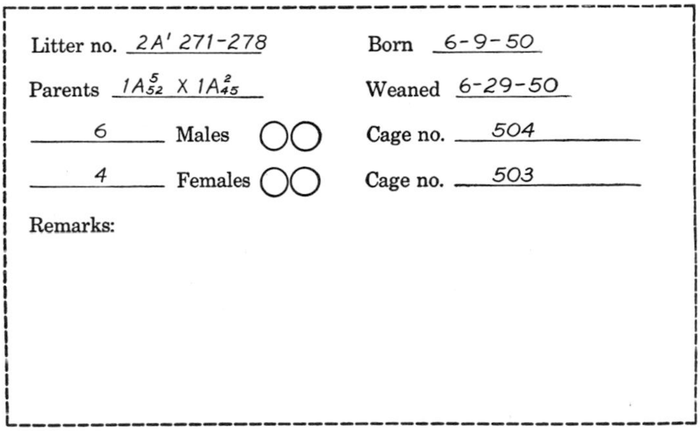

Litter no. _2A' 271-278_   Born _6-9-50_

Parents _1A$^5_{32}$ X 1A$^2_{45}$_   Weaned _6-29-50_

____6____ Males ⭘⭘   Cage no. ____504____

____4____ Females ⭘⭘   Cage no. ____503____

Remarks:

Fig. 20. Card No. 5: Litter card for office file (actual size: 3 by 5 inches).

second section the matings are recorded chronologically, and the outcome of the mating is listed as soon as known. This provides an easy way of discerning the non-productive females and eventually picking out non-fertile males. In the third section are copied all the data from the breeding card. This serves as a check in the effort to avoid error.

Figure 19 shows the form for a card of bright color which is used to indicate the presence of a litter in a cage. Figure 20 demonstrates the "new litter" card, filed in the office of the colony.

GENERAL CONSIDERATIONS

In endeavoring to develop a healthy, uniform colony of rats, important essentials to be considered are:

1. Selection of breeders.
2. Intelligent and regular care.

3. Temperature of colony room 72° F. with 5° variation, humidity 45 to 50 per cent. Temperature of the room for newborn rats should be 78° F. with a relative humidity of 50 per cent, to prevent ringtail.
4. Cleanliness.
5. Diet.
6. Exercise.

These factors will be considered in the course of the chapter.

Selection of Breeders

Fertility is at its maximum when rats are from 100 to 300 days of age. At this time the average number of young in a litter should be above the norm (six per litter). The mothers are usually able to nurse all young that they may cast.

The proper selection of breeding stock depends primarily upon experience, which enables one to recognize the better animals. Breeders should be "streamlined," not plump or fat. They should be free from all external physical defects, and the coat should have a uniform, smooth, and glossy appearance. It is important also to select breeders from families that have a record of high fertility.

Dr. Helen Dean King (1918a, b, c) has shown that brothers and sisters may be mated with no detrimental results, if breeders are selected carefully. Inbreeding of well-selected breeders tends to eliminate undesirable characteristics and to produce a homozygous strain which is most desirable for use in various types of experimental work. The King "A" strain of albino rats are in their 135th generation of brother and sister inbreeding.

About ten generations of brother and sister mating reduce heterozygosity to about 8 per cent. In comparison, breeding of brother and many half-sisters reduces heterozygosity to about 20 per cent in eight generations (Wright, 1922).

To reduce variability, according to Hutt (1945), the biologist should (1) learn the range of variation to be found in the species with which he works; (2) select the most suitable breed or variety for his purpose; (3) if possible, obtain foundation stock of a strain or strains already differentiated and proven suitable for that purpose, preferably inbred; and (4) maintain the colony without introducing new blood and with as much inbreeding as can be practiced without endangering vigor and reproduction.

Handling

The domesticated strains are easily handled and have practically forgotten how to bite in self-defense. The colony caretaker should be interested in the welfare of the animals, and the animals in return will become exceedingly friendly. Rats should not be lifted by the tail, as are mice, but are picked up gently. It is not advisable to use gloves at any time unless handling the wild, untamed forms.

Rats object strongly to noises, and certain types of noises will cause audiogenic seizures (Farris and Yeakel, 1943). The more frequently the animals are handled by the regular caretaker, the tamer they will be. It is advisable to discourage visitors to the colony, as they not only worry the animals, but also may carry in lice, mites, etc., unknowingly on their clothing.

Phenomena of Reproduction

Estrus Cycle. The estrus cycle of the rat occurs usually every 4 or 5 days, and each is characterized by periodic histological changes in the epithelium of the uterus and the vagina.

One of the earliest signs of estrus in the Wistar strain female albino rat is ear quivering, elicited by stroking the animal gently on the head or back. This reaction does not occur unless the rat is in heat. In view of the fact that the rat is nocturnal, this reaction does not occur under normal conditions until the very late afternoon or early evening. The copulatory response test (pelvic digital stimulation which produces lordosis) follows the ear-quivering test. Such a response is sufficient basis to assume that the animal is in heat. When in heat, the female is nervous and apprehensive, darts about, and on being touched braces herself (fig. 21A).

It has been shown by Wang (1923) that estrus cycle changes and activity cycles parallel each other. In our experience (Farris, 1941), the most accurate method for the determination of onset, regularity, and length of heat is by observing the amount of running activity by the turntable method during the estrus cycle. Estrus is evidenced by increased running activity, which begins at the onset. Other signs of heat are observed by examination of the vaginal smear and in the appearance of the superficial genitalia (dry, with radiating ridges and characteristic blue color rather than the moist, pink condition of diestrum).

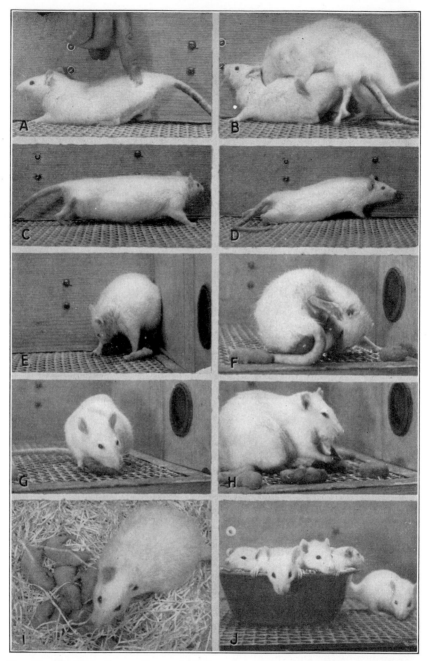

Fig. 21. Stages in the mating and birth cycle of the rat: A. Estrus. B. Mating. C and D. Labor, with visible extensions. E and F. Delivery. G. Removal of amniotic sheath. H. Placenta being eaten. I. The newborn in nest. J. Weaned young.

To secure female albino rats in estrus for experimental purposes, a colony room is darkened completely from 6:00 A.M. through 6:00 P.M. Four 100-watt lights are then turned on automatically and remain on until 6:00 A.M. By this method the rats display more regular estrus cycles, and as a rule come into heat about ½ to 3 hours after 6:00 A.M. on the fourth or fifth day.

Mating. Vigorous Wistar albino males attempt copulation fifteen to seventy times in 15 to 20 minutes, with one to two ejaculations, providing the female is in active heat. Note in figure 21B the typical female lordosis during normal copulation.

Period of Gestation. Careful observations have shown that the gestation period of the albino rat is usually between 21 and 23 days. In a series of precisely timed females, gestation occurred in 22 days, 14½ hours (range 10½ to 21 hours), in the primipara. A flow is evident usually 1½ to 4 hours before the first birth. Gestation may be extended a week or more, however, if a female is carrying a large number of young and suckling a previous litter (King, 1913).

Puberty, Ovulation, Menopause. Both males and females become sexually mature at about 50 to 60 days of age. In Wistar rats, the vaginas are open in about 60 per cent of the females between 35 and 50 days (range 15 to 67 days). According to Dr. Harry Shay (personal communication), the testes descend in 85 per cent of the males between 18 and 31 days (range 15 to 51 days).

Blandau and Soderwall (1941) reported that ovulation was complete by the eleventh hour after the beginning of heat in postparturitional (and postpartum) rats.

Ovulation has been shown to occur in mated rats (Farris, 1942) in about 8½ hours after the start of running activity on the day of heat in the mated females, and in about 10 hours in the unmated females.

The menopause occurs usually when the females are 15 to 18 months old. This is indicated in the activity records of the female, which show irregular cycles and frequently great bursts of activity during the changing period. Postmenopause is characterized by rather regular bursts of voluntary running (Farris, 1945) at 17- or 18-day intervals.

Parturition. The onset of parturition is characterized by rather regular extensions, with the interval shortening from 2 minutes to 15 seconds. During early labor the female stretches (fig. 21C) while walking about the cage. With the approach of parturition, the extensions become more severe (fig. 21D). The female rests on her abdomen, and the hind limbs extend off the cage floor (note left leg). Uterine ripples and contractures are frequently visible on the abdominal wall. During the extension and contracture period the cervix dilates, and the abdominal mass descends. Note the abdominal mass before descending, in figure 21C.

The time required for parturition depends largely upon the age, litter size, and physical condition of the female. It may take but a few minutes or be prolonged for many hours, the usual litter being born within 1 to 2 hours.

Birth of Young. The first fetus is usually expelled by head presentation, and the others follow either by breech or by head delivery. In most cases, the female licks the vulva preceding the delivery (failure of the rat to lick the vulva seems to be correlated with difficult delivery). During delivery she stands in a semi-crouched position (fig. 21E).

Birth Weight. The weight of newborn rats is influenced by several factors: sex, size of litter, physical condition of the mother, and her age. The average birth weight of albino rats in The Wistar Institute Colony is about 5.7 grams, with an average of 8.4 per litter.

Weight and other changes according to the approximate age of the mother are shown in table 1.

TABLE 1

THE EFFECT OF INCREASE OF AGE OF THE MOTHER ON THE NUMBER OF
MALES AND FEMALES BORNE

Age of Mother	Number of Litters	Number of Individuals	Males	Females	Average Weight of Newborn
120–180 days old	50	441	227	214	5.6
180–230 days old	50	488	255	233	5.7
230–290 days old	{ 50	499	212	287	5.9
	{ 50	490	236	254	5.9
290–365 days old	50	388	210	178	5.9

Placenta. The female helps with the expulsion of the placenta by pulling it out with the mouth (fig. 21F). She promptly eats it (fig. 21H), holding it and working it in her forepaws. The placenta, as Hammett has demonstrated, apparently furnishes a growth-promoting substance to the milk.

Cleaning Young. The mother licks the young rat and removes the amniotic sheath (fig. 21G). The mother very seldom pays any attention to the young until the placenta has been eaten.

The Nest. The mother finally assembles the young at the nest site (fig. 21I). She then cleans herself and permits nursing.

Sex Differences at Birth. In newborn rats the sexes can be distinguished by external characters (Jackson, 1912). The male is charac-

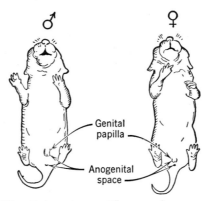

Fɪɢ. 22. External sex differences in young rats.

terized by a larger genital papilla and a greater distance between the anus and the genital papilla than in the female (fig. 22). Changes in the anogenital distance at various ages are shown in table 2.

TABLE 2

Aɴᴏɢᴇɴɪᴛᴀʟ Dɪꜱᴛᴀɴᴄᴇ ɪɴ Yᴏᴜɴɢ Aʟʙɪɴᴏ Rᴀᴛꜱ ᴏꜰ Vᴀʀɪᴏᴜꜱ Aɢᴇꜱ

	Average Anogenital Distance (millimeters)	
Age	Male	Female
Newborn	2.8	1.2
7 days	5.2	2.7
14 days	8.2	4.9
20 days	12.0	7.0
42–50 days	21.0	13.0

Females lack nipples at birth, but show six pairs when they are from 8 to 15 days old. Occasionally supernumerary nipples are present.

Age Determination. The rat is born hairless and blind, with closed ears, undeveloped limbs, and a short tail. Locomotion is effected by wriggling and paddling. The head is always searching, with quick response to olfactory and taste stimuli. The approximate age of a rat may be determined as follows:

	Days
Before birth	
Eyelids of fetus fuse	17
After birth	
Ears open	2½–3½
Incisors erupt	8–10
Young find way to mother	9–11
Eyes open	14–17
Hair obscures genitalia	16
Hair obscures nipples in females	16
First molar erupts	19
Second molar erupts	21
Third molar erupts	35
Testes descend	40
Vagina opens	72
Menopause occurs	450

Life Span. The life span of the albino rat is assumed to be about 3 years, which may be considered as equal to 90 years of human life (Donaldson, 1924).

Under the favorable climatic conditions of California, rats have lived over 40 months (Slonaker, 1912). In regions where there is much humidity with high temperature in summer and extreme cold in winter, few individuals live for more than 2 years.

Domestication. A strain of wild gray Norway rats has been maintained at The Wistar Institute for more than 30 years (King, 1939). These rats are more acutely sensitive than albinos and are easily startled by any quick, sharp sound or movement and by any new odor, human or otherwise. They can be made as docile as albino rats, however, if they are handled frequently while young.

When tamed, gray Norway rats can be handled with comparative ease, and thus are excellent for special research work. The strain is homozygous; the animals are large and vigorous and have proven to be good breeders. They are more resistant to infection than are other strains of Norway rats.

The breeding procedure outlined for the albino rat is satisfactory for gray rats, but because of their gnawing propensities, gray Norway rats should be kept in metal-lined cages.

Foster Mothers. In certain experimental problems treatment of the mother often causes a lack of milk supply essential for the young. Foster mothers may be substituted with marked success, particularly in the albino strain.   King (1939), with reference to the wild grays, says: "Many females would rear litters from other strains if they were barred from the nest until the young had acquired the nest odor. However, alien young were always killed if the mother had access to them as soon as they had replaced her own offspring."

Breeding Methods Used in The Wistar Institute Albino Rat Colony

1. Selected breeders are mated for the first time at between 100 and 120 days of age.   The males and females used are approximately the same age.

2. To obtain a large number of young, two males and six females are placed together in one cage.   The female rats remain there until found pregnant.

3. Each female rat is examined twice weekly for signs of pregnancy.   Pregnancy is indicated by the placental sign, which is a droplet of blood appearing at the vaginal opening, or by a marked increase in weight, approximately 25 to 30 grams, and by an enlargement of the abdomen which is evident at about the fourteenth day of gestation.

4. All female rats found pregnant are isolated in individual cages, and remain there until the litters are born and weaned.

5. Litters are weaned at 21 days of age.   Each member of a litter is then weighed, and those weighing less than 30 grams are discarded.

6. After the weaning of a litter, the mother may be given a rest period of about 2 weeks before being mated again.

7. No stock female is permitted to have more than seven litters. Male breeders are removed from the breeding cages at 12 to 14 months of age.

8. Accurate records are kept for each female and male born and maintained in the colony.

9. A record is made of all males and females that prove to be sterile, and of any abnormalities which may be detected.   Unless such animals are being used in special work, they are destroyed.

10. By regulating the number of animals per litter it is possible to obtain at weaning age an animal of more uniform size and weight. This is done by allowing a standard figure of eight young per litter, and discarding any excess number.   Small litters may be increased to the standard figure by adding the discarded members of another litter born on the same day.

11. For gray Norway rats and their mutants, it is preferable to breed pairs of litter mates.   The female should remain in the same

breeding cage until the litter is weaned, about 28 to 30 days. The male may be left in the cage without harmful results.

DIET

The finest laboratory rats are obtained when a variety of good-quality natural foods is used. Many combinations of foodstuffs would adequately supply their nutritional requirements. One of these combinations as listed below has been used at The Wistar Institute since 1905 in the colony of Dr. H. D. King.

It is known that diets of highly purified ingredients can be prepared in dried checker or cube form, and promote excellent growth and reproduction in rats. This type of diet is calculated on food values measured in terms of proteins, fats, carbohydrates, vitamins, minerals, and water. The cube method offers certain advantages in that it is economical and simple to feed and handle, does not spoil, and permits ease in cage cleaning. However, the diet becomes thoroughly monotonous, as evidenced by the rats' preferential interest in any other type of food, when available.

The dietary requirements of the rat in the form of nutritional essentials are listed in a chapter by Dr. R. H. McCoy in the monograph, *The rat in laboratory investigation* (Farris and Griffith, 1949).

The diet in use in the albino colony of The Wistar Institute is about as follows: An ample supply of dog chow checkers is always available. Milk is supplied four times per week, per individual rat. Approximately 6 oz. of evaporated milk diluted with equal portions of water is fed in individual dishes placed in the cages. Female rats that have weaned a litter are not given milk until they have stopped lactating. Green vegetables are made available twice weekly. Several pieces of raw cubed carrots with green tops are offered, or such items as raw lettuce, spinach, or other greens. Raw pig's liver in inch cubes is also fed twice weekly.

Fresh, clean water for drinking is available to the rats at all times.

In another Wistar Institute colony the following cooked diet is used: The basic daily diet consists of

        ½ pail soybeans
        ½ pail peas
        2 pails whole wheat
        1 scoop salt
       12 lb. fish
        1 scoop whole milk powder
        6 pails corn meal
       20 gal. boiling water

The pail employed is a standard 5-quart one, and the scoop has a capacity of 1 quart.

The food is cooked for about 1 hour, or until the water boils away and the food is fairly dry, in a 50-gal. steam cooker. It is spread on trays to cool. This food can also be put through a meat grinder to form macaroni-like strands. It is then placed on wire screens and dried in a hot room. This makes an excellent dry diet for feeding over week-end periods. It keeps indefinitely.

The basic diet formula is usually varied 3 days per week to include other items such as bone meal, powdered eggs, meat, liver, kidneys, heart, or tomatoes and salmon, in place of the fish.

The quantities given in the formula will feed 900 to 1000 rats. Each cage of 14 rats receives a 1 quart scoop of food daily, and each cage of 9 rats receives a 1-pint scoop of food daily, with double portions on Saturdays. The rats are fed once daily.

The well-supplied storeroom should contain whole wheat, corn meal, split peas, canned tomatoes and salmon, egg powder, whole milk powder, and salt.

As a precaution against the usual brown grain weevils or the dull gray flower moth which occur in dry diets, it is advised to avoid over one week's storage of food, particularly during the summer months. All foods should be stored in metal containers and under dry conditions. It has been customary to sterilize the food before use.

Loosli (1945), in an excellent chapter on feeding laboratory animals, shows a comparison of diets fed to different rat colonies. The following diet devised by Maynard (1930) is reported as being a ready-mixed stock diet for rats which gave satisfactory results during growth and reproduction. The mixture was composed of linseed oil meal 300 lb.; ground malted barley 200; wheat red dog flour 440; dried skimmed milk 300; oat flour 300; yellow corn meal 400; steamed bone meal 20; ground limestone 20; and salt 20. Extra vitamins A and D were supplied twice a week, but no green feed or other supplements were fed. The same mixture has served as a stock diet in the Cornell rat colony for the past 18 years and has been considered adequate.

DISEASES

There are certain diseases of the rat occurring in well-run colonies, and these conditions should be recognized. If they are detected early, specific treatment may be instituted and a cure effected. In standard practice, unless the diseased animal is of particular interest, it should be destroyed.

Pneumonia

Symptoms. The animal has unnatural, noisy, labored, and quickened breathing. The ears and nose are waxy looking. The animal sits usually with its back arched. The fur is roughened and dull white, rather than smooth and glossy. The animal is usually cold to the touch. The nostrils are reddish, often with a bloody discharge, and the eyes are frequently matted. As the rat refuses to eat, there is a loss of weight. Râles are usually heard and felt in the lungs. The rat dies usually from this disease.

For pathological and microscopical examination findings, the reader is referred to the chapter on spontaneous diseases, by Dr. Herbert Ratcliffe, in *The rat in laboratory investigation* (Farris and Griffith, 1949).

Treatment. Unless the animal is valuable experimentally, it is economical to destroy all sick rats. The newer treatments for human pneumonia prove satisfactory.

The incidence of respiratory infections greatly increases when the following conditions exist in an animal colony: (1) the air in the room is stagnant and heavily laden with fumes from the excreta; (2) the debris pans are not changed often enough; (3) the bedding needs changing; (4) the cages need cleaning; (5) the room temperature is unsatisfactory; (6) the atmosphere is dusty.

Middle Ear Disease

Symptoms. Slight tilting of the head to one side, and nose close to the floor, turning away from the midline.

Characteristics. The rat may stagger and lack coordination. When such a rat is held up by the tail, it will usually rotate rapidly.

At postmortem examination, the auditory bulla is usually the source of extensive grayish yellow pus.

Treatment. According to King, the young should be separated at birth, and placed with foster mothers free of this disease.

Paratyphoid

Paratyphoid disease of rats is caused by bacteria of the genus *Salmonella.* This condition is unusual in well-run colonies. The rats may become infected by food or by contact with the excrement of wild rats or mice.

Symptoms. The ears become pale and anemic looking. The eyes become chocolate brown rather than bright pink. The coat is usually rough and dull, and the rat is below standard. Crusts are visible around the nose, and diarrhea is common.

The postmortem and microscopic findings are described by Dr. H. L. Ratcliffe under "Spontaneous diseases of laboratory rats" (*The rat in laboratory investigation*, Farris and Griffith, 1949).

Control and Treatment. According to Slanetz (1948), a practical method of paratyphoid control has been developed in a feces culture method. In order to maintain a low incidence of *Salmonella* infection in breeders it is necessary to test young potential breeders before mating.

It is likely that the infection can be eliminated in rat colonies (Slanetz, 1946) by incorporating streptomycin in drinking water for 7 consecutive days in adequate dosage (100 or more units daily).

Ringtail

This is a condition in which newborn rats soon develop reddened tails that become swollen, and then constricted, with corrugated rings. The tail and occasionally the toes blacken and drop off. Ringtail occurs as a rule in late fall or during the winter season.

Treatment. This condition has been eliminated by providing live steam to the room and maintaining the humidity at about 50 per cent. The temperature in these rooms is kept at 76 to 78° F. Drafts should be avoided, and when the condition is prevalent it is recommended that the metal cages with open wire mesh be eliminated. A wooden cage with solid wooden walls and bottom and a screen top serves well to eliminate drafts immediately after the mother delivers her litter. After the litter is weaned successfully, the rats may be placed back in the usual cages.

Faulty Dentition

Signs and Symptoms. The incisor teeth become abnormally long, interfering with normal feeding, and often extending into the muscles of the lip or cheek.

Treatment. The excessive growth should be clipped with bone forceps.

PARASITES

The parasites of the albino rat are detrimental to growth and breeding and may modify experimental findings. For accurate research, animals with such infection are undesirable, although they apparently live and become adjusted to the infection with little disturbance.

For detailed information relative to parasites, the reader is referred to the fundamental chapter by D. H. Wenrich, "Protozoan parasites of the rat" (Farris and Griffith, 1949), and the fine chapter by H. L. Ratcliffe, "Metazoan parasites of the rat" (Farris and Griffith, 1949).

Included in chapter 16 by W. E. Dove are new methods for the control of laboratory pests and parasites. If the animal quarters are severely infested, it is suggested that reliable and well-established pest-control laboratories, with experienced entomologists, be contacted for opinions.

The parasites may be divided into two groups, the external (ectoparasites) and the internal (endoparasites).

Ectoparasites

The ectoparasites found on the albino rat that are of chief concern are the blood suckers, which include fleas, sucking lice, bedbugs, and certain mites which cause much irritation and scratching, with some loss of blood.

Fleas. According to Ratcliffe (Farris and Griffith, 1949), fleas of several genera—*Pulex, Xenopsylla, Ctenocephalides,* and *Ceratophyllus*—are common parasites of wild rats in various parts of the world. Ova are deposited in the floor of burrows and hatch into worm-like larvae which feed on organic matter. Each molts three times before spinning a cocoon in which pupation occurs and from which the adult emerges. Fleas are troublesome only in untidy colonies in which the bedding remains unchanged for weeks.

CONTROL. Thorough cleaning and insecticidal sprays are effective in controlling this condition. (See chapter 16.)

Lice. The sucking lice (Anoplura) have long, flat bodies, small heads with simple three- to five-jointed antenna, and a simple eye on each side of the head. They are about 1.4 mm. in length. There are three pairs of stout legs of a pearly white color. They are parasitic during their entire lives and deposit their eggs on the hairs of their host (Greenman and Duhring, 1941).

CONTROL AND TREATMENT. Cages should be cleaned thoroughly and disinfected with insecticidal sprays (Pyrenone Emulsion; see page 72). The animals could be powdered with derris, containing 1 per cent rotenone. This may be injurious to young rats. Another effective treatment is to incorporate 5 per cent DDT (dichloro-diphenyl-tri-chloroethane) in a mineral diluent.

Bedbugs (Cimex lecturalius). These are brown, flat, wingless animals, the adults measuring about 3 by 4 mm., the young being smaller. Ova are deposited in the crevices of cages, after taking blood meal. As bedbugs are common parasites of man, it is important that all new caretakers be checked before entrance, to prevent bringing the bugs into the colony. Bedbug infestation is quite uncommon in American colonies.

CONTROL. Pyrenone Emulsion is a harmless treatment and is effective for 3 weeks. A 5 per cent solution of DDT in odorless kerosene, sprayed on walls, cages, etc., until moist, is also a satisfactory treatment. One application lasts for several months (see page 72).

Mites (Acarina). Several varieties of mites may be found on the skin of the albino rat, as indicated by minute red and swollen areas, about 1 to 2 mm. in diameter, giving a generally inflamed appearance. The "scab-mite" penetrates the skin and deposits its eggs as it burrows in, on the edges of the ears and along the tail. The tail appears exceedingly rough and red and may even bleed.

CONTROL. The usual thorough cleaning of the colony and cages is essential. The lesions may be treated with a sulphur ointment, which serves well to cover the bare spots on the rat's tail or ear.

Endoparasites

Of the internal parasites, the tapeworms, or Cestoda, are most frequently observed in the albino rat. This is probably because the encysted larval tapeworm, *Cysticercus fasciolaris,* a form common to rats, is easily visible in the liver of the albino rat, appearing as small ivory-white cysts.

The Nematoda, or true roundworms, and the Acanthocephala, or thorn-headed worms, are also found in the albino rat.

The protozoa found in the intestinal tract of the albino rat are listed by Wenrich (Farris and Griffith, 1949).

Of the blood parasites, *Bartonella muris* is described as occurring in colonies other than that at The Wistar Institute. The rats are subject to an anemia especially marked after splenectomy, and associated with the presence of small coccoid or bacilloid bodies on or in the red blood cells (Hall, 1934). *Bartonella* infections have been controlled with neosalvarsan, 0.014 gm. per 100 gm. of rat, given intramuscularly (Duca, 1939).

Detailed Methods for Control and Treatment of Parasites

All equipment, colony rooms with emptied cages, bedding, feeding dishes, cooking utensils, etc., should be carefully sterilized, preferably in a steam sterilizer with 15 lb. pressure for 10 to 15 minutes. If this method is impractical, the cages should be immersed in a 5 per cent cresol solution for one-half an hour, then thoroughly rinsed in clean running water, and set aside to dry. All bedding, shavings, excelsior, and paper clippings should be sterilized in the autoclave, if possible.

The walls, floors, etc., of the animal rooms should be sprayed until thoroughly wet with a Pyrenone (piperonyl butoxide-pyrethrum) Emulsion, twice the first month, and once a month thereafter, until absolutely free of pests. This substance is advised because of the low toxicity. It is made as follows:

Pyrenone * (refined) R.E.	50–5	20	cc.
Atlox	1045 A	12.5 gm.	
Water to make total of 100 cc.		78.5 gm.	

Use above solution	1 part
Add water (at time of use)	4 parts

* The Pyrenone R.E. 50–5 is standardized to contain 50 gm. of piperonyl butoxide plus 5.0 gm. of pyrethrins per 100 cc. in a deodorized base oil vehicle (Ultrasene—Atlantic Refining Co., Philadelphia, Pa.). Pyrenone is obtainable from U. S. Industrial Chemicals, 60 East 42nd St., New York City. The Atlox is manufactured by the Atlas Powder Company, of Wilmington, Del.

When the cages are emptied and sterilized, the emulsion should be put on them (especially inside) with a small scrubbing brush, paying particular attention to wet down thoroughly the joints, cracks, and area around the water bottle. This coating is effective for 30 days. During this period the ectoparasites on the animals and in the cages are killed.

Severely infested rats should be anesthetized in a glass jar, and under close observation. After light anesthesia, the mites, which come to the surface, are brushed into the ether jar or into a cresol dish. The rats can be dipped into a solution of Pyrenone Emulsion. A larkspur solution can also be used, by painting the liquid over the entire animal with a toothbrush, until it is saturated with the solution. The larkspur solution is made up as follows: 1 part tincture of larkspur seed, 3 parts 90 per cent ethyl alcohol. All lice, including nits and mites, are killed at once. These should be brushed or picked off the rats with forceps. Nits attached to the hairs need not be removed, as the eggs are killed immediately by this treatment. If this drastic treatment is too time consuming, coating the cages with the Pyrenone Emulsion is very satisfactory, but not quite as certain under the circumstances.

Flies and Mosquitoes. It is advisable to maintain a colony free of these pests, and long-period control can be accomplished by spraying all the resting places, such as walls, cages, and lamps, with a solution of DDT in odorless kerosene (Ultrasene—Atlantic Refining Company). (See also chapter 16.)

Cockroaches. Two to three types of cockroaches appear in animal houses, particularly in heated quarters or where there is heavy moisture.

CONTROL. A cockroach spray should be applied twice the first month, and monthly thereafter, in order to destroy the eggs or young that hatch out. Pyrenone Emulsion, described previously, is recommended. Sodium fluoride has proven satisfactory, but a mixture of pyrethrum powder and sodium fluoride, 50–50, is a little faster and probably more effective.

A. J. Lehman, Chief of the Division of Pharmacology of the Federal Security Agency of the Food and Drug Administration, recommends a deodorized kerosene base containing 5 per cent DDT and 2 per cent chlordane. As both of these compounds are poisonous substances, care must be taken to make certain that neither rats nor their food become contaminated. A coarse spray is directed against the walls and floor. The animals and cages are preferably removed from the room during application. (See chapter 16.)

PRACTICAL SUGGESTIONS

1. Select the finest stock for your experiments, and inbreed until the strain is homozygous.

2. Quarantine all new stock until proven clean, before adding to any experimental section.

3. Prevent introduction of infectious diseases and parasites.

4. Proper housing

    A. A well-lighted, well-arranged colony of proper construction is essential.

    B. Control adequately temperature, humidity, and exchange of air.

    C. Provide proper cages, and do not overcrowd the animals.

    D. Have suitable bedding.

    E. Avoid drafts.

    F. Avoid unnecessary noises.

    G. Maintain animals on the finest diet possible.

    H. Supply fresh water.

5. Sanitary practices: Clean debris pans daily. Clean individual cages weekly, and thoroughly disinfect and change cages completely twice a year.

6. Personnel: Select, when possible, individuals who like animals and enjoy handling them, as time permits. A contented animal is the best research animal.

7. Visitors: Admit visitors who have not visited other colonies within 24 hours, thus avoiding chances of carrying insects or parasites on the clothing.

FIG. 23. Shipping box for eight adult rats.

SUGGESTIONS FOR SHIPPING OR IMPORTING RATS

Shipments of animals are made via Railroad Express or Air Express, and are sent express collect or prepaid. Shipments of animals are insured by the Railway Express but not by Air Express.

The shipping crate for rats (fig. 23) is 9 by 10 by 22 inches, with a 6½ by 22 inch screened opening that extends across the top and front of the crate, as shown in the illustration. The movable top slat permits insertion of the rat bedding, food, and water. A maximum of eight adults or twelve immatures is sent in one crate.

For a 3- to 4-day trip, the following food is recommended, to be placed in the crate: 2 heads of lettuce (good size), 3 lb. of potatoes, 2 apples or carrots. For an overnight trip, 3 lb. of potatoes is sufficient. The food should be washed and soaked before being placed in the crate. Water is furnished in a can that is fastened securely in a corner of the crate. The can may be refilled while en route by pouring water through the screen mesh. It is designed with a rim around the mouth, to prevent the water from sloshing out. A card with instructions to water en route is attached immediately above the can location on the outside of the crate.

The crate is clearly labeled in crayon pencil, and also with an addressed typed card. A statement as to the value of the shipment, the number of crates, and the number of rats in a crate must be included.

Procedure for Shipping Animals to Canada

1. Obtain "Shipper's Export Declaration" forms from Railway Express Agency. These forms are free of charge.

2. Fill in two copies of the above form. Although there is space for a notary's seal, notarization is unnecessary. As these forms are subject to revision, it would be wise to check with the Railway Express Agency or the Customs Office.

3. Custom commodity number must appear on this form and also on the shipping crate. To obtain this number, call the Department of Commerce.

4. The "Shipper's Export Declaration" form asks for a license number. This is unnecessary if the animals are to be used for experimental purposes.

5. These forms are sent along with the shipment.

6. It is unnecessary to have the animals examined.

7. For shipments to Canada, special MA invoice forms are necessary. These may be obtained from commercial stationery stores. Make out MA invoice forms in triplicate and mail along with the regular invoice. MA invoices need not be notarized.

8. Make out the usual "Collect" or "Prepaid" express slips.

Shipment of Animals to South America

1. Call airlines to find who will carry the shipment.

Line up itinerary from departure point to delivery point, and inform the consignee.

2. Air express shipments are sent prepaid and cannot be insured.

3. It is necessary to send along with the shipment a certification that the animals are in good health and free from parasites.

4. Shipment is also accompanied by four copies of "Shipper's Export Declaration" form, which is obtainable from the Railway Express or the Custom House.

5. Consular or commercial invoices are not required for Venezuela, but a special declaration is often required in quadruplicate. This form should be made up on letterheads and all copies must be signed. The following information must be given and in the following order:

1. Name and address of shipper.
2. Name and address of ultimate consignee.
3. Date of shipment.
4. Name of point (city) from which shipment is made.
5. Name of point (city) of destination.
6. Name of carrier.
7. Marks and numbers as appearing on each package. (When there is more than one package on one Airwaybill, each package must be numbered differently and consecutively.)
8. Declaration of contents.
9. Gross weight, followed by gross weight in kilograms. Kilograms must be carried to three decimal points. Weight must be written out in words.
10. Value. State value in words, not figures.
11. Name of country of origin.
12. "I hereby declare under oath that the information given in the foregoing document is exact and I will indemnify the carriers for any damages which may arise therefrom," followed by the name of shipper, with signature of person signing for shipper, and title of such person.

It is unnecessary that the foregoing declaration be notarized or certified in any manner other than by the signature of the shipper.

Importing Animals from Foreign Countries

1. Arrangements for the flight are made by the shipper, who notifies consignee as to airline company, flight number, and departure time.

2. Check with the local airport or the company carrying the shipment in regard to flight arrival.

3. Animals should be released by customs either at the port of entry or locally.

4. To have the animals released at the port of entry, obtain the services of a broker in that city through the air transit company.

5. To have the animals released by customs locally, it is necessary to secure a permit for customs release from:

Surgeon General's Office
Department of Public Health
Quarantine Division
Washington, D. C.

This permit must either be in written form or be sent from the Custom House through its teletype system.

6. Have the animals picked up immediately upon their arrival.

REFERENCES

Blandau, Richard J., and Arnold J. Soderwall. 1941. Postparturitional heat and the time of ovulation in the albino rat. Data on parturition. *Anat. Rec.,* vol. 81, p. 4.

Donaldson, Henry H. 1924. *The rat: data and reference tables,* 2nd ed. The Wistar Institute, Philadelphia.

Ducca, C. J. 1939. Studies on age resistance against trypanosome infections. II. The resistance of rats of different age groups to *Trypanosoma lewisi* and the blood response of rats infected with this parasite. *Amer. Jour. Hyg.,* vol. 29, pp. 25–32.

Farris, E. J. 1941. Apparatus for recording cyclical activity in the rat. *Anat. Rec.,* vol. 81, pp. 357–362.

——. 1942. Studies of reproduction in the albino rat. Abst., *Anat. Rec.,* vol. 84, p. 454.

——. 1945. The effect of menopause on the voluntary running activity in the albino rat. Abst., *Anat. Rec.,* vol. 91, p. 273.

Farris, E. J., and Gustav Engvall. 1939. Turntable for exercising rats. *Science,* vol. 90, p. 144.

Farris, E. J., and J. Q. Griffith. 1949. *The rat in laboratory investigation.* J. B. Lippincott, Philadelphia.

Farris, E. J., and Eleanor H. Yeakel. 1943. The susceptibility of albino and gray Norway rats to audiogenic seizures. *Jour. Comp. Psychol.,* vol. 35, pp. 73–80.

Greenman, M. J., and F. L. Duhring. 1941. *Breeding and care of the rat for research purposes.* 2nd ed. The Wistar Institute, Philadelphia.

Hall, P. R. 1934. The relation of the size of the infective dose to number of oöcysts eliminated, duration of infection and immunity in *Eimeria miyairii* Ohira infection of the white rat. *Iowa State Coll. Jour. Science,* Vol. 9, pp. 115–124.

Hutt, Frederick B. 1945. Genetic purity in animal colonies. Chapter in Animal colony maintenance. *Annals N. Y. Acad. Sciences,* vol. XLVI, Art. 1, pp. 5–21.

Jackson, C. M. 1912. On the recognition of sex through external characters in the young rat. *Biol. Bull.,* vol. 23, pp. 171–174.

King, Helen D. 1913. Some anomalies in the gestation of the albino rat (*Mus norvegicus albinus*). *Biol. Bull.,* vol. 24, pp. 377–391.

——. 1918a. Studies on inbreeding. I. The effects of inbreeding on the growth and variability in the body weight of the albino rat. *Jour. Exp. Zool.*, vol. 26, pp. 1–54.

King, Helen D. 1918b. Studies on inbreeding. II. The effects of inbreeding on the fertility and on the constitutional vigor of the albino rat. *Jour. Exp. Zool.*, vol. 26, pp. 335–378.

——. 1918c. Studies on inbreeding. III. The effects of inbreeding, with selection, on the sex ratio of the albino rat. *Jour. Exp. Zool.*, vol. 27, pp. 1–35.

——. 1939. Life processes in gray Norway rats during 14 years in captivity. *Amer. Anat. Mem.* 17. The Wistar Institute, Philadelphia.

Loosli, J. K. 1945. Feeding laboratory animals. Chapter in Animal colony maintenance. *Annals N. Y. Acad. Sciences*, vol. 46, Art. 1, pp. 45–75.

Maynard, L. A. 1930. A diet for stock rats. *Science*, vol. 71, pp. 192–193.

Slanetz, Charles A. 1946. Control of *Salmonella* infections in mice by streptomycin. *Proc. Soc. Exp. Biol. Med.*, vol. 62, p. 248.

——. 1948. The control of *Salmonella* infections in colonies of mice. *Jour. Bact.*, vol. 56, no. 6.

Slonaker, J. R. 1912. The normal activity of the albino rat from birth to natural death; its rate of growth and the duration of life. *Jour. Animal Behavior*, vol. 2, pp. 20–42.

Wang, G. H. 1923. The relation between spontaneous activity and oestrous cycle in the white rat. *Comp. Psychol. Mono.*, vol. 2, no. 6.

Wells, W. F. 1940. Bactericidal irradiation of air. *Jour. Franklin Institute*, vol. 229, p. 347.

Wright, S. 1922. The effects of inbreeding and cross-breeding on guinea-pigs. III. Crosses between highly inbred families. *U. S. Dept. Agr. Tech. Bull.* 1121.

ACKNOWLEDGMENTS

The author wishes to thank George Wahmann & Company, Baltimore, Md., for supplying the following illustrations: figures 5, 7, 9, 12, and 13. Figures 2, 6, 10, 21, and 22 are reprinted from *The rat in laboratory investigation*, E. J. Farris and J. Q. Griffith, Jr. (editors), by the kind permission of J. B. Lippincott Company, Philadelphia.

3. THE CARE OF EXPERIMENTAL MICE

LEONELL C. STRONG

Yale University School of Medicine
New Haven, Connecticut

INTRODUCTION

Where experiments require large numbers of individuals or rapid succession of generations, mice are economical. They are the smallest and fastest-breeding mammals available, since they average only about an ounce in weight, and breed in the third or fourth month of life, with litters of up to a dozen or more young, practically every month.

Wild mice are generally unsuitable for the laboratory, however; they are incurably vicious and prone to escape. Even the young of wild mice reared in captivity soon reveal their heritage.

Domesticated strains are quite tractable by comparison and may even become pets. There are numerous distinct strains, characterized by various colorations, behavior traits, anatomical and physiological peculiarities, and size. Proper choice of strains for an experiment is often a crucial point and should be given careful consideration.

"Fancy mice" are obtainable from pet stores and various dealers advertising in such magazines as *Small Stock Journal* and *All-Pets Magazine*. Mostly they may be classified as "white mice" of unknown origin; other colors are also available, especially orange or "red," and "Dutch" or piebald, and waltzers. Indiscriminate use of such types in the laboratory may give erratic results. "Mouse units" and minimum lethal doses may vary widely.

For standardized genetic material it is best to select among a number of highly inbred strains available from larger laboratories. Such strains, and first-generation hybrids between them, offer a wide range of types, each of extreme uniformity, as far as heredity is concerned.

Two factors appear to have been involved in the origin of the genetic variety encountered in mice: a long period of domestication, with selection of mutants; and species hybridization. The house mouse

79

of Europe differs considerably from the Asiatic species, but crossing can readily occur. The small Japanese waltzer, for example, seems to be chiefly derived from the Oriental species.

The production of the various inbred strains of mice has necessitated many changes in the care of such animals. Before inbreeding was established, almost any selection of good animal food was sufficient to supply the nutritional requirements of mice. One well-known experiment was performed on mice which had received a diet of bread and water! Also the hybrid vigor that was maintained by pen or uncontrolled matings was sufficient to carry on any particular group of experimental mice. However, from time to time, infectious diseases, such as paratyphoid, would be introduced into a colony and the entire group of experimental mice would be wiped out almost overnight. This discouraged many investigators from setting up long-time experiments.

The use of inbred mice has controlled, in proper hands, excessive biological variability but has unquestionably reduced vigor. Many of these animals, even though they may be physiologically weak or delicate, have other admirable characteristics which make possible quantitative research in the various biological sciences. It therefore becomes necessary to protect mice of a highly inbred strain by the application of rigorous procedures for their care. Not all inbred mice are physiologically weak; some of them have high fertility, early sexual maturity, and, having been produced by selection, great longevity. Other inbred mice, such as the JK, require a higher temperature than do mice of other strains; mice of the I and F strains require for their proper maintenance quite a different diet from that of mice of other strains. Many other differences between the strains could also be indicated. Not all laboratories are equipped for the application of all these environmental variables for the maintenance and continuation of all these various inbred mice. It is desirable, therefore, to discuss the general principles involved in the care of a colony of mice and leave it up to the investigator, first, to incorporate variations suitable for the needs of the colony, or, second, to use only those strains of inbred mice that do well under his particular laboratory conditions.

CARE AND HOUSING

In the first place, it is extremely important to maintain a colony of mice by themselves and not in contact with any other species of animals. In the second place, no new animals should be brought into the laboratory from any outside source. If these two ideal conditions cannot be maintained, then the new mice from an outside source should be isolated from the main colony for an adequate period of strict quarantine before being introduced into the laboratory. Most of the difficulties in the maintenance of inbred mice come from the violation of the above two rules. Even though the newly introduced mice do not show any symptoms of disease, they may carry strains of infectious organisms to which they themselves may be partially resistant, but which would introduce diseases into the colony to which the laboratory strains have no or very little resistance. It is the experience of the author that epidemics frequently occur as a result of the introduction of foreign mice.

The inbred mice do well in a laboratory in which the temperature and humidity are both controlled, although these conditions are not absolutely necessary except in those cases where a pedigreed inbred strain has been maintained for many years or generations and its possible loss would be a severe blow to scientific research. It would be too costly to experiment with uncontrolled conditions that could be adequately controlled. Mice do well at any reasonable temperature from 70 to 80° F. The author is acquainted with one laboratory in which the temperature was maintained at 85° F. The important feature is to maintain an even temperature, avoiding radical changes from time to time. In the laboratory of the author the physical equipment for temperature control is not adequate to maintain an absolutely constant temperature throughout the year. Therefore the temperature during the winter is maintained at 72 to 74° F., which is probably a little high for ideal conditions, so as to average these temperatures with the higher ones maintained during the hot summer months, when the apparatus used is adequate only to reduce the temperature of summer heat to 76 to 78° F. The temperature at which mice should be maintained depends upon the type of cage in which they are kept, an item that will be discussed later. If wooden boxes with open-wire-mesh covers are used, then the temperature of the room may be kept lower than when an open-wire-mesh or solid metal cage is used. Another reason for keeping the laboratory moderately warm is convenience to the investigator. If the mice are comfortable, they will not cover the young with nesting material, and it is easier for the attendant to see them. The humidity should be kept between 50 and 60 per cent. This is a very difficult item to control, especially during the dry winter months in which the humidity of the incoming air must be increased by adequate equipment. If the humidity becomes too low, more difficulty with pneumonia or hemorrhagic lesions of the lungs will ensue.

Perhaps the best box for the keeping of mice is a wooden one with an open-wire-mesh cover as illustrated in figure 1. The boxes are made of ½-inch planed knot-free white pine measuring 6 by 12 inches

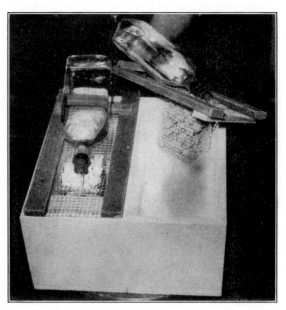

Fig. 1. A wooden box in which mice are kept, showing wire-mesh cover containing a food hopper and a water bottle.

(four pieces) and 6 by 11 inches (three pieces). The cut lumber can be obtained from any lumber yard. When assembled, the box with an open top consists of two equal compartments or cages. They should be painted with two coats of a paint not containing lead or any other material that may be detrimental to the health of the animal. One coat should be a good interior flat paint; the second coat a gloss supreme. The author uses Society White Paint (one coat satin elite, one coat gloss supreme) and has boxes which have been in constant use for 15 years with only an occasional repainting. Some laboratories prefer to dip boxes into either shellac or varnish or some combination of the two. Other laboratories prefer to soak wooden boxes in creosote. In the cover there is inserted a small open-wire hopper in which pelleted mouse food is placed. The water bottle is laid on top of the cover.

A bent glass drinking tube, commonly referred to as a goose-neck, is inserted into the water bottle and held in place by a suitably bored rubber stopper. The flow of water is controlled by a narrowing of the end of the goose-neck which is inserted through the top cover of the mouse box. The tip of the drinking tube in the mouse box should be near the hopper so that the mice have access to the water at the same time that they are eating the pelleted food through the open-mesh hopper. If properly adjusted, there should be no dripping of the water on the bottom of the cage unless the bottle is jiggled.

An ideal breeding box should contain one to three females and one male. Mice under investigation may be kept five or six in a cage. Any more adult animals in a box is conducive to overcrowding, which probably influences the experimental data being obtained. For example, spontaneous breast tumors and spontaneous lesions involving the mucus-secreting cells just anterior to the pylorus can be suppressed or eliminated merely by keeping ten or more adult animals in a box. Breeding mice may be kept together in the same breeding box, or the pregnant females may be isolated into individual boxes.

Boxes in which breeding mice are kept should be changed once a week, while boxes containing adult experimental mice should be changed twice a week. As mice grow older they soil the boxes more, and consequently their boxes should be changed more frequently. All boxes can be lightly scraped (with a rubber spatula preferably) and then soaked for a short time in warm water to which has been added a small amount of creosote or some other standard disinfecting solution. The boxes are then lightly scrubbed with lukewarm soapy water, after which they are rinsed in clear lukewarm water and piled upside down in a warm dry atmosphere for at least 24 hours or until they are thoroughly dry.

Clean sawdust or wood shavings (kiln dried) are then placed in each box to the depth of approximately ½ inch. A small amount of cotton batting is placed upon the sawdust. Some investigators prefer not to use cotton, but it is invariably the rule that when pregnant female mice are preparing to have young they will make a dome-shaped nest of the cotton and will continue to keep their young in it for several days. When cotton is present, many adult mice will also stay under it from time to time. The cotton may also serve as a protection against a radical reduction of temperature when anything goes wrong with the heating apparatus. Grüneberg (1943) advocates the use of hay instead of sawdust, shavings, and cotton. He notes, however, that part of the hay is eaten. Another author has found small nesting compartments of cardboard (ice-cream containers) helpful in breeding, especially for delicate types such as waltzers.

It is commonly stated that paratyphoid inevitably results if mice are kept in wooden boxes. This is a misconception. The author is aware of one colony of mice where the cages were never sterilized. The colony had been in existence for several years, and it was one of the healthiest colonies of mice that he has ever seen. Another colony in which the boxes had actually been soaked in creosote has shown many animals harboring paratyphoid. For emphasis, therefore, it should be kept in mind that a flare-up of paratyphoid is frequently due to the introduction of new mice into a colony from some outside source without adequate protection by strict quarantine. Another cause of paratyphoid may be found in the possibility that some mice in the laboratory have been neglected. This happens occasionally when students leave on vacations without making adequate provision for the care of their animals.

A general plan for a laboratory using a colony of mice kept in wooden boxes on suitable racks made of angle iron is given in figure 2.

Some investigators prefer solid metal cages. The main reason given for the use of these cages is that they can be thoroughly sterilized by steam with or without pressure. One difficulty of keeping mice in solid metal cages is that they do not raise their young well unless the temperature of the room is kept considerably higher than it is when wooden boxes are used. Another disadvantage of solid metal is that the acid urine of mice will corrode the metal and thus produce holes through the bottom of the box. Properly handled, a wooden box has a longer life span than a solid metal one.

It is standard practice among modern nutritionists to maintain experimental animals in screen-bottom cages. In theory this is to prevent coprophagy, a practice which furnishes the animal with an undesirable source of numerous nutritional factors. Cages have been devised which greatly inhibit or completely abolish the animal's movements and thus make it difficult or impossible for the mouse to have access to its urine and feces. So far it has not been possible to maintain animals in good health under such drastic conditions. Whether this is due to the presence of unidentified factors in urine or feces or whether it is lack of exercise has not been determined.

The usual type of screen-bottom cage (fig. 3) used by nutritional investigators is easy to keep clean and has a relatively long life. It is possible to maintain mice in such cages in excellent health, not only during their adult lives, but also during the period of most active growth. By introducing a piece of wire screen of somewhat smaller mesh than the bottom of the cage and placing on this a small quantity of shredded paper, it is possible to maintain lactating females and their litters (for a precise description see Fenton and Cowgill, 1947). Temperatures between 73 and 78° F. have been successfully used

Fig. 2. The arrangement of a laboratory equipped with metal racks, and the wooden boxes for the mice.

with this type of caging setup. For nutritional experiments it is usually necessary to maintain a single animal in a cage. Even so, it is essential to clean the cages once each week and preferably more frequently. A great problem which one encounters with certain types of screen-bottom cages is to provide a constant supply of water. The investigator is confronted with the necessity of compromising between a narrow-bore delivery tube which passes easily through the wire-screen front of the cage and a wide-bore tube which avoids the accumulation of a large air bubble in the tip of the tube. A little trial and error will determine the diameter of tubing and the angle at which such tubing should be bent for best results. If available, a dish-washing machine such as is used in restaurants is of considerable help in cleaning a large number of cages.

Some colonies of mice are maintained in glass jars on top of which are placed open-wire covers containing an open-wire hopper and water

FIG. 3. On the left is a metal rack equipped with suspended metal cages with open-mesh bottoms, with trays containing sawdust or rolled paper. On the right is a balance suitable for the weighing of mice. This system is more suitable for work on nutrition.

bottle. The advantages here are, first, that the glass jar can be sterilized, and, second, that every experimental animal in the laboratory can be seen and examined by the investigator or an animal caretaker in a very short time. A disadvantage is the high breakage rate, particularly when handled by animal caretakers. In 1940 the author tested out a mouse cage made of plastic. This cage had the advantage of glass in that all the animals could be seen within a very short time. It also had the advantage over glass of not being breakable. It had the disadvantage that it could not be sterilized. Some of the newer plastics can now be sterilized, and they are being commercialized. Whether they will be successful for the care of mice over a period of time has not as yet been answered. The initial cost to most investigators is prohibitive.

In practically all the systems the open-mesh hopper for the use of pelleted food has replaced the older method of dropping the mouse food on the bottom of the cage. When food is placed on the bottom of the cage, a certain amount of soiling, especially by urine, ensues. When the hopper is used, there is less contamination with feces and urine, but contamination is not completely eliminated. Certain types of pelleted food inevitably get matted together with urine. When the hoppers are periodically changed, this matted food should always be discarded. The hoppers are then soaked in hot water and dried for some time before being put back into use. The contamination of food in the hopper can be partly eliminated by a hopper having solid metal sides with only the bottom or lower one inch made of open-wire mesh.

NUTRITION

There are numerous pelleted foods for mice on the market. It is very probable that none of them is completely adequate for a colony of mice over a long period of time. The main difficulty is that very few manufacturers will guarantee not to change the ingredients from time to time. The reason given for this is the fact that the manufacturer tries to keep his product at a uniform price to the consumer. When one ingredient becomes scarce and increases in value, some substitute is made. These changes are not always revealed to the consumer even though the use of the product probably enters into the results obtained. Such changes may even interfere with the application of the strict rules of quantitative research in biology, where all variables in the experimental setup should be controlled. This being the case, it is probably wise to supplement any pelleted food with other food materials.

The diet for mice that has been used by the author for many years will be discussed, and then variations from it will be considered. Nurishmix, a product of the Pratt Food Company, Philadelphia, Pa., is placed in the hopper twice a week. A supplement of a mixture of wheat, oats, sunflower seeds, and calf meal pellets is given to the mice (a small handful at least once a week). Mice in the breeding colony also receive a small piece (quarter slice) of enriched bread, soaked in fortified milk to which has been added medicinal cod-liver oil (10 cc.

to a quart). The sliced bread is quartered on one day and kept over-night. The quart of milk is shaken and partly poured over the bread. Then the cod-liver oil is added to the remaining milk, shaken, and poured over the bread. Periodically fresh lettuce thoroughly washed in cold running water is given to the mice. However, the use of lettuce is usually not necessary. Under this regime there is little likelihood of a dietary deficiency. The entire regime is probably not necessary, but, with the maintenance of thirteen highly inbred strains and the development of many new ones, it is unquestionably desirable. It is a matter of record that since the author has built his present laboratory not a single strain of inbred mice has ever been lost. Grüneberg (1943) has avoided the use of supplements of any kind, with excellent results. MacDowell's colony (see Laanes, 1936) uses moist food, given in daily allotments, without any water bottle. The Hagedoorns' (1939) method is similar. The chief objection to such food and feed-ing is extra labor, but it does save space and no water containers are needed.

Modifications of the dietary regime for mice are innumerable. Many laboratories are using Laboratory Chow, Dog Chow, or Fox Chow, all products of the Purina Company of St. Louis, Mo. These are being used with or without supplements. One supplement that the author is acquainted with is a mixture of calf meal powder, wheat-germ meal, and dried yeast held together by a small amount of cod-liver oil. However, when this supplement was replaced by the mixed grain mentioned above, the mice did considerably better. Most mice will do well on any one of a large number of pelleted mouse foods on a short time basis. However, it is well to bear in mind that if the investigator desires his experimental mice to live two years or even more, then a careful checkup of the diet should be made before the investigation is started.

It has been possible in recent years to design highly purified diets (in an approach to the completely synthetic diet) which permit excel-lent growth and even moderate success in supporting reproduction and lactation in highly inbred strains of mice. The use of such purified diets is of primary interest to the nutritionist and to the biologist concerned. A purified ration which is as nearly complete as one can prepare today was reported by Fenton and Cowgill (1947). It has also been possible on the basis of newer nutritional information to prepare stock rations which improve the nutritional status of a mouse colony above that obtainable with most commercially available feeds. Maintenance on such improved rations can considerably increase the weight of the offspring at the time of weaning. Under excellent conditions a 21-day-old mouse should weigh well above 10 grams.

For nutritional research it is probably best to place the diet in a container capable of holding rather small amounts of food designed to eliminate spillage. One device which accomplishes this quite well is the glass ointment jar of 1- or 2-ounce capacity. A hole of ¾-inch diameter is cut into the jar top, and a circular piece of wire screen (4-mesh) is placed on top of the food in the jar so as to be able to sink freely toward the bottom as the food is consumed. The entire assembly is placed in a small glass beaker. Almost all the food spilled by the animal during feeding will fall into the beaker and will thus not be lost from the cage. The jar assembly without the beaker is weighed before being placed in the cage. It is weighed again together with the spilled food in the beaker 24 hours later, at which time a weighed jar with fresh food is supplied to the animal. By this admittedly tedious method fairly accurate data on food consumption can be obtained. Much of the published work up to 1944 on the nutritional requirements of the mouse has been reviewed by Morris (1944).

REPRODUCTION

(With Section on Marking and Records)

The young mice may be weaned at 21 days of age. They actually do better and in fact will live longer if they nurse their mothers from 28 to 30 days. If a pregnant female is separated from the male, a second litter cannot be expected for at least 42 days, since the estrus period that the mouse has shortly after the birth of a litter is not repeated until after her young are weaned. When the sexes are kept together, a second pregnancy may be started within 24 hours of the birth of the first litter. The implantation of the eggs for the second litter is somewhat delayed, so that the second litter, instead of being born in 19 to 20 days, may take up to 26 days. One other advantage of keeping two or three females with one male is the fact that the mice will suckle each other's young, so that if one female is not a good mother her young will be nursed by the other mothers in the same cage. However, when a female is separated from the male before the birth of her offspring, she will usually raise more young per litter than when kept with the male and other breeding females. Grüneberg (1943) unqualifiedly recommends keeping a pair of breeders (male and female) together indefinitely. He states that up to 135 young from a female have been produced this way. The Jackson Laboratory, however, routinely removes pregnant females to maternity wards. Such moving may result in errors of identification sometimes. With a continuation of a pedigreed brother-to-sister mating, it is also easier to keep track of the offspring from a single female, since when two or three females are kept in the same box, they may produce young within a few hours of each other.

Marking and Records. At weaning age the mice should be individually marked and recorded in the ledger. It should be noted that males can be kept together if started young, except JK and perhaps other strains that manifest viciousness and cannibalistic tendencies. Several systems of individual markings have been used, such as (1) touching picric acid or some other colored solution to various parts of the mouse, (2) the amputation of one or more toes, or (3) the use of punched holes and notches on the ears, etc. The use of colored solutions applies only to albino or light-colored mice. This system also has the disadvantage that the distinguishing spots are lost at each molt of the hair and must be, therefore, periodically renewed. Toe amputation and ear punches are permanent. The toe amputation system is restricted by the fact that many mothers are cannibalistic and will amputate tails, toes, and feet of their young during the nursing period. The ear-punch system is by far the best so far devised. Variations of this system are used in different laboratories.

One type that is frequently used was developed as a modification of a system devised by Professor W. E. Castle of Harvard University many years ago. This system recognizes three positions on the ear, the front, the middle, and the back. Numbers 1, 2, and 3 are produced by punching holes through the ear in these three positions with a poultry toe clip which may be purchased at any poultry supply house; numbers 4, 5, and 6 are produced by punching notches in one of the same three positions. The notches may be produced by the same instrument, care being taken that only one-half of the punching point is placed on the edge of the ear. Scissors may also serve. Numbers 7, 8, and 9 were made originally by Professor Castle with a peculiar combination of two notches. The author could never remember these combinations and consequently changed the original system to the following: 7, two notches in the front; 8, two notches in the middle; and 9, two notches in the back of the ear. Tens are punches on the left ear; units on the right ear. Thus, by a combination of punches and notches, the numbers from 1 to 99 can be indicated by a fairly permanent system. Ear punches are subject to error from mutilation.

Where males are together, fighting may lacerate the ears. Skin disease, parasites, and scratching may open punches into notches, and notches into bays or ribbons. However, with a careful check of the serial number of mice, especially on the cards for individual mice, these difficulties are not serious. In fact, some mice that have been ear punched at weaning age will continue to show clear or distinctive ear marks during their entire life span.

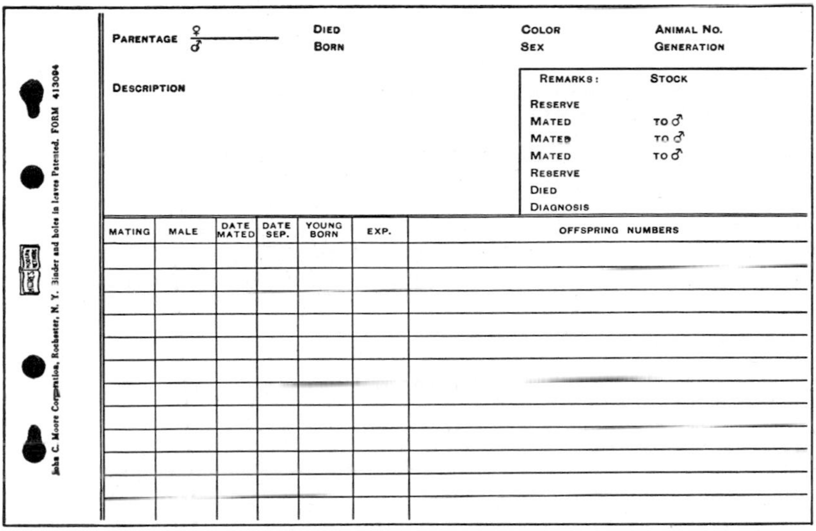

John C. Moore Corporation, Rochester, N. Y. Binder and holes in leaves Patented. FORM 413094

PARENTAGE ♀ ♂ DIED BORN COLOR SEX ANIMAL NO. GENERATION

DESCRIPTION

REMARKS: STOCK
RESERVE
MATED TO ♂
MATED TO ♂
MATED TO ♂
RESERVE
DIED
DIAGNOSIS

MATING	MALE	DATE MATED	DATE SEP.	YOUNG BORN	EXP.	OFFSPRING NUMBERS

FIG. 4. A breeding sheet containing records for mice.

Records on mice vary considerably, according to their purpose. For short-term experiments, such as in nutrition or carcinogenesis, simple lists or card indices may suffice. Much more elaborate systems are needed for long-term genetic work. An ideal arrangement is as follows:

At weaning age each mouse should receive a distinguishing ear mark which is recorded in a permanent ledger, providing this is not done at birth. An individual card with the serial number of the mouse is also made out, on which should be recorded all pertinent data obtained throughout the life history of the animal. Female mice which are used for breeders may also be given a breeding sheet. One of these is shown in figure 4. A third set of records may also be kept for each mouse. Figures 5 and 6 show one card containing data of modifications made upon an original card that was designed by W. S. Murray. The data here are such that they may be transferred to a card-sorting machine with keyed pertinent data. With this system, discussed above, all records are in triplicate, and the chances of losing data on any given individual, even though thousands may be raised in a single year, is very slight.

Several systems of giving serial numbers to mice are in use. Some investigators record the birth of a new litter together with serial numbers in a permanent ledger book. Others give serial numbers at

Date of Birth		Stock		Exp.		Female	
Date of Tumor					Pedigree		
Date of Death				Induced Tumor		Male	
Age at 1st litter							
Days between 1st and 2nd							
2nd and 3rd							
3rd and 4th							
4th and 5th							
5th and 6th							
6th and 7th				Female	Young	Male	Young
7th and 8th							
Age when Tumor appeared							
Position							
Appearance in relation to last litter							
Duration of Tumor in days							
Age at death							
Number of young born							
Number of young born dead							
Number died before weaning							
Number of males							
Number of females							
Number of sex undetermined							

FIG. 5. Records for each mouse can be assembled as shown. Variations from this system can be used according to the needs of the investigator.

weaning age, while in another laboratory serial numbers were not given to the mice until they died. With the third system there is a considerable likelihood of errors creeping into the investigation.

It is generally accepted that for quantitative research in biology, when mice are used and reproducible results are desirable, one or more of the recognized inbred strains now in existence should be selected. Most investigators do not have the time to use this elaborate procedure of recording pertinent genetic data, and in fact for many investigators this is unnecessary. The rate of appearance of spontaneous mutations is somewhere in the neighborhood of 1 in 26,000 mice. When an investigator, therefore, is doing a small investigation over a short period of time, the chances of this mutation process taking place, which may disturb the results he obtains, are not very high. In fact, very few investigators would use these large numbers in a two- or three-year program; consequently, it is only necessary for a few laboratories to continue the laborious process of the continuation of the already established inbred strains and the development of new strains showing new biological characteristics. With time, these pedigreed inbred strains become extremely valuable and if confined to one laboratory may be irreplaceably lost by some unpredictable catastrophe. It is,

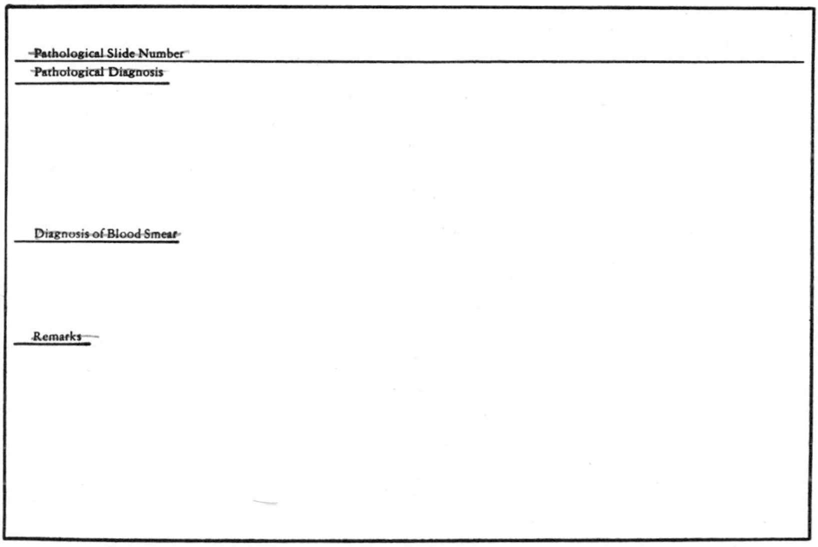
Pathological Slide Number
Pathological Diagnosis
Diagnosis of Blood Smear
Remarks

FIG. 6. This is the reverse side of figure 5, showing further data on each animal.

therefore, desirable to have these strains continued in several places, perhaps under the jurisdiction of some federal agency or cooperative arrangement.

Any investigator, therefore, who desires to obtain mice from one or more of these strains should be able to do so at any time. He should be in a position to continue the colony thus obtained for a period of two or three years. During this time, if possible, but not necessarily, the colony should be continued by brother-to-sister matings. Pedigreed records of individual matings need not be kept. Above all else, if the line is to be continued for any lengthy period it should be done by the non-introduction of mice of any other source. This simplification of the time-consuming process of a pedigreed brother-to-sister mating should not increase biological variability too much, and at the same time many more animals for an immediate investigation could be produced. If this colony, however, is separated too long from the original or ancestral inbred strain, then mutations will eventually produce divergent sublines. For this reason the entire colony should be discontinued, and a new supply from the original pedigreed stock should be obtained. For the sake of using highly inbred mice, too many sublines of any inbred strain should not be continued. Nor should any of them be continued on a too-large basis. One well-known strain exists in thirty sublines, each one of which is distinct and has biological characteristics unique to itself. This is very confusing, even to a geneticist, and it must be even more so to other investigators not trained in this science.

A practice that deserves consideration is the production of first-generation (F_1) hybrids between selected inbred strains. These frequently are far superior to the parental strains in health, fecundity, and longevity. Moreover, the animals of any particular hybrid type show as great uniformity as the inbreds do. Thus the experimenter may contrast "delicate" types with robust ones, or employ hybrid females as foster mothers, and so on. From the genetic standpoint only the first cross is uniform, so that subsequent generations, outside of the science of genetics, are not generally useful and in many instances are quite misleading.

As a general rule it is better to experiment on mice in the laboratory in which they were born and raised. However, when an outside supply of mice is necessary, then care should be employed in permitting the animals to become accustomed to their new environment. As a result of shipment by express, especially for long distances, mice undergo unpredictable physiological changes. For example, the colony of mice of the author has been transported *in toto* upon several occasions. Each time the physiological characteristics of some of the mice have changed. These unexplained changes influence the incidence of spontaneous tumors. When a well-known stock of mice showing spontaneous tumors of mammary origin was shipped to London, the incidence of spontaneous tumors was completely wiped out. However, after several generations of uniform laboratory conditions in London the incidence of spontaneous tumors gradually increased. Recently the Strong CBA stock was shipped from New Haven to Salt Lake City by Dr. T. Dougherty. Upon arrival of the mice in Utah, the lymphocyte count had diminished by 3000 per cubic centimeter. Within a few months in Utah, however, uniform readings of blood counts could again be made.

DISEASES AND PARASITES

Experimental mice are susceptible to many diseases. Some of them such as paratyphoid are very serious. In textbooks on the care of animals, particularly in the fine book of the Jackson Laboratory, *The biology of the laboratory mouse,* there are excellent chapters on the diseases and parasites of mice. Most of these the author has never seen, and if the rule of never bringing mice from an outside source into the laboratory before adequate quarantine is kept by all investigators working in the same laboratory, most of these diseases probably never will be seen. One of the serious pests that perhaps carries many diseases to mice is the common mite (*Lipomyssus bacoti*). It is extremely difficult, in the author's experience, to control this pest. The system of running the cages and racks through the steam chest has completely eliminated the tropical rat mite at the National Cancer Institute. The same procedure also eliminates bedbugs.

The best dusting powder for the control of the common mite consists of one part talcum powder and one part powdered derris root to which has been added a small quantity of the detergent 2:6-dimethylnaphthalene. Care should be taken, however, to sterilize this dusting powder, since most of the cheaper talcum powders on the market are contaminated with bacteria which may produce skin lesions in mice. The dusting powder may be dropped in small quantities onto a group of mice huddled in the corner of a box or placed individually on them near the base of the tail. In the latter case, the fingers should be drawn through the hair on the dorsum toward the head in order to aid the dusting powder in reaching the base of the hairs. If too much dusting powder is used, it creates a dust hazard which becomes a serious problem, particularly in an air-conditioned laboratory. Cockroaches may be controlled by a periodic spraying of fresh pyrethrum. Since the efficacy of pyrethrum diminishes rapidly, it should be replaced every few days. Bedbugs may best be controlled by kerosene; house flies by flypaper or one of the proven innocuous fly sprays. It is perhaps a very serious mistake to spray or paint a laboratory with some of the new insecticides before they are tested adequately on the animals that are being used. All the vermin that are bad for the health of mice can be controlled by methods which have been proven harmless, and therefore, new methods should be avoided until proven advantageous. One should particularly avoid carbontetrachloride and DDT in the mouse laboratory, as these are quite toxic to mice. In the control of vermin it is also unwise to use hollow tile or other building material in which vermin can propagate. Even when the laboratory is disinfected, it is extremely difficult to disinfect the hollow tile, and within a short time the vermin are back.

SUGGESTIONS

A few miscellaneous notes: Most of the odor from a mouse colony emanates from the cleaning and disinfection of boxes. A separate room for this purpose, with a ventilating blower, will minimize odor, which otherwise may be very objectionable.

Certain items are useful in the laboratory: dissecting instruments; disinfectant or boiling arrangements; soap and running water; preservatives for specimens, and bottles; bread pans with wire covers used for temporary holding and carrying; a gas outlet for killing mice, or a bunsen burner; supplies of thumbtacks in various colors; one or more large (5-gallon) glass jars or crocks in which to place mice for observation during weaning, etc.

REFERENCES

Bittner, J. J. 1941. Care and recording. Chapter 13 in *Biology of the laboratory mouse*, ed. by G. D. Snell, Jackson Memorial Lab. Blakiston, Philadelphia.

Fenton and Cowgill, *Jour. Nutrition*, vol. 33, p. 703 (1947).

Grüneberg, H. 1943. The genetics of the mouse. Appendix 2 in *The keeping and breeding of mice for genetical experiments*. Cambridge University Press.

Hagedoorn-Vorstheuvel La Brand, A. C., and A. L. Hagedoorn. 1939. Mouse breeding made easy. *Jour. Heredity*, vol. 30, pp. 147–148.

Keeler, C. E. 1931. *The laboratory mouse*. Harvard University Press.

Laanes, T. 1936. Die Züchtung und Pflege von Laboratoriumsmäusen. *Abderhalden's Handbuch der Biologishes Arbeitsmethoden*, Abt. IX, Teil 7, pp. 593–609.

Morris, H. P. 1944. *Jour. Nat. Cancer Inst.*, vol. 5, pp. 115–141.

4· THE GUINEA PIG

HEMAN L. IBSEN
Kansas State College
Manhattan, Kansas

INTRODUCTION

Guinea pigs (*Cavia porcellus*), called cavies by the fanciers, were originally from South America. Their wild ancestors are still to be found on that continent. The Indians of Peru keep some of the domesticated animals in their huts and use them for food. These stay under the stone sleeping-benches and are prevented from escaping by a low stone barrier placed along the bottom of the door (Castle and Wright, 1916). Some of these animals attain a rather large size (1500± grams), which is probably the result of selection carried on by the Indians.

Guinea pigs have been crossed with at least two wild species (*C. rufescens* and *C. cutleri*). It is only with *cutleri* that hybrids fertile in both sexes are produced, thus indicating that this species is probably the wild ancestor of our present-day domesticated variety (Castle and Wright, 1916).

The following account of the care of the guinea pig is based mostly on my own 34 years of experience with these animals. Many practical breeders, raising them for laboratory purposes, use other methods, but, since most of these men remain in business a relatively short time, it may mean that in some respects their methods are inadequate. Also, they usually keep their animals in unheated buildings, which are maintained at somewhat above the outside temperatures in cold weather by means of the body heat of the animals themselves, and which on the other hand are apt to be too hot in summer.

Our animals have been kept in a laboratory heated to about 70° F. in winter and, when possible, maintained at less than 80° in summer. Experience has shown that temperatures consistently above 90° are unfavorable, causing loss in weight and also abortions.

97

Even under favorable conditions, the mortality sometimes seems to be somewhat higher than one would expect, indicating there is still something to be learned about the care of these animals. There is no question, however, but that the beginner can be saved from some pitfalls by profiting from our experience.

GUINEA PIG VARIETIES

The fanciers classify guinea pigs according to the length and the direction of the hair. Those animals with smooth, short hair (about $1\frac{1}{2}$ inches in length) are called English. This is the most common variety, and apparently is the one best adapted for laboratory use. It comes in many colors, the most common of which is probably tricolor (spotted black, red, and white). Other colors are self black, self chocolate, self red, and self agouti (wild color). These colors may also be combined with white spotting. A second variety is called Abyssinian. In these animals the hair is short, but rough, radiating out from centers and forming what are called rosettes. The latter vary in number, the most common being four. A third variety is called Peruvian. These guinea pigs also have rough hair, but it is long (usually about 6 inches). As a result, the hair tends to protrude over the head as well as over the posterior part of the body. The remaining variety is usually not exhibited at shows and is not given a distinctive name by the guinea pig fancier. However, it corresponds to the Angora variety of rabbits, cats, etc., in having long hair which lies fairly smoothly over the body, and does not have any rosettes.

REPRODUCTION IN GUINEA PIGS

Guinea pigs are unique among rodents in having a relatively long gestation period. Mice and rats carry their young about 21 days, and rabbits about 32 days. With guinea pigs, gestation varies from 62 to 72 days, 68 being the most common. The size of the litter affects the length of the gestation period. Large litters, made up of five or more animals, are usually carried less than 68 days, while litters containing only one individual may be carried the limit, 72 days. Large litters with five or more individuals usually have a low average birth weight (figs. 1A and 1B). Some of the individuals in large litters are weak, and some are born dead. Others are in danger of freezing to death even though the temperature may be approximately 60° F. With large litters the mother is unable to remove enough of the amniotic

A

B

FIGS. 1A and 1B. A female shortly before and shortly after parturition. Her weight on September 13, 1914, was 1318.3 grams. On September 14 she weighed 791.1 grams, a decrease of 527.2 grams. Her litter on September 14 weighed 449.2 grams, or more than 56 per cent of her weight after parturition. The four males in the litter averaged exactly 90 grams, while the weight of the only female was 89.2 grams. Usually there is about a 2-gram difference in favor of the males.

MIDDLEBURY COLLEGE LIBRARY

fluid from the hair with her teeth, and is not able to hover over all her young. A number may therefore freeze to death because of the evaporation of the fluid. Placing the young that are in danger of freezing near a radiator usually helps them to revive. Individuals in large litters are generally handicapped, and some may die before they are a month old. The most satisfactory litters contain three or four animals.

Because of the long gestation period, the fetuses are in an advanced stage of development before they leave the body of the mother. Their eyes are open 2 weeks before they are born (Ibsen, 1928). Their body is covered with hair, and their teeth are so well developed that they can eat solid food from birth.

In a number of cases the mammary glands were removed from the mothers before parturition (unpublished data). The young were left with these females, and most of them lived without receiving any milk whatever. They were handicapped, however, in that it took about 10 days for them to regain their initial weight. Those young receiving their full supply of milk are also somewhat handicapped at first. It usually takes 2 to 4 days for the newborn to regain their initial weight. Those animals that did not receive any milk eventually caught up with the others.

The relatively long period spent *in utero* also has an effect on sexual maturity. Our records show that some females were only about 20 days old when they were bred by their fathers. Under similar conditions a fairly large percentage is bred before 30 days of age. The males attain sexual maturity somewhat later. They are usually at least 70 days old before they are able to become sires, although they attempt copulation at an earlier age.

A healthy female comes in heat a few hours after parturition. By keeping a male in the cage it is possible to have as many as seven successive litters each a gestation period apart. A young female will continue to grow more or less normally under these conditions if she is well fed. However, she will not attain the same weight as a virgin female of the same age. The chief difference is that the breeding female puts on less fat. The estrus cycle is about 16 days (Stockard and Papanicolaou, 1917).

Unless special precautions are taken, it will be found that the average production per female for the colony as a whole will tend to decrease as time goes on. The reason is that an increasing number of females will have become either sterile or, at least, infrequent breeders. If records are kept of the performance of each animal, a sure

remedy can be found. The method we have used is to check the breeding record of each female. If she has not had a litter during the past four months, and on palpation shows no signs of pregnancy, she is discarded. We usually leave two females with each male. If both have apparently stopped breeding, the male as well as the two females are discarded. This method not only increases the number of litters per female, but it also increases the average number of young per litter. It seems to show that regular breeders tend to liberate more eggs.

BIRTH WEIGHT

Animals that weigh 40 grams or less at birth usually are weak, and a fairly large percentage of them die. Those weighing 60 grams or more (up to 140 grams) have a relatively good chance of survival. Sometimes the extremely heavy ones at birth are sluggish and may die. Males on the average weigh approximately 2 grams more at birth than do the females. In fairly large litters (five or more to a litter) there is usually a considerable difference in the birth weights of the individuals. This has been shown to be due to the degree of crowding in the uterus (Ibsen, 1928). Such differences in weight usually disappear as the animals become older.

MATURE SIZE

Guinea pigs increase in body length until they are about 15 months old. Some individuals increase slightly in length after they have reached this age. Although there is usually only a slight increase in body length, there is more variation in body weight. Both the increases and the decreases in weight after the animals have reached 15 months of age are due to either the gain in or the loss of body fat. Because of increases in body fat some animals attain their greatest weight when they are 3 years or older. Others may lose most of their fat and as a result may weigh much less at 3 years than they did at 15 months.

Mature females vary in body length (stretched out) from about 28 cm. to 32 cm. Males vary from 30 cm. to 35 cm. The body weights of mature females range from about 700 grams to 1300 grams. Mature males usually weigh at least 800 grams and may weigh as much as 1600 grams. Fanciers state they have produced males that weigh 1800 grams.

LACTATION

Guinea pigs usually have two mammae. If they have more than two, the extras are not as a rule functional. Thus in the case of litters containing more than two members, the mothers are unable to nurse all of them at one time. Large litters with four or more individuals, including a weakling, may crowd out the latter and cause its death.

My observations (unpublished) have indicated that the lactation period in guinea pigs is about 21 days. Records have also been kept on the amount of milk produced by placing the mother at intervals with the litter in an empty box, and weighing the animals before and after the milking period. As might be expected, females vary in regard to their milking capacity. It was estimated that the best individual examined produced as much milk per body weight per day as a fairly high-producing Jersey cow.

It should be noted that by the time the mother is through furnishing milk to her litter some of the females in the litter are already able to reproduce.

NUTRITION

Guinea pigs differ from other rodents and from mammals in general (except man) in requiring a relatively large amount of vitamin C. This is furnished in fresh green feeds such as alfalfa, clover, and blue grass. The C vitamin is also found in cabbages, lettuce, carrot tops, and sprouted oats. Cabbages should be fed sparingly at first; otherwise some of the animals may go off feed and die. Sprouted oats are usually fed during the winter months, one reason being that they grow best when the temperature is in the neighborhood of 75° F. Our method of sprouting oats will be described in a later section. Both the tops and the roots of sprouted oats are eaten by the animals. However, they do not take to this feed very readily. It is therefore advisable to feed it with other green feed for about two weeks. Young animals especially may be unwilling to eat it, and as a result die of scurvy.

Guinea pigs require a relatively large amount of roughage in their feed. It is therefore of advantage to keep them well supplied with hay. Our animals get alfalfa hay coarsely ground in a hammer mill. It is moistened with water (2 quarts to 6 pounds of hay), the reason being that this prevents the feed from being dusty. Dry, powdery feed is not readily eaten by guinea pigs. Alfalfa is high in calcium and vitamin A. Clover hay is probably satisfactory, but we have not had much experience with it. Prairie hay, however, has been found unsatisfactory, probably because it is low in both calcium and vitamin A.

The "grain" mixture fed consists mostly of ground oats (passed through either a ⅛- or ³⁄₃₂-inch mesh of a hammer mill). If it is fed much coarser, some of the animals will not eat the oat hulls. To every 100 pounds of ground oats are added (and thoroughly mixed in) 11 pounds of a mixture containing the following:

Bone meal	1.5
Skimmed milk powder	1.5
Soybean meal	2.0
Meat tankage	2.0
Wheat shorts	2.0
Cottonseed meal	2.0
	11.0 lb.

In winter especially, 2 ounces of irradiated dry brewer's yeast (9F, 9000 units of vitamin D per 1 gram of yeast, sold by Standard Brands, New York) are added to the 11 pounds of the mixture. It should be mixed in thoroughly because of the extreme potency of the vitamin D content. The vitamin D, the vitamin A in the alfalfa, and other ingredients of the feed, and the calcium in the alfalfa, bone meal, skimmed milk powder, etc., not only aid in bone growth, but build up resistance to colds and pneumonia (Ibsen, 1927).

The proportion of the ingredients in the 11-pound mixture has not been arrived at scientifically, but it is the result of an attempt to add various proteins to the feed without having enough of some of the ingredients (such as meat tankage) to affect the palatability.

The grain ration also is moistened before being fed. We have found that 1 quart of water is sufficient to thoroughly moisten 8 pounds of the feed. The lumps that form as a result of the moisture should be broken up to some extent.

If possible, the moistened feed (both the hay and the grain) should be fed within 12 hours. In hot weather especially there is a possibility of it souring if it is not spread out thin over night.

Although both kinds of feed are moistened, and there is a large percentage of water in the green feed, it is advisable to keep drinking water available to the animals at all times. The water dishes should be kept reasonably clean. Some guinea pigs have a habit of defecating in the water dish, and many of them drop feed into it.

Some breeders state that they can furnish enough water for their animals by means of the feed, keeping it well moistened, and by feeding a good supply of green feed. We have never tried this method. We find that, in the hot summer months especially, the green feed tends to become dry, and that during these months the animals drink a very appreciable amount of water. On very hot days they either come in bodily contact with the earthenware water dishes or place a leg and as much of their bodies as possible directly into the water.

Method of Feeding. Guinea pigs never eat very much at a time, but they eat often. Therefore it pays to keep feed before them most of the time, or about 20 out of every 24 hours. If possible, it is desirable that the feed dishes be empty in the morning. The morning feed should be enough to last the animal all day, and in the late afternoon only enough should be added to keep the animal going most of the night. If there is feed left in the dish in the morning, the morning feed should be light in order to insure it being cleaned up by evening. Keeping the feed dish full all the time is about as objectionable as keeping it empty most of the time.

We have kept no records in regard to the weight of the feed given each animal. One learns to judge by experience. Our method has been to give each animal the amount of grain that can be held between the thumb and the first three fingers. This amount will vary within fairly wide limits, but it is usually sufficient to keep the animal supplied for about a day. The same method is used in feeding the ground hay. A larger volume of hay can be held in this manner.

The green feed is placed directly on the floor of the cage. The number of animals in a cage may vary, and the method used is to give enough so that it will take three or four hours to clean it up. It is surprising how much green feed can be consumed. Some of it will drop through to the pan underneath, but it is a relatively small amount. Lawn clippings would not be very satisfactory as green feed when fed this way because too much would drop through.

Salt spools are hung in each cage, and in such a manner that they readily rotate on the wire through the center. By allowing free rotation there is less salt wasted.

Sprouted Oats. Sprouted oats are usually fed through the winter months when other green feed is not available. Green feed grown outdoors is preferable if it can be obtained. Some breeders with relatively small colonies are able to keep their animals well supplied through the winter months by feeding them green feed obtained free of charge from grocery stores. This consists of the outside leaves taken from lettuce and cabbage and of carrot tops. Care should be taken not to feed anything that is partially spoiled. It is also advisable to wash the green feed before giving it to the animals.

FIG. 2. An all-metal oats sprouter with twenty-eight of its thirty-two pans in place. The rack is so constructed that two additional pans may be placed on the very top, making the total number of pans thirty-four. The pan (or pans) containing the newly soaked oats is placed each day in the upper compartment at the reader's right. Each succeeding day it is moved further down till it reaches the bottom. Then it is moved to the top of the reader's left. The plan is always to have the newer oats above the older oats to avoid contamination from the drippings. If the two topmost places are used for pans containing well-sprouted oats needing artificial light to green up the tops, each of these pans should have a solid-bottomed pan underneath to prevent contamination.

When green feed of this kind cannot be obtained readily or in large enough quantities, it is usually necessary to feed sprouted oats. One objection to the latter is that it usually takes several weeks before the animals eat it readily, and the second objection is that a certain amount of equipment, space, and skill, is necessary to produce it. Also, it grows best in a temperature of about 75° F.

The oats are sprouted in metal pans 2 feet square and 2½ inches high (figs. 2 and 3). Each pan is perforated with round holes about

¼ inch across and about 2 inches apart. It takes approximately 3½ pounds of dry oats to make a full pan of sprouted oats. A pan of sprouted oats ready to feed weighs about 16 pounds, depending upon

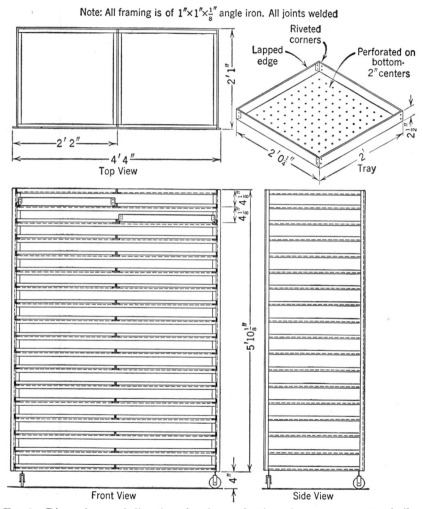

FIG. 3. Dimensions and directions for the production of an oats sprouter similar to the one shown in figure 2.

the amount of water present, and takes approximately 10 days to produce. It will furnish green feed for 1 day for about fifty animals.

It is obvious that if one plans to feed a pan of sprouted oats daily, it becomes necessary to start a new batch each day. The first step is to soak 3½ pounds of the dry oats for 3 or 4 hours in a plentiful supply of water. The excess water is then drained off and the oats placed in a pan, having them cover only about one-half of the bottom. If they are spread over the entire bottom of the pan, they will lose too much of their moisture.

FIRST DAY. After the oats have been in the pan for about 20 hours they are mixed thoroughly by hand, allowed to cover only about one-half of the bottom, and then watered and placed back in the rack.

SECOND DAY. The next day the process is repeated, allowing the oats to cover slightly more of the bottom of the pan.

THIRD DAY. About the same as the second day.

FOURTH DAY. By this time each oat should have a tiny root and the beginning of a top, and would therefore occupy more space. After being stirred thoroughly by hand, the oats should be spread over the entire pan and watered. From this point on they are left undisturbed, and only watered.

FIFTH, SIXTH, AND SEVENTH DAYS. Growth continues, and each succeeding day more water is needed to keep the plants supplied. By the seventh day it becomes necessary to allow more room above the pans because of the growth of the top.

EIGHTH, NINTH, AND TENTH DAYS. The tops should be 1 inch or more in height on the eighth day. At this point we usually place the pan on top of a cage and under a bright light, leaving it there continuously till the tenth day. Each day it should be watered thoroughly. On the tenth day the tops should be a fairly intense green, and should be 4 or more inches long. The roots should be fairly well matted and almost white in color.

Chunks of these sprouted oats are torn out and placed on the floor of the cage. The animals, when they become accustomed to them, will eat the roots as well as the tops. Both contain vitamin C (Bogart and Hughes, 1935). If after 24 hours there are still some sprouted oats left in the cage, they will be found to be dry, and brown in color, and should be discarded.

After the sprouted oats have been fed from the pan, the latter is brushed with a 2 per cent coal-tar disinfectant which is allowed to dry on the pan.

In spite of all precautions one will occasionally get a supply of oats whose roots become moldy or slimy before it is ready to be fed. Our experience has been that such oats have become infected in the field. Up to the present we have found no satisfactory treatment for this condition. Every chemical used that was powerful enough to prevent the infection also killed or at least greatly retarded the oats.

If it is desirable to feed two or more pans of sprouted oats a day, it becomes necessary to modify the above procedure slightly during the first few days. Enough for two pans, or 7 pounds of oats should be placed in one pan for the first 3 days, and on the fourth day half of the contents of this pan should be put into another and left undisturbed, except for watering, from that time forward.

Since producing good sprouted oats is a continual process requiring 10 days, it becomes necessary to follow some plan consistently. The method suggested here is as follows: After the contents of the ten-day pans are fed, the pans, as previously stated, are disinfected and placed in reserve. The nine-day pans are then thoroughly watered and placed in the location previously occupied by the ten-day pans (they are not watered on the day they are fed). The eight-day pans are watered and moved to the nine-day position, etc., until finally there is a vacancy for the pan or pans containing the oats that have been soaking for a few hours.

Fig. 4. A unit containing ten cages. As shown, the bottom of each cage is entirely of ½-inch wire cloth. The plans for this unit are shown in figures 5 and 6.

CAGES

All-metal cages with wire-mesh bottoms are advocated (figs. 4, 5, 6, 7). The first cages we made had ¼-inch wire cloth bottoms in order to prevent the newly born animals having their "elbows" caught in them. This proved unsatisfactory because the feces of adults did not pass through readily, and usually formed solid masses in the floor of the cage. The newer cages are so constructed that the entire floor is made up of ½-inch mesh. It is found that if the animals are born on such floors they usually are able to avoid being caught in them. When they are caught, they can be extricated by bending the lower part of the leg against the upper part. The big advantage in this

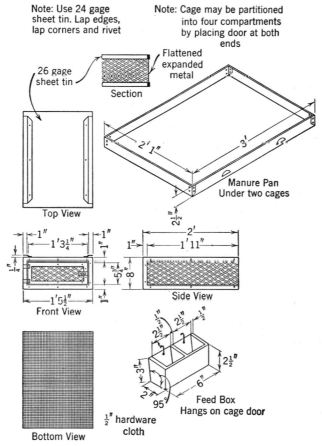

Fig. 5. Dimensions and specifications for building a single cage, a feed dish, and a pan to go under two cages.

type of floor is that only in very rare cases do the feces fail to pass through.

All-metal cages with snug-fitting doors prevent mice from entering. Guinea pigs pay no attention to mice, and the latter, if given a chance, will befoul the feed, making it unpalatable to the guinea pigs.

The feed dishes also are all metal (figs. 5 and 8) and are hung on the door of each cage, thereby making it more convenient to put the feed into them.

Two types of cages are used. One has small compartments for single individuals (figs. 8 and 9). There are four such compartments to a cage, with a door on two sides, and ten cages to a unit. The

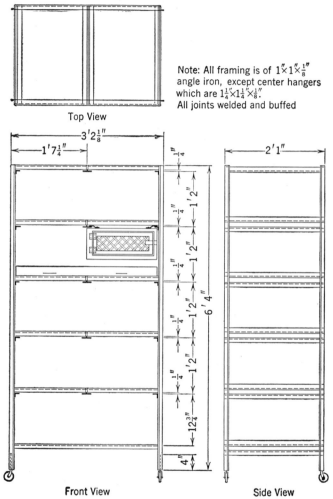

Note: All framing is of $1'' \times 1'' \times \frac{1}{8}''$ angle iron, except center hangers which are $1\frac{1}{4}'' \times 1\frac{1}{4}'' \times \frac{1}{8}''$.
All joints welded and buffed

Top View

Front View   Side View

FIG. 6. Specifications for building a metal rack to hold ten guinea-pig cages.

other type of cage is the same size as the first, but it is not divided into compartments and has only one door. There are either five (fig. 7) or ten (figs. 4 and 6) such cages to a unit. Each of these cages is large enough for four females, or for a male mated to two females. If the litters born in these cages are not too large (five or more), they are left with their mothers until they are about two months of age. In this type of cage two feed dishes are hung on the door.

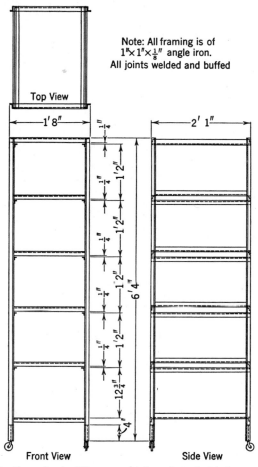

Note: All framing is of $1'' \times 1'' \times \frac{1}{8}''$ angle iron. All joints welded and buffed

Top View

Front View   Side View

FIG. 7. Specifications for building a metal rack to hold five guinea-pig cages.

DISEASES

If properly fed, guinea pigs seem to be resistant to most diseases, especially to those that attack the lungs and nasal cavities (Ibsen, 1927). Proper feeding, however, may aggravate intestinal-tract diseases. Such a disease was introduced into our laboratory with some stock bought about six years ago. It was not recognized as such until it had spread through the entire colony of about 400 animals. At its

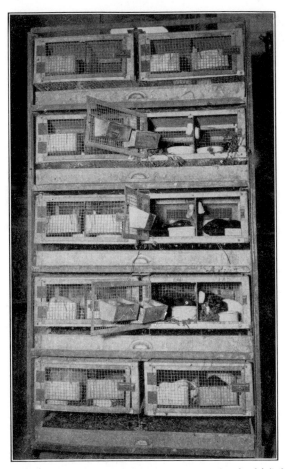

Fig. 8. Photograph of a unit containing ten cages each of which has been subdivided into four compartments, making a total of forty compartments to a unit. Some of the doors were left open in order to show the metal feed dishes hanging on the doors. The feed dishes are subdivided into two parts, one for the "grain" and the other for the ground hay. Salt-spools are shown hanging on the partitions between the compartments. Two different sizes of earthenware water dishes are in evidence. The smaller size is preferable, but the larger size has to be used in cages where the occupants have the habit of turning over the smaller dishes. Green alfalfa can also be seen on the floors of some of the compartments. This type of cage is used mostly for males.

height as many as ten animals would be found dead in one day. All showed the same symptoms, an inflamed intestinal tract and the presence of tremendous numbers of the protozoon *Balantidium*. All seemed to die suddenly. It was not established that *Balantidium* was the causative organism. The feces of all affected animals remained hard. To prevent the spread of the infection all the cages were modified to the extent that the bottoms were entirely covered with ½-inch mesh wire-cloth, and were frequently brushed with a 2 per cent coal-tar disinfectant. Several remedies were tried with only partial success. By far the most satisfactory is a patent medicine recommended for controlling coccidiosis in poultry, Ren-o-sal (Dr. Salsbury's laboratories, Charles City, Ia.). It is added to the water used for wetting the feed, and has been administered continuously for a number of years without any noticeably bad effect from the arsenic. No deaths are occurring from the disease.

Deaths due to other causes (Eaton, 1949) happen sporadically in our colony, but there is no evidence that any of these causes is dangerously contagious.

Fig. 9. End view of the cage shown in figure 8. The animals may seem crowded, but with this type of cage they remain clean and healthy.

SHIPPING GUINEA PIGS

At ordinary temperatures there should be no difficulty in shipping guinea pigs, provided the shipping crate has a solid bottom covered with about an inch of wood shavings. Each animal should have just enough room so that it is able to stand comfortably. The crate should admit plenty of air. Enough food should be placed in the crate to last until the animals reach their destination. Carrots with their tops attached will furnish both feed and water. A small amount of hay is also advisable. One good feature about guinea pigs, so far as shipping is concerned, is that though they are rodents, they will not attempt to gnaw their way out of the shipping crate.

In very cold weather there should be slightly more wood shavings on the floor of the crate. The animals can be crowded to a greater extent. In order to help keep them warm a large quantity of prairie hay should be stuffed into the crate. If the hay is alfalfa or clover, they may eat it. Besides, prairie hay is more fluffy than the other kinds, and the animals are more willing to stay under it for protection from the cold. Carrots, without their tops, are more suitable for feed in very cold weather.

In hot weather with the temperature in the neighborhood of 90° F. a different procedure should be used. The wood shavings should be soaked in water for several hours and then drained and placed two inches thick on the floor of the crate shortly before the animals are placed in it. Each animal should be allowed more room. No hay should be put into the crate, and a fairly plentiful supply of carrots with their tops removed should be used for food. We have found from experience that a large percentage of a shipment will die in very hot weather if the animals are kept entirely dry and covered with hay.

CHARACTERISTICS

Guinea pigs are unique in a number of respects. Unlike other rodents, such as rats and mice, they are unable to climb. This is convenient in that it tends to keep the sides of the cages clean. Because of their short, small legs and comparatively heavy bodies, they find it difficult to jump. Therefore it is possible to keep them in pens with fences only about 2 feet high without danger of their escaping. For the same reason they can be safely carried in open-top wire baskets about 18 inches high.

They are timid in their relations with man, some more so than others. The very timid ones will make desperate attempts to free themselves when held, and as a result one's hands may be scratched. A few in desperation may even bite if they are given the opportunity. The less timid ones will be relatively docile if they are held correctly. There is a tendency for the beginner to hold them too tightly, under which conditions all of them will squirm (figs. 10 and 11).

The males especially are not timid in their relations with each other. If two adult males are placed together, they will attempt copulation with one another. Apparently they cannot distinguish the sexes by smell. After a relatively short time one will resent the attentions of the other and will fight him off. This leads to a battle, and one of the two is always vanquished. From then on, he is the underdog. The posterior part of his back will be partially devoid of hair, and the skin will be covered with small cuts. If records have been kept of his weight, it will be found that he weighs considerably less than he did at the time he was placed with the other male. For this reason males that are not being used for mating are maintained in individual cages (fig. 8).

Females usually do not fight each other, but there are exceptions. Occasionally one will attack the others in the cage, causing them to squeal and to make desperate attempts to escape. The tumult will subside as soon as someone looks into the cage, but the guilty indi-

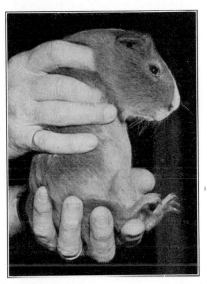

Fig. 10. A satisfactory method of holding a guinea pig—the two-hand method. The animal rests comfortably on the lower hand. Squeezing tightly with the upper hand should be avoided.

vidual can be detected as a rule by the presence of hairs in her mouth. Experience has shown that this sort of thing is not temporary. One way to remedy it would be to remove the offender and place her in an individual cage. The other method, which is the one we use, is to break off with a pair of pliers the two lower teeth of the offender.

If a cage contains several females, it will be found that they usually resent a new arrival if it also is a female. One method of keeping peace is to saturate the belly and back of the new arrival with a coal-tar disinfectant. The smell acts as a repellent. The treatment also should be effective in preventing the spread of disease. Similarly, if a female is placed in a cage containing a male and another female, the latter will resent the intrusion. The coal-tar treatment is effective also in this case. Another advantage is that it delays and calms down the attentions of an overzealous male.

Guinea pigs are noisier than other laboratory rodents. They squeal when they think they are to be fed. They are very sensitive to sound and hear the slightest rustle of green alfalfa, etc. If they are very noisy at feeding time, and especially at other times, it is usually a sign

Fig. 11. Even a large animal may be held in one hand. It is resting on the hand. The holder's thumb is partially visible above the animal.

that they either are being underfed or have an inadequate ration. If a bunch of keys is rattled, there is a sudden scurrying, followed by a dead silence. If the rattling is repeated, it is followed by a low murmuring from what seems to be the entire colony. One gets the impression that the low murmur is intended as a warning signal.

There is a popular belief that guinea pigs will kill wild rats and can therefore be used for rat eradication. Quite the reverse is true. Rats will kill young guinea pigs by biting a hole in the throat and, apparently, sucking the blood. Mature guinea pigs are attacked in the same way, but they do not die. Dogs, especially terriers, will kill guinea pigs if given the opportunity. Cats may eat the young animals, especially if they themselves are nursing a litter.

REFERENCES

Bogart, Ralph, and J. S. Hughes. 1935. Ascorbic acid (vitamin C) in sprouted oats. *Jour. Nutrition,* vol. 10, pp. 157–160.

Castle, W. E., and Sewall Wright. 1916. Studies of inheritance in guinea-pigs and rats. *Carnegie Inst. Wash. Pub.* 241.

Eaton, Orson N. 1949. The guinea pig. *U. S. Dept. Agr.*

Ibsen, Heman L. 1927. Cod-liver oil for snuffles in rabbits and pneumonia in guinea-pigs. *Science,* n.s., vol. 66, pp. 509–510.

——. 1928. Prenatal growth in guinea-pigs, with special reference to environmental factors affecting weight at birth. *Jour. Exp. Zool.,* vol. 51, pp. 51–91.

Stockard, Charles R., and G. N. Papanicolaou. 1917. The existence of a typical oestrous cycle in the guinea-pig with a study of its histological and physiological changes. *Amer. Jour. Anat.,* vol. 22, pp. 225–283.

5· BREEDING AND CARE
OF THE SYRIAN HAMSTER
Cricetus auratus

SAMUEL M. POILEY *

National Institutes of Health
Bethesda, Maryland

INTRODUCTION

The golden hamster (*Cricetus auratus* Waterhouse), a comparatively new addition to the family of laboratory animals, has already demonstrated its usefulness in some phases of medical research. Since 1931 it has been the subject material for a great variety of scientific investigations, and results have been good, bad, and indifferent. Its reactive versatility has been found to be limited, although gratifying progress was reported in several fields.

However, newspaper and magazine articles written for layman consumption have been overly optimistic in many cases and have led to fairly widespread belief, on the part of the average commercial breeder, that the market is unlimited. It has been erroneously assumed that this animal will replace the guinea pig, because of not only its high productivity but also its susceptibility to infection. Breeding has been reported as sporadic and seasonal by many observers, and susceptibility to a goodly number of infections as not noteworthy. Although these animals have proven satisfactory in quite a few types of investigational activities, the quantities of hamsters used for individual experiments do not approach the numbers of other species of laboratory animals required for medical research.

Serial passages and other techniques, now in primary stages, may subsequently be adapted to hamsters but, at the present writing, must not be considered too optimistically. The diverse nature of medical research and the variations in the physiological mechanisms of the

---

* We are indebted to Mr. Ernest P. Walker of the National Zoological Park, Washington, D. C., for the hamster photographs.

many genera preclude the use of but one member of the order of mammalia for scientific purposes.

To judge by the bulk of mail and telephone inquiries received, the market at the present time is glutted with hamsters. Many individuals have succumbed to the advertisements, invested in breeding stock and equipment, and are now considerably perturbed because they can realize but a small return on their original outlay. It is therefore considered advisable for the sensible entrepeneur first to explore the salability of his product, then develop outlets and adjust his breeding schedule accordingly.

On April 9, 1839, Mr. G. R. Waterhouse (*Annals of Natural History*, 1840) at a meeting of the Zoological Society of London, Rev. F. W. Hope presiding, exhibited a new species of hamster from Aleppo, Syria. He stated that it was of a species lesser than the common hamster and was remarkable because of its coloring. His description follows:

The fur is moderately long and very soft, and has a silk-like gloss; the deep golden yellow coloring extends over the upper parts and sides of the head and body, and also over the outer side of the limbs; on the back the hairs are brownish at the tip, hence in this part the fur assumes a deeper hue than on the sides of the body; the sides of the muzzle, throat, and under parts of the body are white, but faintly tinted with yellow; on the back, and sides of the body, all the hairs are of a deep gray or lead color at the base; and on the under parts of the body, the hairs are indistinctly tinted with gray at the base. The feet and tail are white. The ears are of moderate size, furnished externally with deep golden-colored hairs, and internally with whitish hairs. Mustaches are of black and white hairs mixed.

The skull differs from *Cricetus vulgaris* in not having the anterior root of the Zygomatic Arch produced anteriorly in the form of a thin plate, which in that animal, as in the rats, serves to protect an opening which is connected with the nasal cavity; the facial portion of the skull is proportionately longer and narrower. In size there is much difference, the skull of *Cricetus auratus* being one inch and six lines in length, and ten lines in breadth, measuring from the outer side of the Zygomatic Arches.

It may be added that there is a patch of reddish brown hair on each side of the chest, separated by the very pale grayish white hair covering the greater portion of the ventral surface. The hair covering the ventral surface of the trunk is sparse as compared to the density of the pelage covering the sides and dorsal surface. The forelegs each possess four toes and a rudimentary thumb, whereas the hind legs each have five well-developed digits. The eyes are prominent and black in color. The ears are large, erect, very dark in pigmentation, and practically nude. The cheek pouches, figure 1, are diverticulate rather than invaginate, and are unfurred, being lined with an epidermoid type of mucous membrane. When in a relaxed position they are approximately 1 inch in depth and can be distended by pouched food to a distance of approximately 2 inches.

The skin is very loose, and folds may be pulled several inches from the trunk. The tail is scant, 10 mm. in length, flesh colored, and very sparsely covered with short whitish hairs. Differentiation between adult males and females is rather simple and may be ob-

FIG. 1. Female hamster with cheek pouch distended by pocketed food.

FIG. 2. Male hamster; note protruding sex glands.

served without handling. The male bodily contour presents an elongated or pointed aspect posteriorly, figure 2, because of the fact that the testicles are carried posteriorly rather than pendant. The females are larger and heavier, adult does weighing approximately 165 to 180 grams and measuring 17 to 18 cm. in length; and adult males weighing approximately 150 to 170 grams and measuring 15 to 17 cm. in length. The *Cricetus* differs from the *Mesocricetus* in that the former possesses eight mammae, whereas the latter may have from fourteen to twenty-two mammae.

Before 1930, accumulated information relating to the golden hamster was cursory, and but very little more concerning its activities in its natural habitat has been acquired up to the present time. It is generally assumed that this rodent makes its burrows in corn and grain fields, hoarding enormous quantities and causing appreciable damage to crops.

In April of 1930, I. Aharoni (Bruce and Hindle, 1934) of the Department of Zoology, Hebrew University, Jerusalem, succeeded in capturing a mother and twelve young from a burrow eight feet deep. Previously the only known specimens extant were the skin and skull of a female found by Waterhouse and a specimen preserved in alcohol at the Beirut Museum in Beirut. The group found by Aharoni was brought to Jerusalem several months later, and four months after the hamsters arrived at the aforementioned city (Doull and Megraill, 1939), the first litter was born in captivity.

In 1931 several specimens were brought to the National Institute for Medical Research, London, England, by Dr. S. Adler of the Microbiological Institute of Jerusalem. These animals formed the nucleus of their colony, and subsequently shipments were made to various parts of the world. The hamster first made its appearance in the United States in 1938, through consignments to the Rockefeller Institute in New York and the Western Reserve University in Cleveland. The foundation stock of the colony at the National Institutes of Health at Bethesda, Md., was received from the Western Reserve University. This was composed of two groups of animals in two shipments. The first group, consisting of two males and four females, was received on April 24, 1940, and the second group, comprising three males and four females, on May 17, 1940. These animals, under controlled breeding adjusted to the needs of the laboratories at the National Institutes of Health, have produced numerous progeny. It might be proper at this point to note that all present-day golden hamsters are descendants of the original family of thirteen hamsters found by Aharoni.

Lydekker (1915) classifies hamsters of all species among the Sigmodontinae, a subfamily of murine rodents which includes the hamsters of the Old World as well as a large number of South American genera of rat-like mammals. North American forms are the "chisel-teeth" (Cahalane, 1947) such as *Onychomys, Sigmodon hispidus,* and *Peromyscus.* The golden hamster is a close relative of the common European hamster (*Cambridge natural history,* 1902) (*C. frumentarius*) and the Chinese striped hamster (*C. griseus*).

NOTE. In this text and in many scientific writings, the golden hamster is commonly referred to as *Cricetus auratus.* However, because of the number of mammae, correct nomenclature is inclined towards *Mesocricetus auratus.*

HOUSING

The type of caging available for laboratory animal breeding is usually influenced by the kind of construction materials to be used, number of animals to be housed per unit, space requirements, individuality of the operator, and, last but not least, funds available. The cheapest building materials are those provided by nature for free-ranging mammals, i.e., the Good Earth and that which grows thereon. The aforementioned, however, is impractical for indoor use, and man again is forced to improvise. The superiority of natural burrows over man-made shelters is debatable, and, moreover, available statistics do not indicate that the former are preferable. It has, nevertheless, been demonstrated that caging conditions simulated to represent natural burrows have resulted in satisfactory reproduction (Schneider, 1946).

Through succeeding generations, the animal becomes acclimated, and after orientation an evolution in caging takes place, thereby simplifying housing. The adaptability of the hamster hastens this process and permits the fabricator to adapt his construction to a variety of materials, such as wood, metal, asbestos, or compositions.

The various species of mammals require caging systems that are peculiarly suited to their individualities. Wild animals or animals that are not many generations descended from their uninhibited ancestors require quarters containing a darkened or semi-darkened section that will give them some feeling of security. There are almost as many types and sizes of cages in use as there are animal colonies. Laidlaw (1939) reported the use of a cage measuring 15½ by 9 by 8½ inches. Bruce and Hindle (1934) have successfully raised hamsters in cages whose dimensions were 20 by 15 by 6 inches and which contained a dark portion 15 by 5 by 5 inches.

Upon arrival of the first group of hamsters at the National Institutes of Health they were placed in a two-compartment cage (Poiley, 1941), one compartment of which was darkened. A small wooden box was installed for nesting. This cage is similar to the one designed by The Wistar Institute (Greenman and Duhring, 1923) for caging white rats. A wooden cage was selected because it was felt that animals are much more comfortable in quarters of this nature. Variations in temperature are not readily transferred through wood.

However, this cage was found impractical because of the hamsters' predilection to gnawing wooden objects. After expending time and energy in recapturing escaped animals, and costly repairs, it was decided to resort to galvanized sheet metal. A simplified metal cage, of the same dimensions as the original cage, was adopted, and the hamsters thrived therein. Galvanized sheet-metal nest boxes were also provided but were discarded after a short trial period. It was observed that these metal boxes tended to "sweat" during excessive humidity, and many of the females were casting and rearing their litters outside of the nest boxes. The nest boxes were discarded for all time, and the darkened portion of the cage eliminated. With the admittance of light to all parts of the cage and the ability to survey happenings outside, it was found that the animals became more tractable and displays of irritability became rare.

A shortage of breeding cages, at one time, necessitated the use of a nutritional type of cage for maternity purposes. These cages had hardware cloth for bottoms, and although nesting materials were provided, the females invariably devoured their young. After too many sad experiences of this nature, it was decided to discontinue the use of nutritional cages and a small rabbit cage measuring 14 inches high, 14 inches wide, and 16 inches deep was used for breeding and rearing purposes in order to conserve space. White pine sawdust was used as bedding, and it was also determined that nesting materials were unnecessary. The does would unconcernedly scoop out a depression 3 inches in diameter, towards the rear of the cage, deposit their litters therein, and apparently very happily go about the business of raising a family without benefit of cover.

Two factors influenced the decision to discard the rabbit cage as a breeding unit. Primarily, the height of the cage was too great in proportion to the dimensions of the animal. Secondarily, the door was hinged on the side and constructed of wire. Because of the climbing propensities of the hamster, very often, when the door was hurriedly opened, an animal would ride out and drop to the floor. After a period of trial and error, the design of cage shown in figure 3 was adopted. This cage is of the shoe-box type but, as an added feature, contains a pan. It measures 14 inches wide, 7 inches high, and 22 inches deep. The cage pan width and depth are essentially those of the cage but reduced sufficiently so as to allow a comfortable fit. The height of the cage pan is 2 inches. Ventilation is provided by the perforations through the sides and the hardware cloth top. The latest models are constructed of 0.064 gauge, ½ hard aluminum. At the present time, they have been in use only two months, and complete information is unavailable as to the durability of this metal when used for this purpose. Theoretically, it should outlast galvanized sheet metal. The perforations in the sides also facilitate watering, as any one of them may serve as the entrance for the glass tube.

Feed receptacles are not provided, as the hamster will pouch food particles and deposit them wherever it so desires. This cage is very convenient for all purposes in that it may serve as a breeding cage,

maternity pen, or stock cage. In the last case, eighteen young hamsters or twelve adults may be accommodated very comfortably. The
cage top may be removed or slid back for inspection of contents.

Figure 4 shows a section of these cages in position. A floor space 40
inches wide by 22 inches deep is all that is required to contain a rack
of 16 cages housing 192 to 288 hamsters. This spacing includes
allowances for water bottles and aisles between racks.

Fig. 3. Latest model of hamster cage.

Environmental temperatures exert some effect on warm-blooded
animals (Mills, 1945). Recommended year-round temperatures for
hamster raising range from 70 to 80° F. It is suggested that optimum humidity be in the neighborhood of 50 per cent or equivalent
to the humidity recommended for human comfort. With air conditioning, the above conditions can be controlled. However, air conditioning has its drawbacks also. It is possible that animals raised
in an air-conditioned building and transported elsewhere may be
exposed to extremes of temperature for a significant period of time.
The precipitation of respiratory infections, in this event, becomes very
probable. It is therefore speculated that a temperature lock may be

the solution to this problem. Nevertheless, hamsters have been raised for 7½ years in the latitude and longitude of Washington, D. C., without ill effects, and it is concluded that air conditioning is not altogether essential.

The animal quarters should be provided with adequate lighting equipment in order to facilitate the work and lessen fatigue for the

Fig. 4. Method of stacking cages in cage racks.

attendants. Overtired personnel are incapable of exhibiting the patience required for good animal care.

NOTE. After 18 months' use, it was decided that aluminum cages were impracticable because of the hamster's tendency to gnaw holes in this material. However, this metal displays greater corrosion resistance than does galvanized sheet metal.

HANDLING

The golden hamster is, comparatively speaking, a tractable animal, although occasionally indulging in obstreperous actions no doubt calculated to impress the observer with its ferocity and freedom from inhibitions, figure 5. Biting is an animal-like reflex action akin to the human being's instinct to jump when startled by an unexpected noise. Oftentimes, this mammal may not find it necessary to go to such extremes, but may resort to verbal chastisement and warning.

Although these deplorable tendencies are the exception rather than the rule, they nevertheless serve to alienate the hamsters from the affections of many persons. An intimate acquaintance with *Cricetus auratus* will result in more amicable relations and some understanding of the reasons for these acts. Its antisocial characteristics may be

attributed to irritability caused by temporary discomfort in the gastrointestinal tract, unclean or uncomfortable bedding, or malaise. It may be caused by nervousness due to noise or excitement in the animal quarters, too much handling, or handling by many strange caretakers. Irritability may also result from onset of estrus, uncongenial cage mates, gestation, or postparturition.

FIG. 5.   Female hamster exhibiting her teeth.

One of the recognized proper techniques in the rearing of animals is, obviously, to allow certain designated personnel to care for certain designated groups of animals. Familiarity with the caretaker eliminates sensitivity to a great extent, and he in turn will experience considerably less difficulty in handling and caring for the animals. The animal breeder must view his charges sympathetically and adapt his procedures to their psychological makeup. He must be familiar with the types of noises which startle these rodents, their idiosyncrasies, and their reactions to methods of handling.

Many commercial raisers recommend the use of glass jars, tin cans (Hayner, 1946), small transfer cages, or gloves for the handling or inspection of the hamster. The above methods, however, are not satisfactory, for the animals do not have the opportunity to become accustomed to the scent and feel of the human hand. Furthermore, the hamster may at times bite the gloved hand with no discomfort to the handler and the act may pass unnoticed. These biting proclivities may easily become habitual and result in future lacerations when the animals are inadvertently held without gloves.

One of the intangible factors in good animal care is the ability to suppress man's inherent fear and dislike of the genus *Rodentia*. Animals detect fear in man through their sense of smell, since fear engenders a mild perspiration in man. This scent will cause them to become irritable and antagonistic. When aroused, they scamper about the cage, occasionally turning on their backs, baring their teeth, uttering their characteristic cry, and acting in a wholly pugnacious fashion. Even under these circumstances, it is possible to pick them up barehanded.

The following technique is observed by the experienced caretaker. The hand is placed in approximately the same horizontal plane as the animal's torso, and is then slowly moved towards the animal, finally grasping it firmly but gently about the body and removing it from the cage. The animals dislike being handled too often or too roughly. At the National Institutes of Health's animal colonies, hamsters are always manipulated without gloves or other artificial means.

The hamster may also be picked up with a bare hand by another method. The animal is allowed to climb up the side of the cage. During the course of its climb or when it reaches the top, it may be handled without difficulty. For inoculation purposes, the hamster is placed on one's chest with the right hand, and the loose fold of skin at the base of the skull is grasped by the thumb and forefinger of the left hand. The animal can now be restrained with ease. If the injection is to be performed by the holder, while the thumb and the forefinger of the left hand are occupied with the loose fold of skin, the fourth finger encircles the left hind leg, which is then clamped between the third and fourth fingers. The right hand is used for the operation. If a two-man team is to perform the work, the manipulation will be correspondingly simplified. Hamsters will very seldom bite when being held. If they do so, it may generally be attributed to food odors adhering to the handler's skin. Even under these circumstances, puncture wounds are very rare.

The gentling of animals has a two-fold purpose. Primarily, careful handling of breeding stock proves its worth in the production of numerous and healthy progeny. Secondly, a docile animal is appreciated by the research worker, for very often a nervous animal may succumb from shock when being injected, occasioning a loss of valuable time and material. It is noteworthy to bear in mind that the handling of unruly animals injected with infectious material is a hazardous occupation.

BREEDING

Nature has endowed wild mammals with certain physiological mechanisms designed for adaptation to environmental conditions. Domestication negates the need for adjustments because of climatic conditions, security from predators, and absence of forage. Reproductive functions also alter appreciably when animals come under the jurisdiction of man, and monestrus mammals become seasonally or wholly polyestrus. The physical characteristics are influenced by improved nutritional methods intended not only to promote growth but also to increase reproduction and facilitate lactation. Systemic maladjustments disrupt the reproductive rhythm, subsequently resulting in abortions, stillbirths, and malformation of offspring.

The estrus cycle in viviparous mammals varies according to their generic order. Man, primates, the equine group, cattle, etc., differ from the rodents in that the latter are multiparous, whereas the former are but very seldom so. Although the duration of the estrus cycle varies according to the species of mammal, the functional activities are essentially similar. Although the practical animal breeder very often deems it unnecessary to concern himself with an intimate knowledge of the underlying physiological functions of reproduction, it is nevertheless advantageous to possess some understanding of the basic principles. The individual who pursues animal breeding and raising as a vocation, understanding these principles, will find that his knowledge will prove beneficial in the production of a gratifyingly thrifty herd.

Nature has blessed or burdened the female with the bearing and raising of offspring; therefore the functions of the female reproductive tract are of prime importance. The production of ova and the consequent adaptation of the physiological mechanisms for the expected pregnancy are practically alike for all species save the rabbit, cat, and ferret (Dukes, 1935). However, in the golden hamster, certain differences in the genital tract have been observed during recent investigations.

Graves (1943) reports that the egg of the golden hamster differs from that of the rat by its large size and abundance of coarse yolk granules. He also points out certain essential differences in the blastocyst and morula stage as compared with that of the rat. Deanesly (1938–1939) observes that the estrus cycle continues throughout the year, although matings are fewer from October to

February. She also notes that the epithelium of the cervix and upper vagina is unusual in that it abruptly becomes cornified about 1 cm. from the vulva. At this point the vaginal wall forms two lateral ventral pouches normally filled with layers of cornified epithelium. It is therefore difficult to trace stages in the cycle by means of vaginal smears, because of constant growing and sloughing of material in the pouches. Sheehan and Bruner (1945) have also encountered difficulty in tracing cyclic changes by standard methods and have found it expedient to observe the following procedure: A sample is obtained from the upper portion of the vagina by means of a pipette moistened with physiological saline. The specimen after being fixed with equal parts of ether and 95 per cent alcohol is stained by the Shorr (1941) method, passed through 70, 95, and 100 per cent alcohol, cleared in Xylol, and mounted in clarite. The four stages of the cycle, i.e., proestrus, estrus, metestrus, and diestrus, can then be noted. Shorr (1941) states that cornified elements stain a brilliant orange-red; non-cornified elements stain a green, which is deeper in younger cells and paler in older cells. Leucocytes, erythrocytes, bacteria, and spermatozoa can be readily observed.

Peczenik (1944) has observed the effects of sex hormones on supposedly sterile females. By single injections of diethylstilbestrol and chorionic gonadotrophin it was possible to induce pregnancy in eleven out of eighteen sterile females. Females younger than ten months of age, when so treated, produced two or three healthy litters which subsequently proved to be fertile when bred at sexual maturity. When older females were injected and bred, they produced but a single litter, none of which survived more than a few days.

Selle (1945) selected ten females aged 21 days for mating with sexually mature males. He found that they copulated at the ages of 27, 28, 29, 30, 31, 38, and 42 days, with the exception of one animal. The female that copulated at 28 days of age gave birth to five young at 44 days of age. He ascertained that the gestation period generally occupied 16 days, although some of the animals were 2 to 5 hours short of 16 days. Observations at the National Institutes of Health indicated that sexual maturity occurred at the age of 1 month, i.e., from 28 to 31 days. In contrast to Selle's report, it has been ascertained that gestation may endure for 16 days plus as many as 6 hours.

Copulation occurred at night and at any time after 7:00 P.M. E.S.T. and followed a definite pattern of behavior. For the first 5 minutes or so after the female was introduced into the male's cage, activity was confined to running about the pen with intermittent mutual examination of genitalia, figure 6. When the female finally decided that the male was acceptable to her, she crouched and awaited his attentions. Some females have been seen to remain passive for as many as five to six mountings by the same male. The majority of copulations, according to observations, occurred between 9:00 and 10:00 P.M. E.S.T. Sheehan and Bruner (1945) report that copulation occurred after 8:00 P.M. C.S.T. in Iowa. Selle (1945) states that hamsters will copulate after 6:00 P.M. P.S.T.

FIG. 6. Getting acquainted. Male hamster with back to camera; female facing camera.

Routine procedure at the National Institutes of Health includes the mating of hamsters at the age of 6 weeks, and no deleterious effects have been noticed. The vaginal smear technique is not used for detection of estrus because the size of the colony would render this task too cumbersome. For simplicity, the female is placed in the male's cage for a period of 1 week. At the end of this time, she is removed to another cage where she remains for 16 days. If the female does not give birth by the end of this time interval, she is remated. Inasmuch as pedigree records are maintained for these animals, each hamster is identified by a cage number. For the first 9 days after birth, the family group is allowed to remain undisturbed except for feeding and removal of dead animals. After the aforementioned length of time, cage cleaning is resumed. In this manner, cannibalism and disinterest towards offspring on the part of the mother have been practically eliminated.

The young animals are weaned at the age of 3 weeks and separated according to sex. The mother is not allowed a rest period and is immediately placed with a male for mating. Resting females, it has been ascertained, tend to develop deposits of adipose tissue and under these circumstances exhibit a lack of libido. During the early life of this colony, many females were dissatisfied with preselected males and incompatible mates were slaughtered by them. It was often found expedient to breed a young female with an older male. Subsequent generations, however, have reconciled themselves to matings with any male.

The literature cites conflicting reports concerning breeding seasons, maximum breeding ages, and consecutive litter sizes. From its inception, all data pertaining to the N.I.H. colony has been carefully recorded. These facts are presented in the tables included in the text.

Several observers report that breeding ceases in October and is not resumed until March. Table 1 has been compiled for a period of $7\frac{1}{2}$ years and shows the effects of seasonal variations in positive and negative matings. During the existence of this colony, production has been adjusted to fluctuations in demand by the laboratories. For peak demand, unrecorded females from stock have occasionally been bred in order to increase supply. The figures in the tables, therefore, do not present a picture of all animals raised in this colony. They are, however, based on a pedigreed group of 400 females and 160 males, which constituted the bulk of the animals used for production.

The scarcity of data for the last six months in 1943 and the first five months in 1944 was not caused by the animals' failure to breed. It is merely an indication of the occasional sporadic nature of the demand for hamsters by the laboratories, as influenced by the type of investigational activity. From June, 1940, until June, 1943, these animals were in great demand and were being tested for their susceptibility to all types of infections. During 1945, 1946, and 1947, it was definitely decided that hamsters were to be considered good reactors for certain research problems, and their use for these purposes has increased progressively. The figures in table 1 indicate that some of the early difficulties encountered because of seasonal variation have been overcome. The year 1947, in particular, demonstrates the adaptability of the Syrian hamster to entirely different climatic conditions after several generations. This hamster colony is not being raised in air-conditioned surroundings and at times is exposed to temperature changes of 20° F. in a short space of time, and humidities as high as 85 per cent.

This colony has never exhibited a complete failure to breed at any time. However, the percentage of unsuccessful matings during the

TABLE 1

Seasonal Influences upon Matings *

(A comparative study of the effects of environment upon the breeding habits of Syrian hamsters)

Month	1940		1941		1942		1943		1944		1945		1946		1947		Total	
	P	N	P	N	P	N	P	N	P	N	P	N	P	N	P	N	P	N
Jan.			2	8	10	9	25	15	1	4	9	7	13	15	18	25	88	83
Feb.			2	5	22	8	12	12	2	5	12	6	17	8	15	18	82	62
March			3	3	34	2	18	12	0	0	12	8	28	7	28	16	124	48
April			12	2	37	6	17	3	0	0	13	7	16	5	19	11	114	43
May	3	0	8	3	12	6	12	4	4	6	15	6	32	18	21	16	104	59
June	3	0	17	3	48	9	8	1	8	5	28	20	23	18	32	10	167	66
July	4	2	17	8	43	10	7	0	5	5	26	5	27	20	48	17	177	67
Aug.	0	6	18	5	26	19	3	1	7	3	19	11	26	6	46	16	145	67
Sept.	3	0	15	6	25	16	2	1	7	0	25	12	32	8	75	10	184	53
Oct.	0	5	11	25	11	62	1	2	3	2	17	24	26	13	34	14	103	147
Nov.	2	3	6	18	9	42	1	3	10	19	5	25	12	17	55	21	100	148
Dec.	1	9	14	9	20	38	0	3	9	7	23	7	16	14	80	32	163	119
Total	13	25	125	95	297	227	106	57	56	56	204	138	278	149	471	206	1551	953

* P indicates impregnation; N indicates unsuccessful matings.

TABLE 2

Effects of Seasons and Selective Breeding upon Average Litter Size

Month	1940	1941	1942	1943	1944	1945	1946	1947
Jan.		* 8.00	7.36+	7.67−	0	6.75	7.08	7.33+
Feb.		* 7.50	7.29+	6.27+	0	4.92−	7.86+	6.67−
March		9.67−	6.94−	7.17−	0	6.50	7.11−	8.21−
April		9.34−	7.48+	5.69−	6.40	6.92−	8.09+	6.88−
May		6.88−	6.63+	8.06−	7.67−	7.41+	6.57+	8.79−
June	* 6.67	9.35+	6.25	* 7.00	7.38−	7.35+	8.87+	10.15−
July	* 5.75	7.65−	6.13+	* 3.20	6.80	7.92−	9.45	9.11−
Aug.	0	8.06−	5.76−	* 8.67−	7.29−	8.75	9.59+	9.17+
Sept.	* 5.33+	6.73+	5.77−	* 4.0	7.29−	7.18+	8.03−	8.96
Oct.	0	6.55−	6.00	* 4.0	7.20	10.00	9.60	9.20
Nov.	* 3.50	7.00	7.17−	* 3.0	7.29−	10.20	9.64−	8.47+
Dec.	* 7.50	8.71+	8.14+	0	6.78−	8.14+	7.94−	8.75

* These figures insignificant because of small numbers involved.

early years of its life increased from October through February. Other observers (Laidlaw, 1939; Black, 1939; Peczenik, 1944) have stated that the animals are in a semi-hibernating condition during these months and, consequently, do not breed. Environmental conditions have apparently altered the picture, and indications point to a fairly constant production rate throughout the year.

The figures in table 2 have been assembled to show the seasonal variations and the fact that litter size may be controlled through selective breeding. Peczenik (1944), Laidlaw (1939), Bruce and Hindle (1934), Black (1939), and Doull and Megraill (1939) have noted seasonal lapses in breeding. The N.I.H. colony paralleled these results during early years, but, at the present time, results show improvement. Offspring of females that bore and raised young from November to March were selected for subsequent breeding stock. Their performance records were studied as to number of viable young produced per litter, and the quantity was also considered as an influencing factor. The litter averages tabulated for the years 1946 and 1947 demonstrate that the selection technique adopted for this colony was satisfactory.

Many hamster raisers have complained about the disheartening tendency of hamsters to devour their young or to lose offspring during the nursing period. According to available records, the majority of deaths occur during first litters and result from apparent lactational failure. Females very rarely lose consecutive litters in this manner. Deaths in later litters may more often be attributed to external disturbances rather than lactational failure. Individual deaths in litters have never been observed. The young are never disturbed from birth until 9 days of age; consequently, litter size at birth is not tabulated. It is therefore assumed that dead babies are consumed by the mother.

Cannibalism has never occasioned much loss in this colony, with the exception of the year 1942, as shown in table 3A. A shortage of proper caging facilities necessitated the use of wire-bottomed nutrition cages for maternity purposes. Female hamsters apparently were dissatisfied with these quarters and demonstrated their irritation by eating their young. With an improvement in housing conditions, the incidence of cannibalism was reduced. The phenomenon of cannibalism is not necessarily confined to all litters borne by one mother and may occur in any litter. It cannot be attributed to any one factor and may be precipitated by any one of a series of external influences.

Non-routine noises, strange caretakers, inadequate quarters, and unsanitary conditions may be causative agents. It is, however, considered advisable to discard females that habitually include their offspring as part of their rations.

Table 3B shows the breakdown of eaten and dead litters as to numerical order; as previously stated, the majority of deaths occur in first litters. The high incidence in 1945 was caused by an attempt to foster rapid production through the use of 1-month-old females for breeding purposes. The disastrous effects resulting from this procedure have caused an abandonment of further attempts to use animals of this age for mating. The practice of discarding females after the fifth litter is routine procedure except for selected individuals. The latter were retained for longevity studies and the accumulation of data relating to litters subsequent to the fifth one.

TABLE 3A

Litter Mortality during Nursing Period

	1940	1941	1942	1943	1944	1945	1946	1947	Grand Total
Litters born	14	125	249	96	56	166	259	278	1243
Litters eaten	1	1	35	5	1	0	0	0	43
% litters eaten	7.14—	0.80	14.06—	5.21—	1.79—	0	0	0	3.46—
Litters died	2	15	2	1	1	31	19	12	83
% litters died	14.29—	12.00	0.80+	1.04+	1.79—	18.67+	7.34—	4.32—	6.68—

TABLE 3B

Breakdown of Dead and Eaten Litters according to Rank *

Litter Order	1940		1941		1942		1943		1944		1945		1946		1947		Totals	
	E	D	E	D	E	D	E	D	E	D	E	D	E	D	E	D	E	D
First	1	2	0	6	1	1	1	0	1	1	0	16	0	4	0	3	4	33
Second	0	0	0	3	2	1	1	0	0	0	0	6	0	2	0	1	3	13
Third			0	2	4	0	2	1	0	0	0	3	0	2	0	3	6	11
Fourth			0	1	6	0	1	0	0	0	0	2	0	6	0	0	7	9
Fifth			1	1	1	0	0	0	0	0	0	0	0	2	0	0	2	3
Sixth			0	0	5	0	0	0	0	0	0	2	0	3	0	0	5	5
Seventh			0	2	8	0	0	0			0	1	0	0	0	4	8	7
Eighth					4	0	0	0			0	0	0	0	0	1	4	1
Ninth					4	0											4	0

* E: litters eaten; D: litters died.

Table 4 depicts the variations in average size of consecutive litters. Before 1944, the optimum number per female was considered to be but four litters. Selective breeding has increased this figure to five litters per female, and the sixth litter may often be considered to be profitable. Many animal breeders discard the hamsters at the age of 9 months, thereby losing the advantages that could be derived from a longer breeding life. During the years 1940 through 1943 the first three litters were most productive; in 1944 this figure was increased to four litters; and the fifth litter was added in 1945.

TABLE 4

AVERAGE SIZES OF CONSECUTIVE LITTERS

Litter Order	1940	1941	1942	1943	1944	1945	1946	1947
First	5.25	6.97+	6.64+	5.39−	6.35	6.58+	8.36−	9.81+
Second	5.33+	8.97−	7.43+	8.17+	7.64−	7.98−	8.62−	9.51−
Third		7.47−	7.46−	7.33+	7.45−	8.10−	8.85+	10.33−
Fourth		7.90−	6.88−	6.42−	7.38−	7.70−	8.69+	9.78−
Fifth		7.38−	5.92+	4.00	6.80	8.00	7.88−	7.83−
Sixth		9.00	7.22+	5.40	6.00	6.50	7.13+	6.50
Seventh		1.00	5.64−	4.00		5.67−	6.82−	4.58+
Eighth			5.40	4.00		6.25	5.00	2.50
Ninth			3.00					

Table 5A has been tabulated for the purpose of assembling data on the approximate longevity of the Syrian hamster. The figures assembled herein relate to animals used for breeding purposes only. Some investigators have reported a life expectancy varying from 1 to 2 years with breeding span lasting from 9 to 12 months. They also report that no viable young were born after the age of 1 year, and data contained in previous tables demonstrate the above assertions to be erroneous.

Fertile females 18 months of age have been encountered, although this phenomenon is not very common. The hamster seems to be a fecund animal, the percentage of fertile males and females being well above 90 per cent. It is therefore probable, though not always possible, to obtain six healthy and worth-while litters per female before the females reach the age of 11 months.

Table 5B has been tabulated for the purpose of demonstrating the fairly high percentage of fertility in this colony.

TABLE 5A

LIFE EXPECTANCY OF SYRIAN HAMSTERS

Age (months)	1-2	2-3	3-4	4-5	5-6	6-7	7-8	8-9	9-10	10-11	11-12	12-13	13-14	14-15	15-16	16-17	17-18	18-19	19-20	20-21	21-22	22-23
Females died	2	6	5	9	9	9	11	6	3	10	10	9	6	5	4	2	3	4	1	1	1	0
% females died	0.50	1.50	1.25	2.25	2.25	2.25	2.75	1.50	0.75	2.50	2.50	2.25	1.50	1.25	1.00	0.50	0.75	1.00	0.25	0.25	0.25	0
Males died	1	9	6	2	4	6	3	4	6	3	5	3	5	2	3	2	2	4	1	1	1	1
% males died	0.63	5.63	3.75	1.25	2.50	3.75	1.88	2.50	3.75	1.88	3.13	1.88	3.13	1.25	1.88	1.25	1.25	2.50	0.63	0.63	0.63	0.63

TABLE 5B

INDICATED FERTILITY IN A HIGHLY INBRED COLONY

	Females	Females (% of total)	Males	Males (% of total)
Breeders discharged at 11 months of age	116	71.00	86	53.75
Sterile animals	30	7.50	8	5.00

BEHAVIOR

The behavior of animals is influenced by their environmental reactions as controlled by climate, foods and feeding conditions, housing and surroundings. Their behavior in the wild state is in accordance with the laws promulgated by mother nature. When they come under the jurisdiction of man, his doctrines, and his fumblings, their behavior patterns vary, and may follow diverse courses at variance with, but also influenced by, their instincts.

In its natural habitat, the hamster builds its nest at the end of a burrow 8 feet deep, lining its quarters with vegetative matter found in the vicinity. In the laboratory, it adapts itself to metal cages, although retaining its instinct for excavating by energetic scratchings in the corners. For bedding and nesting material, it adjusts its needs to the sawdust, straw, or other material furnished by its keeper. Although an abundance of feed is provided by the caretaker, it still retains the urge to hoard food, and will gather the rations provided and store them in a secluded spot. The food is conveyed via the cheek pouches and ejected by using the front paws. In the wild state, hamsters go into semi-hibernation during parts of the year and fail to reproduce. Under domestication, their reproductive cycles alter and breeding is possible throughout the year.

The gestation, growth, and development of the hamster are so rapid that the young ones can be seen eating solid food 8 days after birth. The mother carries food to the nest, and the young animals nibble it even though their eyes have not, as yet, opened. Controlled temperatures obviate the necessity of remaining in the nest, and they are soon scampering about the cage. Until 14 days after parturition, the watchful mother will put them back into the nest if they should scramble out or be dragged out while hanging on her nipples. In common with all rodent mothers, the hamster maintains excellent care of her offspring. In times of seeming danger she will hide her babies in her cheek pouches.

The Syrian hamsters, as the distaff members of the human species will tell us, are "cute" and are surprisingly entertaining while at play, figure 7. They wrestle with each other until they are quite large. Play consists of rolling each other over and playful biting. They seldom draw blood, for the biting takes place on the loose folds of skin on the back and sides. Males very seldom fight, and have never been observed mutilating each other's sexual organs, as do squirrels, rabbits, mice, and many other mammals.   Females will attack males and other females.

Hamsters are good climbers, especially on hardware cloth.   They will seldom retrace their steps, but will either fall, jump, or scramble down.   They are not as resilient as mice when falling, however.   A fall of 4 or more feet to a concrete floor will stun, if not kill, a hamster; whereas after the same fall, a mouse will scamper away.   Hamsters delight in hanging from an inverted position, but do not seem to be able to do so for more than a minute.   The forelegs seem to be con-

FIG. 7.   A litter of eight hamsters 17 days old.

siderably stronger than the hind legs, for the latter are seen to lose their hold first.   They make but very little progress in the inverted position, but just remain suspended.

They seem to be fascinated by an extraordinary activity in their vicinity, and will line up at the side of the cage closest to the operation. Changing one's shoes will cause them to sit up on their haunches and remain perfectly still and watchful until the operation is completed. Mother hamsters may be enticed from their nests merely by scratching with one's finger on the floor of the cage.

Most of their feeding is done at night, although young hamsters up to the age of 3 weeks will eat during the day.   They are sound sleepers, may be picked up in a comatose state, and remain asleep for several seconds in one's hand.   On a bright sunny day, they may become lively about 1:00 P.M.; on a dull cloudy day, they do not begin to stir until about 3:00 o'clock in the afternoon.   During the morning or early afternoon, one animal may occasionally sit up and look about, but it will soon curl up with the rest of its fellows.

Hamsters delight in receiving fresh bedding.   They will scratch through it and eventually try to push all the bedding to one end of the cage. They dig by scratching with the forepaws and kicking the loose material to the rear with their hind legs. For sleeping purposes, a depression about ½ inch deep and 3 inches in diameter is made at the top of the heap and the animal curls itself up therein. Where straw or hay is furnished, hamsters will burrow into it.

Cricetus auratus can also adapt itself to extraordinary noises, although manifestly sensitive to it for the first few days. Window exhaust fans are used for additional ventilation during hot weather. When these fans are first started each season, the animals are very irritable for 2 or 3 days. After this brief period, they seem to become accustomed to the noise. At the end of the season, when the fans are no longer in use, they again become irritable for 2 or 3 days until they accustom themselves to the change.

NUTRITION

The role of nutrition, and the beneficial effects of the individual vitamins, with respect to the well-being of human beings, are fairly well known and generally understood. Nevertheless, the existence of additional factors is suspected, and intensive research in this direction and investigations in connection with known vitamins are proceeding at a brisk rate. Whereas the endpoint of all scientific research is intimately concerned with man, the latter does not readily lend himself for laboratory material. It is therefore necessary to enlist the aid of members of the animal kingdom for test purposes.

Mammals in the wild do not adjust their diets for systemic needs, but feed upon those rations which are readily available and palatable, and can be obtained without undue exposure to predators. The examinations of stomach contents of collected specimens reveal the type of plant or animal life that composed their last meals, but approximate quantities of the various ingredients cannot be estimated. When foraging, animals are not particularly concerned with the freshness of their foods and vitamin contents vary accordingly. Times of famine can cause material alterations in the nutritional makeup of their diets and result in physiological changes. Animals that are being used for laboratory research cannot be subjected to intermittent changes in their diet. One of the prime reasons for the use of the laboratory-raised animals is the elimination of variables such as diet, climate, housing, and care. The most important of these is diet, and it is the duty of the conscientious animal husbandman to select an adequate ration for his charges that will result in high production and good performance in the laboratory.

When man first domesticated mammals, he was content to allow them to graze at will and was well satisfied with production. As grazing areas were reduced and population increased, the need for greater production became all-important. It was suggested that more nutritive rations were required, and the field of animal nutrition came into being. This field has subsequently been expanded to include the small laboratory animal. While a great deal has been accomplished, many factors are still unknown, for biology cannot as readily be manipulated as can a machine of steel. The expansion of medical research has led to the acquisition of new species of mammals (Ranson, 1941) with their attendant differences in feeding methods.

The Syrian hamster, one of the aforementioned newcomers, is still presenting problems of its own. Although inadequate nutrition may not be the only cause, nevertheless, a well-fed animal has a considerably better chance for survival. In the early days, every raiser had his own feeding methods, mainly based on "trial and error." In recent years, however, some progress in the nutrition of the hamster has been reported and should show some beneficial results.

Laidlaw (1939) reported that he varied the diet every day but fed the hamsters a ration consisting of raw green food, beans, wheat, dog biscuit containing meat, bread, maize meal, and an extract of yeast.

Bruce and Hindle (1934) report that they fed the hamsters a diet consisting of puppy biscuits, monkey nuts, mixed grain, raw carrot, green stuff, and fresh cow's milk. They also supplemented their rations during the winter months with vitamin concentrate of A, D, and B.

Sheehan and Bruner (1945) report a diet of balanced dog chow checkers daily, leafy vegetable three times per week, and cod-liver oil several times per week during the winter months.

Several investigators have reported on the nutritional requirements of the hamster, with conflicting conclusions. Hamilton and Hogan (1944) report that the simplest satisfactory diet includes vitamins A, D, E, K, thiamine, riboflavin, pyridoxine, and pantothenic acid. Omission of E causes death; omission of K causes irregular rate of growth, but animals reach maturity in nearly normal time. Few of their animals bore a second litter, and it was concluded that the hamster requires at least one unrecognized vitamin for reproduction. Routh and Houchin (1945) reported that the hamster was good subject material for the demonstration of vitamin deficiencies. Experimental work was undertaken with thiamine, riboflavin, pyridoxine, pantothenic acid, and nicotinic acid. They concluded that the omission of any one of the above factors would cause loss of weight and death.

Cooperman and associates (1943) reported that a synthetic diet containing the six crystalline members of the B group, which was adequate for the rat, was inadequate for the hamster. The addition of p-aminobenzoic acid and inositol increased the survival time of the hamster. They also demonstrated that biotin was needed and the inclusion of vitamin C was unnecessary for hamster nutrition. In direct contrast to Routh and Houchin (1945) they maintained that nicotinic acid was unnecessary. Houchin (1942–1943) reported that the hamster was suitable for studying muscle changes in vitamin E deficiency. Young hamsters placed on high-fat diets that were deficient in E died in 3 weeks. Those placed on low-fat diets died within 2 weeks. Deaths occurred suddenly with very little loss in weight. Therapeutic doses of alpha-tocopherol produced rapid recovery.

These publications indicate that good nutrition for the Syrian hamster must include certain vitamins and vitamin factors. These elements must be contained in foods that are palatable and can be readily assimilated. The reported experimental investigations have been conducted with synthesized diets which are expensive and wholly inadequate for large-scale feeding. The animal husbandman must, perforce, resort to ingredients that can be obtained in the open market and which are available throughout the year. Limiting the ration to as few ingredients as possible will oftentimes decrease feeding costs and render the undertaking profitable.

The basic ration in use at the National Institutes of Health consists of the pellet that is also fed to guinea pigs and rabbits. This pellet is milled according to N.I.H. specifications and consists of the following:

15% Pure oat groat
30.5% Pulverized no. 1 whole wheat
40% Alfalfa leaf meal, U. S. no. 1 sun cured, with a carotene content of not less than 60 gamma per gram
13% Soybean oil meal, high temperature processed and containing not less than 44% protein
0.25% Irradiated yeast, Fleischmann brand, or equal
0.25% Sodium chloride, iodized
1% Calcium carbonate

To every ton of the above mixture, there is added 8 ounces (by weight) of cold pressed wheat-germ oil.

The finished product at the time of delivery shall have a carotene content of not less than 25 gamma per gram, shall not test less than 45 per cent green color, and shall have a guaranteed analysis of not less than 17.5 per cent protein, not less than 2.5 per cent fat, not more than 8.5 per cent fiber, and not less than 58.0 per cent carbohydrate.

The pellets are fed ad libitum and are supplemented with green feed. Each hamster receives 5 grams of carrots, one-sixth of a 1½-inch apple, and approximately 10 grams of kale per day. Water is always available, although the intake is very low. Before January, 1941, the basic ration consisted of a commercial dog chow, and it was observed that the animals did not consume it with evident relish. When the rabbit ration was substituted, an immediate upsurge in intake was noted.

Although this ration may not be the optimum, it has nevertheless proven beneficial up to the present time. The alfalfa, carrot, and kale are excellent sources of vitamins A and C. The wheat and wheat-germ oil furnish vitamin E, and the irradiated yeast seems to supply sufficient D. The grains and yeast supply thiamine, and a large intake of apple, kale, and grain products provides riboflavin. The soybean oil meal is a good source of protein, thiamine, and riboflavin. Niacin and pyridoxine are obtained from the yeast and grain. Pantothenic acid is obtainable from the kale. Excellent sources of inositol are leaves of plants and cereal grains such as are included in this ration. p-Aminobenzoic acid is present as a natural constituent of yeast. Alfalfa, vegetables, and yeast are good sources for biotin. The alfalfa also supplies vitamin K. The lactation vitamin L_2 is obtained from baker's yeast, folic acid from alfalfa and yeast, and choline is present in wheat.

It is felt that the low incidence of infection is due to the high intake of vitamins A and C obtainable from the adopted ration.

The growth curve, table 6, covers the period of 1 year, and its construction is based upon the average weights of 25 male and 25 female hamsters. One male hamster died on the 120th day, 2 died between 160 and 180 days, and 3 more died between 180 days and the end of the weighing period.

In the case of the females, 2 died before 120 days, 3 died between 120 and 150 days, 2 died between 150 and 180 days, 1 died between 180 and 270 days, and 2 died between 270 and the end of the weighing period. The chart indicates that males and females, on the average, gain in weight at approximately the same rate up to the 53rd day of

TABLE 6

GROWTH CHART OF THE GOLDEN HAMSTER

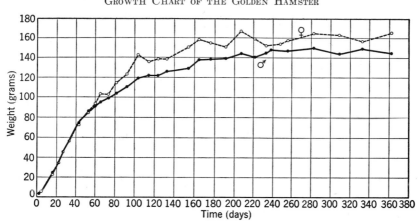

age. From this point the curves diverge and the females gain at a more rapid rate until approximately the 100th day of age. This preponderance of weight in favor of the females seems to continue ad infinitum.

SANITATION

Successful animal breeding and raising are dependent primarily on three essential elements; i.e., first-rate foundation stock, good nutrition, and sanitation. The first and second conditions are not too difficult to fulfill, but the success or failure of the colony is dependent upon the degree of sanitation achieved. Too often, investments of cash, labor, material, and time have been rendered worthless through careless maintenance of animal quarters. Rigid supervision of untrained personnel and indoctrination into the fundamentals of sterility should be a prerequisite for the handling of animals.

It is essential that animal breeders and attendants be informed of the harmful effects resulting from expectorations into feed bins, bedding materials, and animal cages. Individuals harboring respiratory infections should be excluded from the animal rooms. The breeding and raising rooms should be isolated from quarters housing infected animals. The golden hamster is a fairly hardy rodent, but, as has been demonstrated in many laboratories, it is also capable of succumbing to pathogenic organisms.

A variety of bedding materials can be used, all types being in a raw state, i.e., as by-products of some agricultural or industrial process.

Consequently, they have been exposed to contamination from animal excreta, human sputum, insect, and/or vermin infestation. Bedbugs, roaches, and mites (Smith, 1931; Harvey, 1947) are very often encountered in supposedly clean bedding. Some laboratories sterilize all bedding before use as a routine procedure, but this phase of sanitation would be too cumbersome and costly in large installations. The operators of large animal colonies must, therefore, resort to periodic sterilization of cages and equipment and adhere to a set routine of control through insecticides.

At frequent intervals, all cages are placed in a steam chest, receiving streaming steam at atmospheric pressure for 1 hour. Upon removal from the sterilizer, they are washed with a high-pressure water hose (150 lb. per square inch), allowed to dry, and then sprayed inside and out with a DDT solution. After spraying, the cages remain vacant for a 24-hour period before being reoccupied by the hamsters. Cage racks are also sterilized, washed, and sprayed with DDT. The ceilings, walls, and floors are also subjected to spraying with DDT at predetermined intervals. Because of a liberal use of DDT solution, this colony is entirely free of vermin. Incidental equipment, such as water bottles, stoppers, tubes, feed bins, dust pans, and brooms, is also sterilized and washed, but not sprayed. It has been determined that mites are unaffected by DDT, and other methods of control through insecticidal spraying must, perforce, be observed.

The degree of sanitation may be increased by liberal applications of disinfectants, if sterilizing equipment is not available. There are many good disinfectants on the market, and coal-tar derivatives have been proven economical and satisfactory. If materials are hand washed, they should first be soaked in detergent solutions in order to remove adhering greasy residue from fecal material. Cages and cage racks should receive liberal applications of disinfectant. If it is impractical to hose the floors of the animal quarters, then the scrub water should contain detergent and disinfectant.

A mixture of sulphur, rotenone, pyrethrum, and talc in the form of a dust has achieved satisfactory results, but this method causes some discomfort to the animals, specifically involving the respiratory tract. Young animals up to the age of 5 to 6 days have failed to survive after dusting. Liquid insecticides composed of a mixture of pyrethrum concentrate and deodorized kerosene will not cause as much disturbance, but final results are not too satisfactory. Many insecticides will not destroy eggs, and very frequent spraying is then required. It has been found advisable, during an infestation of mites, to clean cages at more frequent intervals, with gratifying consequences. Routine procedure in colony maintenance includes cage cleaning at least once per week. The moth-eaten appearance of the pelage of the hamster and the accompanying dermatitis are oftentimes due to the presence of mites. This can be obviated by applications of a heavy suspension of flowers of sulphur in carbon tetrachloride. This treatment has also been found effective for the eradication of the rabbit ear mite.

Rodent fecal contamination of feeds and bedding is another possible source of infection. Wild rats and mice are also the hosts for insect vectors of many organisms. The construction of the animal quarters should be such as will eliminate the entrance of wild rats and mice. The optimum in inside wall construction is that it be composed of glazed tile, with rounded corners at the junction of walls, ceiling, and floor. Mice, as a group, are very susceptible to DDT, and contact through the foot pads will kill them after a short interval (Konst and Plummer, 1946). Rats do not react as favorably to the aforementioned compound and must be controlled through trapping and poisoning. The most effective rodenticide has been found to be the chemical combination called 1080. However, 1080 is highly potent for all animals and can be used only under restrictions.

Some evidences of intestinal parasites have been observed. The hamster has been shown to be a favorable host for the cestode *Hymenolepis nana* and the nematode *Syphacia obvelata*. Larsh (1946) states that the development period of *Hymenolepis* is much shorter in the hamster as compared to the white mouse. Its length at 11 days was much greater in the hamster than in the mouse. Stunkard (1945) found a fairly high incidence of infestation in a group of hamsters and became alarmed over possible consequent parasitical invasion to the animal caretakers. Subsequent fecal examinations (Watson, 1947) of these individuals proved negative, however, although no special precautions had been observed in handling the hamsters. The animal attendant, therefore, need not be apprehensive on this score, if he follows simple sanitary methods.

At the National Institutes of Health, the death rate in the hamster colony has been very low, because of the standardized policy of changing breeding animals at the age of 10 months. The colony is gauged to produce only the quantities of animals required for research, and the stock animals are issued before reaching the age of 2 months.

Pearson and Eaton (1940) note that the Syrian hamster develops a virus pneumonia that is usually fatal in from 6 to 15 days. Horsfall and associates (1940, 1946a, 1946b) have determined the presence of this virus in many laboratory animals. This organism is one of the several that are latent in the respiratory tract but may become virulent when bodily resistance is lowered through inoculation with some other medium.

The experienced caretaker is the one individual who is best equipped to detect abnormalities in the animals. Being familiar with their normal physical aspect, he should be quick to detect listlessness, anorexia, dull eyes, rough pelage, and a generally emaciated appearance. For precise determinations of causes, pathological and bacteriological examinations are indicated. Occasionally a group of hamsters in a stock cage has been observed to excrete loose stools. Autopsies revealed no lesions or nodules typical of enteriditis or the *Salmonella* group of infections. Organs gave a macroscopical appearance of normality, and bacteriological examinations were negative. This symptom has been attributed to a gastrointestinal upset caused by ingestion of decayed vegetable matter. Several laboratories reported death and autopsy of animals on test. This, however, would not apply to breeding colonies, for the picture is complicated by inoculation with infectious material, receiving deficient diets, or precipitation of latent infections through systemic weakness caused by any of the above-mentioned methods. In any event, it may often prove disastrous to attempt to cure sickly breeding stock. As long as the animals remain alive, they can act as potential agents of infection for the balance of the colony. Abnormal individuals should be destroyed immediately and incinerated, as should also be done with soiled bedding and waste materials.

The laboratory worker desires, as subject material, an animal that will react favorably to numerous infectious agents, but be germ free itself; an animal capable of acquiring induced immunity but not having immunity in itself. Animals in the wild state are exposed to all forms of organisms and parasites and acquire immunity; otherwise they must, perforce, cease to exist. By depriving the laboratory animal of the opportunity to become immune by contact, the animal breeder increases its potentialities as concerns susceptibility and must constantly be on guard, through sanitation, to prevent occurrences of epidemics. It is only through the methods and degree of sanitation maintained that the animal caretaker materially assists the medical researcher in the elimination of tangible complicating factors that may influence this work.

LABORATORY USES

There seems to be a widespread belief among commercial hamster raisers, laymen, and scientific people unfamiliar with the animal that this rodent is susceptible to all pathogenic organisms, intestinal parasites, etc., and is indispensable for laboratory research. The following discussion and references are intended to indicate some of the uses and limitations of the hamster for experimental purposes.

The European and Asiatic hamsters were used for laboratory work before the Syrian hamster was acquired, and some confusion concerning identification has resulted. The Chinese striped hamster (*C. griseus*) was first used as a laboratory animal in 1919. The Syrian hamster made its appearance in the laboratory in 1937, when Adler (1937) thought he had at last obtained an animal that would respond to human leprosy. Dharmendra and Lowe (1940) were unable to reach the above conclusion, although conceding that the bacilli can live and multiply for some time in the original implant. They found no evidence of a progressive infection. Doull and Megraill (1939) concur with Dharmendra.

Arnold (1942) observes that carious lesions may be produced in the molar teeth of the Syrian hamster fed a corn meal diet. These animals at the present time are being widely used for these studies.

Lennette (1941) has demonstrated the susceptibility of the Syrian hamster to the viruses of St. Louis and Jap. B. encephalitis. Both viruses multiply well in the brain, and the titer is comparable to that of the mouse.

Wile and Johnson (1945) found that syphilis in the Syrian hamster runs a course similar to that which is found in mice. The infection is inapparent in that the dark field is negative, silver impregnation stain shows no spirochetes, and Kahn test is negative. Wile used the Nichols strain of *Spirocheta pallida*.

Anderson and Goodpasture (1942) conclude that the Syrian hamster can be used successfully for serial passage of the virus of mare abortion.

Smadel and Wall (1942) report that the virus of lymphocytic choriomeningitis produces an internal systemic infection of *C. auratus* with few if any clinical or pathological signs of disease.

According to Griffith (1941), the Syrian hamster is susceptible, both by subcutaneous inoculation and by feeding, to infection with the bovine, avian, and vole types of bacillus. The bovine type is the most virulent, followed by the human type, and both result in generalized tuberculosis. The vole type develops slowly, producing lesions which do not undergo necrosis or caseation. The avian type is the least virulent, although the bacilli multiply in glands and organs. Macroscopic tuberculosis lesions are rarely encountered.

Taylor and Parodi (1942) report that the Syrian hamster may be used in place of the ferret for the identification of influenza A virus in throat washings. A throat washing containing type B influenza virus produced an immune response in the hamster.

Morton (1942), Larson (1944), and Larson and Griffiths (1945) find that the Syrian hamster is susceptible to *Leptospira canicola* and *L. icterohaemorrhagiae.* Eaton, Martin, and Beck (1942) report that the Syrian hamster is susceptible to viruses of meningopneumonitis and lymphogranuloma venereum.

Gordon and Seton (1941) have used the Syrian hamster for research in connection with scabies and report satisfactory results.

The Syrian hamster has been used for research in connection with leishmaniasis and has proven to be good subject material (Goodwin, 1944; Van Dyke and Gellhorn, 1946; Harrison and Fulton, 1946).

The Syrian hamster has responded favorably to infection with schistosomiasis (Stunkard, 1946; Cram, Jones, and Wright, 1945). *Schistosoma mansoni* has developed well in the hamster and has been carried through ten successive generations.

The field of nutrition as concerns research with hamsters is covered in the section titled "Nutrition."

Research at the National Institutes of Health indicates that the hamster is susceptible to rabies, and the titer is comparable to that of the white Swiss mouse (Habel, unpublished material).

Cricetus auratus is susceptible to the Lansing mouse-adapted strain of poliomyelitis (Armstrong, unpublished material; Plotz, Reagan, and Hamilton, 1942).

Studies have been made of the cellular constituents and values for hemoglobin, erythrocytes, etc., of the hamster's blood (Stewart, Florio, and Mugrave, 1944; Rose et al., 1946).

The hamster does not react satisfactorily for investigations in connection with psittacosis (Davis, unpublished material).

Larson (1945) and Lillie and Larson (1945) report favorably upon the use of the golden hamster as subject material for tularemia investigations.

There are indications that the Syrian hamster is satisfactory subject material for investigation involving histoplasmosis (Habel, unpublished material).

REFERENCES

Adler, S. 1937. Inoculation of human leprosy into Syrian hamster. *Lancet,* vol. 233, p. 714.

Anderson, K., and E. W. Goodpasture. 1942. Infection of newborn Syrian hamsters with the virus of mare abortion. *Amer. Jour. Path.,* vol. 18, p. 555.

Annals Nat. Hist. or *Mag. of Zool. Botany, Geol.,* vol. 4, p. 445 (1840).

Armstrong, C. Unpublished material. National Institutes of Health.

Arnold, F. A., Jr. 1942. The production of carious lesions in the molar teeth of hamsters (*C. auratus*). *Pub. Health Repts.,* vol. 57, no. 43.

Ashbell, R. 1945. Spontaneous transmissible tumours in the Syrian hamster. *Nature,* vol. 155, p. 607.

Black, S. H. 1939. Breeding hamsters. *Internat. Jour. Leprosy,* vol. 7, no. 3.

Bruce, H. M., and E. Hindle. 1934. The golden hamster, *Cricetus "mesocricetus" auratus* Waterhouse. Notes on its breeding and growth. *Proc. Zool. Soc. London,* p. 361.

Cahalane, V. H. 1947. *Mammals of North America.* Macmillan, New York.

Cambridge Natural History. 1902. Vol. 10.

Chaney, M. S., and M. Ahlborn. 1943. *Nutrition.* Houghton-Mifflin.

Cooperman, Jack M., Harry A. Warsman, and C. A. Elvehjem. 1943. Nutrition of the golden hamster. *Proc. Soc. Exp. Biol. Med.,* vols. 52–53, p. 250.

Cram, E. B., M. F. Jones, and W. H. Wright. 1945. A potential intermediate host of *Schistosoma mansoni. Science,* n.s., vol. 101, no. 2621, p. 302.

Crabb, E. O. 1946. A transplantable 9,10 dimethyl 1,2 benzanthracene sarcoma in the Syrian hamster. *Cancer Res.,* vol. 6, pp. 627–636.

Davis, D. Unpublished material. National Institutes of Health.

Deanesly, R. 1938–1939. The reproductive cycle of the golden hamster. *Proc. Zool. Soc. London,* ser. A., vol. 108, p. 31.

Dharmendra, and J. Lowe. 1940. Attempts at transmission of human leprosy to Syrian hamsters. *Indian Jour. Med. Res.,* vol. 28, p. 61.

Doull, J. A., and E. Megraill. 1939. Inoculation of human leprosy into the Syrian hamster. *Internat. Jour. Leprosy,* vol. 7, no. 4, p. 509.

Dukes, H. H. 1939. *The physiology of domestic animals.* 4th ed. Comstock Pub. Co.

Eaton, M. D., W. P. Martin, and M. D. Beck. 1942. Susceptibility to the viruses of meningo pneumonitis and lymphogranuloma venereum. *Jour. Exp. Med.,* vol. 75, p. 21.

Ellerman, J. R. 1941. *Natural history,* vol. 2.

Emmons, C. W. Unpublished material. National Institutes of Health.

Evans, H. M., and G. O. Burr. 1927. The antisterility vitamin—fat soluble E. *Mem. Univ. Calif.,* vol. 8.

Goodwin, L. G. 1944. The chemotherapy of experimental leishmaniasis. 1. The spleen as index of infection in the Syrian hamster. *Trans. Roy. Soc. Trop. Med. Hyg.,* vol. XXXVIII, no. 2.

Gordon, R. M., and D. R. Seton. 1941. A histological comparison of the effects of certain drugs on scabies, as studied in rodent infections. *Annals Trop. Med. Parasit.,* vol. 35, no. 2, p. 247.

Graves, A. P. 1943. *Cricetus auratus* Waterhouse during the first 9 days. Abst., *Anat. Rec.,* vol. 87, no. 4, p. 17.

Greenman, M. J., and F. L. Duhring. 1923. *Breeding and care of the albino rat for research purposes.* Wistar Institute of Anatomy and Biology, Philadelphia.

Griffith, A. Stanley. 1941. Further experiments on the golden hamster (*C. auratus*) with tubercle bacillus and the vole strain of acid-fast bacillus (Wells). *Jour. Hyg.,* vol. 41, p. 260.

Gye, W. E., and L. Foulds. 1939. A note on the production of Sarcomata in hamsters by 3:4-benzpyrene. *Amer. Jour. Cancer,* vol. 35, p. 108.

Habel, C. Unpublished material. National Institutes of Health.

Hagan, W. A. 1943. *The infectious diseases of domestic animals.* Comstock Pub. Co.

Halberstaedter, L. 1940. Tumors produced in hamsters by benzpyrene. *Amer. Jour. Cancer,* vol. 38, pp. 351–358.

Hamilton, J. W., and A. G. Hogan. 1944. Nutritional requirements of the Syrian hamster. *Jour. Nutrition,* vol. 27, p. 213.

Harrison, C. V., and J. D. Fulton. 1946. The effect of treatment on the spleen of the golden hamster (*C. auratus*) infected with *Leishmania donovani. Brit. Jour. Exp. Path.,* vol. 27, no. 1, pp. 4–8.

Harvey, A. E. C. 1947. Care and maintenance of animals and arthropod vectors of disease. *E. African Med. Jour.,* vol. 24:1, pp. 58–71.

Hayner, Al. 1946. *Successful hamster raising.* Bruce Humphries, Boston, Mass.

Horsfall, F. L., Jr., and Edward C. Curnen. 1946a. Studies on pneumonia virus of mice. I. The precision of measurements in vivo of the virus and antibodies against it. *Jour. Exp. Med.,* vol. 83, p. 25.

Horsfall, F. L., Jr., and Edward C. Curnen. 1946b. Studies on pneumonia virus of mice. II. Immunological evidence of latent infection with the virus in numerous mammalian species. *Jour. Exp. Med.,* vol. 83, p. 43.

Horsfall, F. L., Jr., and R. G. Hahn. 1940. A latent virus in normal mice capable of producing pneumonia in its natural host. *Jour. Exp. Med.,* vol. 71, p. 391.

Houchin, O. B. 1942–1943. Vitamin E and muscle degeneration in the hamster. *Fed. Proc. Amer. Inst. Nutrition,* pp. 191–192.

Kelsall, Margaret A., and Edward D. Crabb. 1947. Lymphocytic infiltration and hemoporesis in the liver of tumor-bearing hamsters. *Amer. Jour. Clin. Path.,* vol. 17, no. 12, p. 925.

Kent, G. C., Jr., and Hal Weathersby. 1946. Mating and the oestrus smear picture in the golden hamster. *Anat. Rec.,* vol. 96, no. 4, pp. 74–75.

Konst, H., and P. J. G. Plummer. 1946. Studies on the toxicity of DDT for domestic and laboratory animals. *Can. Jour. Comp. Med.,* vol. 10, pp. 128–136.

Laidlaw, Sir Patrick. 1939. Maintenance of the golden hamster. *Internat. Jour. Leprosy,* vol. 7, no. 4, p. 513.

Larsh, J. E., Jr. 1946. A comparative study of hymenolepis in white mice and golden hamsters. *Jour. Parasit.,* vol. 32, no. 5, pp. 477–479.

Larson, C. L. 1944. Experimental leptospirosis in hamsters (*C. auratus*). *Pub. Health Repts.,* vol. 58, pp. 522–527.

——. 1945. The susceptibility of the golden hamster (*C. auratus*) to tularemia. *Pub. Health Repts.,* vol. 60, no. 29, p. 839.

Larson, C. L., and J. J. Griffiths. 1945. A comparison of the effect of penicillin and immune serum in the treatment of experimental leptospirosis in young white mice and in hamsters. *Pub. Health Repts.* 317, vol. 60, part I, nos. 1–26.

Lennette, E. H. 1941. Susceptibility of Syrian hamsters to viruses of St. Louis and Jap. B. encephalitis. *Proc. Soc. Exp. Biol. Med.,* vol. 47.

Lillie, R. D., and C. L. Larson. 1945. Pathology of experimental tularemia in the golden hamster (*Cricetus auratus*). Reprint no. 2665, *Pub. Health Repts.*, vol. 60, no. 42, pp. 1243–1253.

Lydekker, R. 1915. *Wild life of the world*, vol. 1.

Maynard, L. A. 1937. *Animal nutrition.* McGraw-Hill, New York.

Mills, C. A. 1945. Influence of environmental temperatures on warm-blooded animals. *Annals N. Y. Acad. Sciences*, vol. 46, Art. 1, p. 97.

Morton, Harry E. 1942. Susceptibility of Syrian hamsters to leptospirosis. *Proc. Soc. Exp. Biol. Med.*, vol. 49, p. 566.

——. 1943. The use of the Syrian hamster as a laboratory animal. Abst., *Jour. Bact.*, vol. 45, p. 194.

Pearson, H. E., and M. D. Eaton. 1940. A virus pneumonia of Syrian hamsters. *Proc. Soc. Exp. Biol. Med.*, vol. 45, no. 2, p. 677.

Peczenik, O. 1944. Actions of sex hormones on oestrus cycle and reproduction of the golden hamster. *Jour. Endocrinology*, vol. 3, no. 2.

Plotz, H., R. Reagan, and H. L. Hamilton. 1942. Transmission of the murine strain of poliomyelitis to the Syrian hamster. *Proc. Soc. Exp. Biol. Med.*, vol. 51, pp. 124–126.

Poiley, S. M. 1941. A practical rat breeding cage. *All Pets Mag.*, p. 89.

Ranson, R. M. 1941. New laboratory animals from wild species. *Jour. Hyg.*, vol. 41, no. 2, p. 30.

Reed, Charles A. 1946. Possible establishment of the hamster in the United States. *Amer. Midland Naturalist*, vol. 35, no. 3.

Rose, C. L., et al. 1946. Cellular constituents and chemistry of the hamster's blood. Reprint, *Proc. Indiana Acad. Science*, vol. 55.

Rosenberg, H. R. 1945. *Chemistry and physiology of the vitamins.* Interscience Publishers.

Routh, J. I., and O. B. Houchin. 1945. Some nutritional requirements of the hamster. *Fed. Proc. Amer. Inst. Nutrition*, pp. 191–192.

Schneider, H. A. 1946. On breeding "wild" house mice in the laboratory. *Proc. Soc. Exp. Biol. Med.*, vol. 63, pp. 161–165.

Selle, Raymond M. 1945. Hamster sexually mature at 28 days of age. *Science*, n.s., vol. 102, no. 2654, p. 485.

Sheehan, J. F., and J. A. Bruner. 1945. The care, breeding habits, and vaginal smear cycle of the laboratory hamster. *Turtox News*, vol. 23, no. 4.

Shorr, E. 1941. New technique for staining vaginal smears; single differential stain. *Science*, vol. 94, pp. 545–546.

Smadel, J. E., and M. J. Wall. 1942. Lymphocytic choriomeningitis in the Syrian hamster. *Jour. Exp. Med.*, vol. 75, p. 581.

Smith, Wilson. 1931. *The breeding, maintenance, and manipulation of laboratory animals*, vol. 9, pp. 236–263. *A system of bacteriology in relation to medicine, in Great Britain.* Medical Research Council.

Stewart, M. O., L. Florio, and E. R. Mugrave. 1944. Hematological findings in the golden hamster, *C. auratus. Jour. Exp. Med.*, vol. 80, no. 3, pp. 189–196.

Stunkard, H. W. 1945. The Syrian hamster, *Cricetus auratus*, host of *Hymenolepis nana. Jour. Parasit.*, vol. 31, p. 151.

——. 1946. Possible snail hosts of human schistosomes in the U. S. *Jour. Parasit.*, vol. 32, no. 6.

Taylor, R. M., and A. S. Parodi. 1942. Use of hamster (*C. auratus*) for detection of influenza virus in throat washings. *Proc. Soc. Exp. Biol. Med.*, vol. 49.

Van Dyke, H. B., and A. Gellhorn. 1946. Chemotherapeutic studies in exp. leishmaniasis. *Fed. Proc.*, part 2, vol. 5 (1), p. 209.

Watson, J. M. 1947a. Helminths infective to man, in Syrian hamster. *Brit. Med. Jour.*, Oct. 19, 1946, p. 578; Memoranda, *Trop. Dis. Bull.*, vol. 44, (2), p. 222.

——. 1947b. Syrian hamsters used for studies on kala-azar in Wellcome Laboratories. *Trop. Dis. Bull.*, vol. 44, (2), pp. 222.

Wile, V. J., and S. A. M. Johnson. 1945. Experimental syphilis in the golden hamster. *Amer. Jour. Syph., Gon., Venereal Dis.*, vol. 29, no. 4, p. 418–422.

Williams, R. R., and T. D. Spies. 1938. Vitamin B_1 (thiamin) and its uses in medicine. Macmillan, New York.

6· THE RABBIT

PAUL B. SAWIN
The Roscoe B. Jackson Memorial Laboratory
Bar Harbor, Maine

INTRODUCTION

As a laboratory animal the rabbit has a wide range of uses, for some of which it is peculiarly adapted, because of special characteristics. For example, it has long been recognized as one of the more important animals for immunological and serological investigations. The lower molecular weight and solubility of the serums give it a greater precipitating power than is found in the horse or chicken, which is sometimes used for antiserum production. In addition the body size, which is such as to provide adequate amounts of serum for most work, and the easy accessibility of the large marginal ear veins for injection and withdrawal of blood are distinct advantages.

The fact that the doe ovulates only after copulation has made it possible to time embryos and the onset of many physiological processes with much greater accuracy (within 1 hour) than is possible with other laboratory mammals. In this connection the rabbit has been particularly valuable in the development of the Friedman test for pregnancy, the technique of artificial insemination, and numerous basic studies of the physiology of reproduction. It has also been found the most suitable animal for experiments on the physiology of mammary development.

By the application of special techniques, the eye and ear have provided excellent opportunity for observation under the microscope of various types of physiological and pathological processes, as they proceed in almost their normal environment.

Genetically, twenty-eight specific single gene variations are known which tag ten of the twenty-two chromosomes of the rabbit. Sixteen of these genes are distributed in six chromosomes as shown below, with cross-over percentages as indicated.

Chromosome			
I	*c	y	b
	0	14.4	42.8

Chromosome			
IV	a	dw	w
	0	14.7	29.9

II	du	En	l
	0	1.2	14.3

V	br	f	an
	0	28.3	36.8

III	r^1	r^2
	0	17.2

VI	e	at
	0	26.2

* Gene Symbols

a = non-agouti	dw = dwarf
ach = achondroplasia	e = non-extension (yellow)
an = absence of antigen A	En = English spotting
at = atropinesterase	f = furless
ax = ataxia	r^1 = rex 1, short hair
b = brown (chocolate)	r^2 = rex 2, short hair
br = brachydactyl	r^3 = rex 3, short hair
c = albino	s = satin
d = dilute	v = Vienna
du = Dutch spotting	w = wide band
l = long hair, angora	y = yellow fat

Five genes, dilute (d); Vienna spotting (v); the blood groups, H_1, H_2, vs. h_0; another, H_6, vs. h_6; and satin (s), so far as is known at present are inherited independently of the other six groups. Ataxia is probably associated with group I.

Of these mutants achondroplasia, ataxia, brachydactyl, dwarf, furless, atropinesterase, and the blood differences are comparable with similar physiological variations in man, and should be useful in the study of fundamental processes involved.

Similarly, a considerable body of knowledge is now available about the inheritance of differences in adult body weight and the normal growth gradients, as determined by study of variations in internal structures, asymmetry, and early embryological processes in six inbred races, which should be fundamentally useful in future studies of normal and abnormal growth. In this respect the rabbit has certain advantages, since many of the variations in type and conformation are the same as in man and larger domesticated animals but on a more simplified scale.

CARE AND HOUSING

The prime considerations in housing rabbits for laboratory purposes are those which will facilitate the experimental work and at the same time maintain the animal in its best health and physical condition.

TABLE 1

A List of the More Common Rabbit Breeds Used for Laboratory Purposes and Their Outstanding Characteristics Based on Standards of the American Rabbit and Cavy Breeders Association *

Breed	Adult Body Weight (pounds) ♂	♀	Ear Length (inches)	Colors	Phenotype	Type	Bone	Distinctive Characters
English Angora	5–7		3–4	Albino	cc	Cobby, compact	Medium small	Long hair
French Angora	7–8		4½	Albino	cc	Medium long	Medium	Long hair
California	8–9		4½	White, black extremities	$c^H c^H$	Medium long	Small	
Champagne de Argents	9–11	9½–12	4½–5	Silver, black undercolor		Medium	Small	Breadth
Cream de Argents	9	10	4½–5	Silver, yellow undercolor		Medium	Small	Breadth
Checkered Giants	11+	12+	6	Black Blue	$Enaa$ $Enaadd$		Medium	Colored spots on white background
Belgian Hares	8		5	Gray, agouti	Aww	Long, racy	Fine	Body type
Viennas	9–10	10–11	5	Blue	$aadd$	Heavy-weight, cobby	Medium	
Beveren	9	10	5	Blue White	$aadd$ $aaddvv$	Tapering, mandolin	Heavy	Blue eyes
Chinchilla	6–7½	6¼–8	4½	Gray, agouti	Ach^3	Medium long, chubby	Medium	Fur quality
Dutch	4½		3½	Varied	$du^d du^w$	....	Medium	White belt
English	6–8		4	Varied	En	....	Medium	Colored spots
Flemish	14+	15+	6–6½	Varied	Varied	Long	Heavy	Size
Havana	6		3¾–4	Chocolate brown	$aabb$	Cobby	Medium fine	
Himalayan	3½		3½	White, black extremities	$c^H c^H$	Snaky	....	Markings
Lops	10	11	8–9	Varied		Heavy	Heavy	Ear length
New Zealand	9–11	10–12	4–5½	Red or white	ee or cc	Blocky	Medium	....
Polish	2¼	2¾	2¼–2½	White or colored	cc or C	Short	Fine	Small size
Rex	7	7½	4–4½	Varied	$r^1 r^1$	Blocky	Medium	Short hair

* 5941 Baum Blvd., Pittsburgh 6, Pa.

For example, such things as freedom of observation are particularly important in behavior studies but are minor considerations in serological work. In other studies special equipment may be necessary, as in activity or nutritional investigations, and provision must be made to meet those needs. For general purposes the essential considerations are those which primarily facilitate normal biological processes. These include (1) cages adequate in size to permit normal exercise of the individual and constructed of materials which may be maintained in

Fig. 1. All-metal general-purpose cage. (Courtesy of Roscoe F. Cuozzo, University of Maine.)

a sanitary condition with a minimum of time and labor; (2) quarters in which a reasonably uniform and moderate temperature can be maintained at all seasons of the year and with complete freedom from drafts and dampness. A moderate exposure to sunlight is desirable, particularly in the winter months, but is not essential; (3) facilities for supplying necessary feed and pure water and the removal of waste materials with a minimum of labor. For sanitation the all-metal cage is probably the best, and several cages are on the market. Figure 1 shows a type of cage which is inexpensive, simple in construction, easy to keep clean, and well adapted for all general laboratory purposes, particularly where only a few cages are in use. It consists of top, bottom, and sides of $\frac{1}{4}$- or $\frac{5}{8}$-inch mesh hardware cloth and floor of pressed steel with $\frac{5}{8}$-inch openings through which fecal material can drop into a galvanized sheet-metal pan. Both floor and refuse pan are removable. It is of a size which can be easily submerged in a tank for cleaning or sterilization, which is particularly important in studies involving pathogenic organisms. In this case a series of cages are mounted in tiers upon wooden racks.

Figure 2 shows a similar type of cage but smaller in size (12 by 18 inches) which is in general use in laboratories where rabbits are purchased for immediate use for periods of short duration.

FIG. 2. Holding cage for experimental work of short duration. (Courtesy of Norwich Wire Works, Inc.)

FIG. 3. Collapsible tiered units, including six cages with hay rack and feed trays. Note the hand nozzle used in filling water bottles.

Figure 3 shows a type of cage available before the war which is particularly advantageous in conserving space, since each unit includes six cages each 21 by 30 by 16 inches in double tiers with feed trays and hay racks between. Floors, refuse pans, feed trays, and racks are removable, and the entire assembly can be set up or dismantled in 10 to 15 minutes. Sterilization in either a tank or steam chamber is easily accomplished.

For growing rabbits, or in experimental breeding investigations, much larger cages are essential in order to get full normal development. Templeton, Ashbrook, and Kellogg (1946) recommend approximately 8 square feet of floor space for the smaller breeds, 10 for the medium breeds, and 15 for the giant breeds; the cost of installation of all-metal cages is high. Where space is not too great a factor or where convenience in observation is of prime importance, single-tiered hutches are preferable, and figure 4 shows a type of cage of economical wood and wire construction. This particular type of cage has an advantage in raising young stock, since the nest boxes are below the floor, which prevents premature emergence of the young and facilitates their return to the nest in cases where they do succeed in early escape.

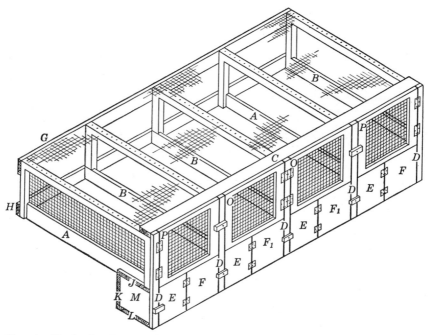

FIG. 4. Single-tiered cage of wood and wire construction. Note the nest box below the floor. (Courtesy of George O. Lilyestrum, Paxton, Mass.)

Feed and Water Receptacles. In cages which are not equipped with metal feed trays, earthenware crocks are quite satisfactory, and there are various kinds on the market. Where possible the crock should have a curved inner surface and lip which tend to prevent the animal from scratching the feed out of it. These crooks may also be used for watering, but a more satisfactory method from the point of sanitation is the water bottle and rack (figs. 3 and 5). For the rabbit, water bottles are best equipped with straight rather than curved glass or metal (brass or copper) nipples and rubber stoppers. Nipples should be 2 inches in length with approximately ⅛-inch opening at the utility end. Water bottles can be filled by carrying to a nearby sink, using a metal milk-bottle carrier. Where any number of cages are involved, considerable time and labor are saved by using a rubber hose of convenient length fitted with a hand nozzle such as is in common use on knapsack pressure tank sprayers. (See fig. 3.)

Where it is planned to maintain the colony size by means of its own reproduction, some form of nest box not only insures better care of the young by the mother but also facilitates observation and any desired record keeping in connection with the young. The entire

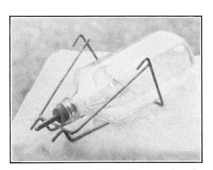

FIG. 5. Construction of water-bottle rack.

FIG. 6. A convenient type of nest box—wooden construction with metal lining of exposed edges to prevent gnawing.

litter may be easily removed to a work table, scales, or other facilities for observation and returned with a minimum of disturbance. For all breeds except the extremely large Flemish, a box (10 by 18 by 7 inches) such as pictured in figure 6 has been found highly satisfactory. This size can be adjusted slightly in either direction to facilitate entrance or removal through cage doors, etc. Wood is preferable to all metal, because of its insulating properties. However, edges, corners, and other exposed surfaces should be protected with metal to prevent destruction from the natural tendency of the rabbit to gnaw. The same precaution applies to cages of any other material.

To obtain the normal growth rates in rearing of young stock, feeders which keep a constant supply of clean grain or pellets before the young are highly desirable, if not essential. Several of these are described by Templeton, Ashbrook, and Kellogg (1946). The principles involved consist of (1) an overhead storage chamber of size proportionate to the number of rabbits and length of time they are to be supplied without refill; (2) a lower feeding chamber of just sufficient size to permit the insertion of the head of the largest rabbit in the colony; (3) an adjustable slot which can be set, according to the maximum size of the pellet or whole grain being used, so as to permit of the fall of a few pellets each time the feeder is slightly jarred; (4) sufficient lip on the lower chamber to prevent entrance by the young rabbits just out of the nest box and wastage by scratching of feed from the hopper.

REPRODUCTION

Biological Considerations

Ovulation. The rabbit belongs to a group of mammals which do not ovulate spontaneously, but only in response to some sort of sexual stimulus, either concomitant with copulation or in response to exogenous gonadotropic hormones. Because of the sensitivity of the ovulatory mechanism, occasionally females may stimulate each other to the point of ovulation. Such ovulation results in the doe becoming pseudopregnant and therefore sterile for approximately 16 days. Does which have no sexual stimulation may remain in heat for as long as a month at a time, but if kept too long, often blood breaks into the ovarian follicles, resulting in pigmented spots on the ovary which eventually atrophy and are resorbed. Ovulation occurs about 10 hours after mating (Heape, 1905; Pincus, 1936).

Estrus. It was formerly supposed that the doe remained in "constant" estrus throughout the breeding season, or at least during that season in which rabbits breed best (Marshall, 1922; Hammond, 1925; Hammond and Walton, 1934). A single crop of follicles was supposed to remain intact for the season. It is now known that follicles come and go in cycles of about *15 or 16 days,* one set degenerating while another set is developing (Smelser, Walton, and Whetham, 1934; Pincus, 1937). It is probably in the transitional period, while the new set of follicles is growing and the old set is retrogressing, that the doe lacks interest in the male (Hartman, 1945). At the height of receptivity of the female, the vulva becomes characteristically swollen and purple in color. Outward manifestations are restlessness, nervousness, and rubbing of the chin and head on the side of the cage or other objects.

Gestation Period. The gestation period of the rabbit is approximately 31 days with a range of 28 to 36 days (Rosahn, Greene, and Hu, 1934; Wing, 1945).

Puberty and Menopause. Small races become sexually mature and produce viable young from matings as early as 4 months, whereas the large Flemish rabbits seldom produce fertile matings before 9 to 12 months. Fall-born rabbits usually mate earlier than spring born. Fertile matings have been secured in Chinchilla females of 6½ years. Individual differences within the breed are well known.

Parturition. Records of parturition terminating 3503 pregnancies recorded at 4-hour intervals over a period of 9 years (table 2) show that two-thirds of all litters are born between 5 A.M. and 1 P.M. and only 8.0 per cent between 9 P.M. and 5 A.M.

TABLE 2

PERCENTAGE OF PARTURITION OCCURRING DURING 4-HOUR INTERVALS

5–9 A.M.	9 A.M.–1 P.M.	1–5 P.M.	5–9 P.M.	9 P.M.–1 A.M.	1–5 A.M.
35.9%	32.5%	11.6%	11.8%	7.6%	0.4%

Birth Weight. This is influenced by the race, size of litter, and physical condition and age of the mother.

Sex Differences at Birth. In males the sexual aperture is round and may be protruded by gentle pressure of the fingers on either side. It is usually much more anterior than the longitudinal slit-like aperture of the female, which seems to be attached close to the anus at the posterior end. Nipples are present at birth in both males and females, and are no criterion of sex.

Fostering. Newborn rabbits may be handled at birth and transferred to foster mothers with little difficulty within 1 or 2 days of birth. Progeny which are younger rather than older than the doe's own progeny are usually more readily accepted.

Nest Building. Loosening of hair in does at parturition parallels involution of the corpus luteum, but some additional factor is necessary for excitation of nesting instincts. The urine of pregnant women will induce nest building (Tietz, 1933).

Weaning. Young rabbits will nurse approximately 6 to 8 weeks but usually little milk is obtained after the sixth week, and the young should be removed from the mother not later than 8 weeks. She may be rebred at once. When the young are raised on self-feeders, it has been found practical to wean at 6 weeks. If the litter is not large, the doe can often be rebred at that time. If, on the other hand, her physical condition is poor, she should be allowed to rest until she has regained her normal weight, vigor, and vitality. Does may be bred as early as 14 days after parturition. Kellogg (1936), in extensive studies designed to determine the effect of breeding rabbits that are suckling young, found no measurable harmful effect until the fourth litter.

Fertility. The number of fertilized ova which develop normally up to the time of birth, in one case at least, has been shown to be a maternal (Hammond, 1928) rather than a fetal character. It was also shown by Hammond (1934) that its causes can be separated into (1) small number of eggs shed, and (2) atrophy of fetuses during uterine development.

Copulation. Males vary greatly in their sexual drive and in the readiness with which they will mate. In general it is best to keep the sexes separate. Mating is most easily achieved when the doe is placed in the cage of the buck. Reversal of this procedure may lead to a savage attack and possible injury to the buck. Bucks are often slow in performing service in a strange cage, and respond best in familiar surroundings. If the buck is virile and the doe is in estrus, mating will occur at once. Successful mating is characteristically indicated by the buck falling on his side. One mating is sufficient. If it is difficult to get the doe to accept service, she may be restrained as recommended by Templeton, Ashbrook, and Kellogg (1946). The rabbit is also peculiar in that it will often copulate during pregnancy and pseudopregnancy but at this time it does not ovulate, and young are not produced as a result of such a mating (Marshall and Hammond, 1946).

Fertilization. At ovulation the ova gather in the fallopian tubes in a mass surrounded by the follicle cells, which fall away as sperm pass through them. They are reached by the sperm in approximately $1\frac{1}{2}$ to 3 hours after ovulation. Several sperm may enter the zona pellucida but only one, apparently, enters the egg. The second polar body is given off 45 minutes or longer after the sperm enters, and the pronuclei are subsequently found (at the earliest 3 hours after ovulation).

Artificial Insemination. This is easily performed in the rabbit. Because of the preciseness with which ovulation can be timed, it has been used extensively as the experimental animal in developing techniques. The most satisfactory method of collecting semen is by means of an artificial vagina developed by Macironi and Walton (1938); details of construction and use are supplied by Lambert and McKenzie (1940) together with certain modifications.

With a little experience in handling the instrument, collections can be made with any buck that is accustomed to the presence of the operator. Repeated collections of semen may be made from vigorous bucks, and as much as 6.5 cc. in exceptional cases may be obtained. Semen can be shipped by mail quite satisfactorily if kept in glass tubes under liquid paraffin in a thermos flask at 50 to 60° F. Young have been produced from semen kept at this temperature for 7 days. At body temperature the life of the sperm is much shorter (12 hours), as is also the case when it is kept in ice (60 hours).

Breeding Practice

Methods in Use at the Roscoe B. Jackson Memorial Laboratory

1. Does selected for breeding are placed under observation at 4 to 8 months of age, depending upon adult body size. Races under 2500 gm. achieve adult body weight at 4 months; 2500 to 3000 gm. races at 5 months of age; and those over 3500 gm. at 8 months. Matings are made as early as does first come in heat thereafter.

2. Does are examined again 15 to 17 days after mating, at which

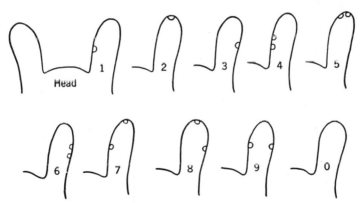

FIG. 7. System of notching right ear for identification of young in the nest. As drawn, the head of the animal is pointing away from the reader.

time pregnancy may be indicated in one of several ways: whining and other objections to attentions of the buck, material increase in body weight since mating, or by palpation. Non-pregnant does will usually remate at this time.

3. Two days before the due date of parturition nest boxes lined with hay, straw, excelsior, or other similar materials are placed in the cage of the doe.

4. All such does are then observed daily until the thirty-third or

thirty-fourth day, or until the litter is born. All litters are recorded at birth.

5. All young are notched in the right ear for identification at birth, using numbers from 0 to 9, as shown in figure 7. This together with sex, coat color, and other characteristics has been found adequate to maintain identity until weaning. However, all young are regularly tattooed at one month, each with its proper consecutive ear number recorded at birth in the ledger book.

6. All litters are weaned between the sixth and eighth week.

7. Usually a rest period of 2 weeks is allowed before a doe is rebred or until the next estrus period. Occasionally a doe whose previous litter was small and is therefore in good physical condition is bred at weaning.

8. No stock female is allowed to have and raise more than four litters a year.

9. All animals of inbred races are weighed at 21 days, which supplies information as to milk production, at 30 days, and at 30-day intervals until of breeding age.

Colony Records. Two types of record forms have been found of practical use in colony maintenance: (1) a running record of births, including weaning, growth, mortality or disposal, and descriptive records; and (2) breeding record cards (3 by 5) which include all matings, and data pertinent to reproduction and racial improvement. The first form is set up in a standard bound ledger book of 300 to 500 pages whose ruling is supplemented to provide columns for all desired information. Horizontal lines are drawn at the time of recording to separate the members of one litter from another. The columns from left to right contain (1) sex and ear number; (2) birth date (recorded only once for each litter); (3) genetic formula or other descriptive information; (4) nipple number; (5) birth weight; (6) mortality or disposal; (7) cage number, and (8 and 9) parentage (female first).

Form No. 2 has spaces at the top of the card for (1) sex and number of the animal; (2) descriptive information; (3 and 4) parentage (female first); (5) cage number; (6) race or cross; (7) nipple number; and (8) birth date. It also has columns for matings, including weight at mating, male used, date of mating, date when litter is due, date of birth, page in ledger where litter is recorded, and the ear numbers of the young derived from the mating. Other columns may be and are used for recording data on special projects.

These cards are kept in a working file with card headings as shown below, and the cards are rotated as each operation is completed.

1. Does ready to be bred.

2. Retest file (cards of does bred and to be tested at 15–18 days).

3. Days of the month—includes cards of does retested at the 15-day period and filed under the date the litter is due. In this file, cards may be easily found each day for which nest boxes should be supplied and where litters are due.

4. Newborn—includes cards for does with young under 1 month of age.

5. Tattooed—includes cards of mothers in front and those of young immediately behind these cards. These are grouped and arranged in order of the ear numbers of young until weaning, at which time the mothers' cards are removed to the breeding file and the young's cards to the weaned file.

6. Weaned—includes all young kept for growth or development studies or breeding. Other headings may be included according to the experimental program of the investigator. In fact one of the principal advantages of this system is the flexibility which permits of expansion of the colony or of a research program without great modification of the record system. It further enables the investigator to inventory easily and quickly and ascertain the status of his stock at any time.

DEVELOPING OF BREEDING STOCK

Although in recent research with other species, particularly with mice and rats, emphasis has been placed upon the development and use of inbred animals, researchers utilizing the rabbit have been satisfied to employ the heterogeneous stocks of uncertain origin obtained from local breeders. The reason for this undoubtedly is in part due to the lack of such inbred strains and in part to the cost in time and effort necessary to develop such stocks in an animal of the size and reproductive capacity of the rabbit. Those investigators who have made serious attempts to develop inbred races of rabbits appear to be unanimous in the conclusion that the domestic rabbit is one of the more difficult species to inbreed successfully. Presumably the long period during which it has been continuously outbred under domestication has been instrumental in accumulating an unusually large number of recessive lethal or semi-lethal genes whose segregation makes inbreeding a costly and hazardous procedure. At the present time there are six races at the Jackson Memorial Laboratory, which have been inbred at least as close as parent to offspring for better than ten, several of them as many as twenty, generations. Experience indicates that successful development of such races is primarily dependent upon rigid selection; first, in the foundation stock; and second, within each new generation, with special emphasis upon the factors contributing to the fertility, vigor, and reproductive capacity of the race, as well as upon any special characteristic of significance to desired investigations.

These factors include:

1. Early sexual maturity.
2. Regularity in breeding of both male and female.
3. Regularity of conception.
4. Litter size or fecundity.
5. Milk capacity of the mother.
6. Maternal instincts of the mother.
7. Growth rate.

Any consideration of the nature and means of measuring these characteristics emphasizes the importance of producing, rearing, and maintaining large numbers of individuals, particularly females, over a substantial portion of their reproductive period.

Litter Size. This is of primary importance, since it provides a very necessary abundance of material for selection of each of the other characters. Hammond (1934) has shown that litter size is controlled by at least two major factors: (1) the number of eggs produced at each ovulation, and (2) the relative number of eggs or fetuses which atrophy during prenatal development. The first is a multifactorial character influenced to a considerable degree by the environment, season of the year, etc.

In one case at least, selection within a race being maintained by parent-offspring mating has been successful in increasing the mean litter size from five to eight over a period of 6 years. In this particular race there is good evidence to show that small size of the first litter of a particular doe is a good indication of small litter size throughout the complete reproductive period of that mother.

To what extent *early sexual maturity* and *regularity in breeding and conception* may be gene influenced in the rabbit has not been determined. It is well known that the larger breeds such as Flemish Giants and New Zealands mature much later (9 to 12 months) than the smaller-sized breeds such as the Polish, which may be sexually functional by 4 months of age. This difference is associated with the difference in growth rate. However, within breeds raised under the same environmental conditions precocious individuals are also well known, and members of at least one inbred race, for their size, are typically slow in becoming sexually responsive, which indicates that other factors, presumably differences in functional maturity of the endocrine glands, are also important. Temperature and molting are major influences which affect the regularity of breeding. The decline in the number of successful matings in September and October can be attributed in a large measure to the shift of the resources of the rabbit to hair production. It is also a well-known fact that rabbits in the southern states will produce fewer litters per year, partly because of failure to breed or to conceive during the hot months of the year. In this respect the male is usually more susceptible than the female, and the large races are affected more than are the smaller-sized races. This could undoubtedly be counteracted by installation of air-conditioning equipment. Rabbits housed outdoors in the colder latitudes likewise may fail to breed or conceive during the winter months, although in the author's experience a cold temperature does not affect reproduction in a vigorous stock.

In all these situations marked individual differences are recognized, and under extreme environmental conditions, selection with respect to these characters is highly profitable.

Milk Capacity. This factor is strongly influenced by environment; unless great care is exercised under conditions of hand feeding, it is difficult to secure the full manifestation of this character, and thus an accurate basis for selection is impossible. It is highly desirable that in developing laboratory stock the mother and young be self-fed. *Maternal instincts* of the mother have been given relatively little attention. However, differences in nest building and in attention to and care and protection of the young are easily recognized, and there is reason to believe that they have a hereditary background. Tendencies to abandon or destroy the young by eating, or to build and line inadequate nests, particularly in cold weather, or to foul the nest are characteristics to be avoided. They are prevented to some extent by fortification of the diet with either vitamin E as wheat-germ oil or by the addition of 5 to 10 per cent of whole wheat grain. Individual differences are also noted in this respect, and hybridization usually tends to prevent its manifestation.

Other factors are of importance for special types of investigations. In investigations in immunology, for example, the size of the ear, the size and distribution of the blood vessels (particularly veins) of the ear, and the relative ease with which the flow of blood can be started or stopped are characteristics of primary importance in selecting or developing laboratory stock, and in studies involving a skin reaction or sensitization certain types of pigmented skin make satisfactory reading of the test impossible. Similarly for ocular transplantation studies blue or brown eye color is much more satisfactory than the pink or red of the albino.

SANITATION

General Procedure. The two most important factors in establishing and maintaining environmental conditions favorable to the health of the rabbit are (1) cleanliness and disinfection to prevent the development of disease, and (2) early recognition and elimination of diseased animals once disease makes an appearance in the colony.

Regular and frequent (preferably daily) removal of all wastes, including manure, unused feed, and soiled nesting and bedding material, is absolutely essential in either solid or wire-floored cages. Such accumulations are not only esthetically disagreeable but they attract flies and other insects which are possible vehicles of infection. Wastes should be removed from the premises each day, or if that is impossible they should be stored in closed containers which should be thoroughly cleaned out and disinfected each time they are emptied. All equipment, including water bottles, crocks, feed trays, and nest boxes, should be periodically washed with hot soapy water and rinsed with a chlorine or other suitable disinfectant solution or steam.

For prevention of disease too much emphasis cannot be placed upon early recognition of disease symptoms. Inactivity, loss of appetite, absence of fecal material, rough coat, dull and partly closed eyes, lowered or unusually high temperature are easily recognized by the experienced caretaker, and such animals should be removed and isolated from the rest of the colony. Special isolation cages, equivalent in number of about 1 to 25 of the regular cages of the colony, kept in quarters entirely separated from the remainder of the colony, are desirable. All discharges should be liberally treated with a 3 per cent compound solution of cresol, and all dead should be disposed of immediately, preferably by incineration. Diseased or dead animals should not be allowed to come in contact with the attendants' clothing and thus spread disease to other members of the colony. All new stock brought into the colony should also be isolated for 10 days before coming in contact with the colony. Likewise, all precaution should be taken to control wild rodents and to prevent access of dogs or cats.

Equipment. Periodic cleaning and disinfection of water crocks, water bottles, feed trays, nest boxes, and if possible hutches are necessary precautions in preventing disease, particularly where rabbits are frequently moved about from one cage to another within the colony. When possible, it is well to have extra water containers, feed troughs, and removable floors which can be kept clean and ready for replacement of regular equipment at a minute's notice.

Proper cleaning consists in washing in hot soapy water and rinsing in clear water, to which a chlorine or other suitable disinfectant has been added. After rinsing, all equipment should be stacked where it will drain dry, or better still where it may be exposed to the sun. All metal parts, feed trays, and cages should not be exposed to corrosive disinfectants. After being washed and thoroughly rinsed they can be best disinfected by boiling or with live steam. Several types of steam chamber have been developed of a size to take cage units such as pictured in figure 3.

Where possible, nest boxes and hutches are best disassembled and disinfected as separate pieces to prevent the persistence of disease germs in crevices and cracks. All dried fecal matter should be removed by scraping before washing, as the presence of organic matter may impair the efficiency of the disinfectant used.

Disinfectants. The coal-tar disinfectants are probably the most easily obtained and commonly used. Satisfactory disinfective solutions of these preparations should have a phenol coefficient of about 5. These are relatively efficient for every organism except possibly coccidiosis.

Ordinary lye is a very good disinfectant and has the advantage that it is odorless. It is injurious, however, to metal, enamelware, and painted or varnished surfaces.

Quicklime is also good but loses its disinfecting capacity when exposed to air.

The blowtorch or fire gun, formerly considered an effective means of disinfection, has been found of little value, and its continued use is detrimental to both wood and metal equipment.

Of recent years certain new chemical disinfectants such as high-molecular alkyldimethylbenzylammonium chloride have been developed, which appear to be highly effective, odorless, and harmless to the rabbit.

The value of sunlight as a disinfectant agent should not be minimized.

Prevention and Control of Infectious Disease. Disease is in a sense a natural phenomenon which can never be completely eliminated.

Prevention is much to be preferred to treatment and possible cure. Purposeful and intelligent practice will usually keep disease at a low level. However, too much emphasis cannot be placed upon the importance of securing stock possessing a high natural resistance to disease, of observing good practices of nutrition, and of providing dry and draft-, vermin- and wild-rodent-proof quarters and plenty of access to sunlight if not attended by excess heat, all of which help to maintain the rabbit's resistance at a high level. The following summary of the causes, symptoms, treatment, and prevention of the common diseases of the rabbit has been adapted and modified from a leaflet by E. A. Lund published by the Bureau of Animal Industry, Agricultural Research Administration.

Specific Diseases

A. Of the Skin.

1. EAR CANKER.
   a. *Cause*—one of two similar forms of ear mite.
   b. *Symptoms*—shaking of head, scratching of ears, yellow, white, or reddish scales on surface of the skin, in and around the opening at the base of the ear, with scaly crusts progressing upward on the inner surface in advanced cases.
   c. *Treatment*—apply a few drops of any vegetable oil, repeating once a week until cleared. A mixture of 1 part iodoform, 10 parts ether, and 25 parts vegetable oil is recommended in stubborn cases.
   d. *Prevention*—very contagious. Clean and disinfect all hutches used by or near infected animals, observe or treat all nearby animals for at least 30 days, and do not introduce infected stock into a colony.
   e. Torticollis or wry neck may be a complication.

2. SKIN MANGE.
   a. *Cause*—mites.
   b. *Symptoms*—reddened, scaly skin, loss of hair, yellow crusts of dried blood serum as a result of persistent scratching or biting.
   c. *Treatment*—clip hair around affected parts, wash with soap and water, and apply salve of 1 part flowers of sulphur and 3 parts lard. Repeat in 4 to 6 days. Compare symptoms with those of ringworm.

3. RINGWORM.
   a. *Cause*—a fungus or mold-like organism.
   b. *Symptoms*—patches of scaly skin with numerous red, pin-head elevations around hair follicles. Hair breaks off or is shed from older patches. Usually starts on head or feet.
   c. *Treatment*—clip, shear, and wash as with mange. Apply tincture of iodine.
   d. *Prevention*—destroy all affected animals and burn or bury; thoroughly disinfect all exposed cages and equipment; use gloves and clothing that can be boiled after handling animals; avoid infection of hands and face through cuts or other abrasions; prevent introducing to colony by animals from unknown sources. Positive identification depends on microscopic examination of skin and identification of fungus. Not all animals or persons are equally susceptible.

4. SORE HOCKS.
   a. *Cause*—bruised or chafed areas which may or may not become infected. Probably due to poor density and quality of hair on the surface of the foot and in some cases to posture and quality of bone, which results in unequal distribution of weight in the feet. Exceptionally nervous animals are more susceptible.
   b. *Symptoms*—tenderness, favoring of feet in movement, and tendency to tread or lift the feet when sitting.
   c. *Treatment*—extreme cases are difficult to treat and should be disposed of. Daily applications of Absorbine Jr. followed with salve, preferably with lanolin base, have been found beneficial. If infected the areas should be dusted with sulfanilamide before applying salve.
   d. *Prevention*—keep hutch floors clean, dry, and in good repair (free from sharp wires or other rough surfaces). Protect animals from disturbing influences which induce stamping. Do not use affected animals for breeding or animals from parents which were affected.

B. Of Moist or Mucous Membranes.

1. URINE BURN.
   a. *Cause*—bacterial resulting from splashed urine which soils skin and results in chapping.
   b. *Symptoms*—inflammation in the region of the anus and genital organs.

    c. *Treatment*—apply salve, preferably with lanolin base. Because of possible confusion with vent disease 1 part calomel to 3 parts lanolin may be advisable. Mercuric or blue ointment is also effective.

2. VENT DISEASE.

    a. *Cause*—bacterial.

    b. *Symptoms*—inflammation of delicate membranes of sex · organs, usually covered with yellow or brownish crusts. Swelling, bleeding, and discharge of pus may occur.

    c. *Treatment*—treat daily with mercuric or blue ointment.

    d. *Prevention*—examine all parents before mating. Avoid use of infected animals.

3. SORE EYES.

    a. *Cause*—usually bacterial.

    b. *Symptoms*—rubbing of eyes, watery or pus-like discharge which mats and scalds the fur and may cause it to fall out.

    c. *Treatment*—treat with ophthalmic solution of sulfathiazole or sulfadiazine 3 to 4 times daily; Argyrol or silver oxide ointment are helpful. Tobacco juice made by steeping tobacco leaves is sometimes effective. Deep-seated infections are difficult to treat, and animal should be discarded unless skilled attention may be given.

    d. *Prevention*—protect animals from drafts and excessive dust, chemicals in paints and sprays, smoke and other fumes and irritants. Isolate affected animals.

4. INFECTED (SCABBY) NOSE OR LIPS.

    a. *Cause*—bacterial, dust, colds and other nasal irritants.

    b. *Symptoms*—inflammation, swelling, chapping, and cracking of nose or lips with formation of yellow, brown, or red crusts.

    c. *Treatment*—local medication is ineffective. Penicillin intramuscularly (10,000 units) at 4-hour intervals is usually successful.

    d. *Prevention*—avoid use of dusty feed, prevent drafts.

C. Of the Mammary Glands.

1. CAKED UDDER.

    a. *Cause*—lack of adequate drainage of the glands.

    b. *Symptoms*—congested, firm, inflamed, and feverish breasts. Advanced stages become hard, devoid of hair and may crack open and emit cheesy exudate.

c. *Treatment*—reduce highly nitrogenous concentrates, allow plenty of exercise, and if possible drain the gland by nursing with own or foster young. Massage to start milk flow and rub down with salve or lanolin.

d. *Prevention*—do not wean young from heavily milking doe too suddenly. If litter is lost, rebreed at once. Remating several days before weaning may help to reduce milk flow. Protect doe from disturbing influences which interrupt proper nursing.

2. MASTITIS.

a. *Cause*—bacterial, usually staphylococcus or streptococcus, sometimes contagious.

b. *Symptoms*—appetite and activity poor, excessive thirst, abnormally high temperature. Nipples and breast become red or purple.

c. *Treatment*—in early stages while temperature is still high (but not over 100) use penicillin, 20,000 units in first and second intramuscular injections; 10,000 thereafter at 4-hour intervals. Discontinue feed and protect from temperature changes. In advanced stages destroy the animal and burn or bury deeply.

d. *Prevention*—avoid bringing affected animals into the colony. Burn all nesting material of infected animals and thoroughly disinfect all nest boxes and other equipment which may have become contaminated. Do not transfer young from an infected doe to a healthy animal.

D. **Of the Respiratory System.**

1. COLDS OR SNUFFLES.

a. *Cause*—any of several types of bacteria and possibly viruses.

b. *Symptoms*—continued sneezing, rubbing nose with front feet, nasal discharge, matting fur on inner side of front feet.

c. *Treatment*—reduce concentrates, and feed plenty of fresh greens. Sunlight and exercise are beneficial. "Nose drops" are usually ineffective, and sometimes are instrumental in spreading infection to bronchii and lungs.

d. *Prevention*—provide plenty of draft-free ventilation, sunlight, and exercise. Increase resistance by selective breeding. Avoid crowding.

2. PNEUMONIA (particularly common in nursing does and very young).

a. *Cause*—bacterial, not usually highly contagious.

b. *Symptoms*—loss of appetite, excessive thirst, labored breathing—often with nose held high. High temperature falling rapidly several hours before death. Diarrhea may occur in last few hours. Advanced stages show bluish coloration of eyes and ears.

c. *Treatment*—successful treatment depends upon early detection. Reduce concentrates and provide greens. Keep at a uniform temperature. Penicillin intramuscularly in hind legs is specific for several of the responsible bacteria. Some types cannot be cured and late cases are fatal.

d. *Prevention* (same as for colds). Pneumonia is usually a secondary disease caused by organisms which become effective in an animal weakened by other diseases.

E. Of the Digestive System.

1. BUCK TEETH.

a. *Cause*—abscesses, injury or abnormal growth of jaws in relation to each other so that continually growing incisors do not occlude.

b. *Treatment*—none, except trimming which will prolong life of the individual.

c. *Prevention*—selective breeding.

2. LIVER COCCIDIOSIS.

a. *Cause*—a protozoan parasite.

b. *Symptoms*—poor appetite, loss of flesh, enlargement of liver, manifest by palpation in living animals.

c. *Treatment*—none. Mild cases are self healing but scars remain in liver. Severe cases are fatal.

d. *Prevention*—use feeding and watering equipment which cannot become contaminated with droppings. Keep hutch floors clean and dry, remove manure frequently.

3. INTESTINAL COCCIDIOSIS.

a. *Cause*—one of four specific one-celled organisms multiplying in the lining of the intestines.

b. *Symptoms*—tendency to soft droppings or diarrhea, poor appetite, loss of flesh, hunched position with hind feet far forward. Mucus may be voided. "Pot belly" on recovery. Pneumonia often a secondary result.

c. *Treatment*—none. Disease runs its course in 5 to 14 days. Resistance may be developed as result of previous attacks.

d. *Prevention*—same as for liver coccidiosis.

4. MUCOID ENTERITIS (SCOURS OR BLOAT).
   a. *Cause*—unknown, probably not contagious.
   b. *Symptoms*—loss of appetite, excessive thirst, hunched posi-
      tion, eyes dull, coat rough, hair erect, abdomen often
      bloated, grinding teeth; dead show much water and gas in
      stomach and intestines. Temperature usually normal or
      subnormal. Intestines show hemorrhage, are empty or
      filled with fluid or mucus. Greatest incidence is near end
      of sixth week; developing and mature stock may be af-
      fected.
   c. *Treatment*—no recommendations.
   d. *Prevention*—no recommendations.
5. TAPE OR BLADDER WORMS.
   a. *Causes*—larval stage of dog tapeworm.
   b. *Symptoms*—absent externally.
   c. *Treatment*—none.
   d. *Prevention*—keep dogs away from premises or at least
      away from feed, water, bedding, and utensils.

F. Miscellaneous.
1. PARALYSIS.
   a. *Cause*—injury to spinal cord, by displaced vertebra.
   b. *Symptoms*—hind quarters drag and are functionless be-
      cause of internal hemorrhage and permanent damage to
      cord.
   c. *Treatment*—none.
   d. *Prevention*—prevent disturbing influences, particularly
      night prowlers and visitors that startle animals.
2. WRY NECK.
   a. *Cause*—infection of organs of balance of inner ear.
   b. *Symptoms*—head twisted to one side, often resulting in loss
      of balance and continued rolling over.
   c. *Treatment*—none. Damage is already done when first
      symptoms are observed.
   d. *Prevention*—eliminate ear mange and other infections.

NUTRITION

The problems of nutrition and feeding methods are of major
importance to the general health and well being of the stock. The
rabbit's digestive system, which is as much as 19 to 26 feet in length,
is adapted to relatively great bulk of matter of low nutritive value.

Not only is there great surface area but it is notable for the enormous development of the caecum. In the wild state the rabbit normally feeds upon large amounts of fresh or dried herbage, including legumes, grasses, cereal grains, and weeds of various sorts. During its natural breeding season, spring-time and summer, there is abundant growth of young green vegetation which is rich in protein, minerals, and vitamins (especially vitamin A) and being soft, succulent, and tender is easily digested. Rabbits do best on a diet composed of these natural sources of food, but in the laboratory particularly it is not always possible to supply any quantity of such feed, and usually it is more economical of funds and labor to make use of the commercially prepared rabbit feeds which are now on the market in pelleted form.

Templeton (1945) recommends rations for dry does, herd bucks, and developing young which provide the following proportions of dietary elements.

	Per Cent
Protein	12–15
Fat	2–3.5
Fiber	20–27
Nitrogen-free extract	43–47
Ash or mineral	5–6.5

This ration is probably a good standard ration for most laboratory animals which are not being worked very hard.

For pregnant does and does with litters he recommends a greater proportion of protein as follows:

	Per Cent
Protein	16–20
Fat	3–5.5
Fiber	14–20
Nitrogen-free extract	44–50
Ash and mineral	4.5–6.5

The above proportions have proved most economical; the upper limits in each give better results than the lowest, and there is no danger in feeding higher levels of protein than recommended providing the ration is adequate in all other ingredients.

Throughout this country, these elements can be secured from the cereal grains plus a certain amount of dry, preferably legume hay, supplemented by a small amount of green feed or root crops. For most rabbits a good standard ration is two parts of whole oats or barley, two parts of whole wheat, and one part of soybean, peanut, or linseed cake in the pea-size or pelleted form, plus all the best-quality legume hay which they will consume, with carrots or green feed twice a week. The last are not absolutely essential in non-pregnant or non-lactating animals. Stock rabbits have been successfully maintained by Hogan and Ritchie (1934) on whole oats and alfalfa plus about 75 cc. of whole milk for lactating females.

Buckwheat and the grain sorghums are palatable and where available may be substituted for barley. Corn, particularly the hard part of the seed, is usually wasted. Ground, rolled, or milled grains are less desirable for feeding and when stored become less palatable and decrease in nutritive value. Even when the best-quality legume hay is also fed, whole grains do not furnish a sufficient quantity of protein and should be supplemented with some one of the high-protein supplements, such as soybean. Such supplements should be fresh, as distinguished by their nutty odor or flavor, and preferably should be in the pelleted form. For rabbits these pellets should be $\frac{3}{16}$ inch in diameter and $\frac{1}{8}$ inch long.

Many laboratories today are resorting to proprietary pelleted rations, of which there are a number in the market. These rations have several advantages, particularly if they include the proper proportion of ground alfalfa hay. In the first place, since all the ingredients are fine ground and thoroughly mixed, each morsel is exactly like every other, and thus there tends to be less wastage resulting from the rabbit's scratching out part of the feed to find the particular kind it desires. Secondly, if the hay is included, the labor of feeding is reduced by about half. However, on such rations young growing rabbits seem to be more susceptible to bowel troubles such as bloat and diarrhea, and in wooden cages are more apt to gnaw the cages in an effort to supplement the roughage content of their ration.

Experiments carried out at the United States Rabbit Experiment Station at Fontana, California, indicate that rabbits which have access to hay in addition to a pelleted ration are less susceptible to impaction and less likely to develop the fur-eating habit, and does so fed are not so likely to eat the bedding material in the nest boxes.

Legume hays, alfalfa, clover, soybeans, etc., for rabbits should be fine stemmed, leafy and green in color, and free from mildew, mold, or dust. The grass hays are much less palatable even when in the best of condition, and there is considerable wastage.

Many rabbits are kept without access to green feed, but a small amount once or twice a week helps to keep them in health and vitality. Legumes are best, but carrot tops, lawn clippings, and numerous weeds such as dandelion and plantain are quite satisfactory. Lettuce and cabbage are apt to cause loose feces and should be avoided in any quantity. During the winter months, root crops such as turnips, beets, and carrots are usually the most easily obtained and economical. Of these, carrots are much the best. Any green feed that is spoiled or moldy should be avoided, and any excess should not be left on the cage floor to spoil and become contaminated with fecal material.

Plenty of fresh, clean water should be kept available at all times and particularly in summer. Salt also should be available at all times. Most of the proprietary foods contain the necessary $\frac{1}{6}$ pound per 100 of concentrates, but if natural grains and hay are fed this should be provided in lumps or spools attached to the side of the cage.

Relatively little definite information is available concerning the mineral and vitamin requirements of rabbits. Hogan and Ritchie (1934) state that "the nutritional requirements of the rabbit are distinctly different from those of the rat or chick and the difference is only in a minor degree due to need of bulk." Rabbits were reared with a considerable degree of success on a simplified ration containing 15 per cent yeast. Bishoff and Sansum (1932) have maintained rabbits for a time on alfalfa alone. When barley was the only food, rabbits developed fatty liver and nephritis.

Vitamin A is apparently essential in the diet. Vail (1944) reports a spastic paralysis, "head-down disease," in domestic rabbits of California, traced to deficiency of vitamin A.

It has not been conclusively established that rabbits require all the known members of the B complex, but it appears certain that they need two or three vitamins which are not now available in crystalline form (Loosli, 1945).

Biotin deficiency can be created in rabbits by feeding rations rich in egg white, which results in a characteristic dermatitis due to a neutralization of biotin by the egg white fraction (Lease, Parsons, and Kelly, 1937; Du Vigneaud, Melville, and Gyorgi, 1940; and Gyorgi et al., 1941).

Niacin (nicotinic acid) is an essential growth factor for rabbits fed a purified diet (Wooley and Sebrell, 1945). Absence causes a loss of appetite, a drop in body weight, and severe diarrhea 24 to 28 hours before death.

The possibility that the rabbit may require vitamin C has been considered by several workers, but apparently none except Findlay has found any reason to believe that this requirement actually exists. Hogan and Ritchie (1934) have reared several generations of rabbits on rations in which vitamin C could not be detected and are convinced that at most provision of vitamin C is a minor problem.

Vitamin D appears not to be essential in the diet; in fact cod-liver oil, which is a standard corrective for D deficiency, is definitely known to have unfavorable influences, eventually resulting in muscular dystrophy. This, however, can be cured or prevented by the feeding of α-tocopherol (Epstein and Margolis, 1941). In the writer's experience 5 to 10 per cent of whole wheat or the equivalent of wheat-germ oil is a corrective and preventive from destruction and particularly eating of young at parturition. The minimum requirement of α-tocopherol is given by Epstein and Margolis (1941) as probably 0.32 mg. per kilogram of body weight.

Rabbits lacking vitamin A become less resistant to anthrax and pneumococcus, and it is probable that, like other animals suffering from pronounced vitamin deficiencies, they are much less resistant to disease generally; since, as indicated above, there is a certain interrelation between the necessary amount of vitamin D and α-tocopherol, it is not improbable that other similar relationships exist.

Manganese deficiency in rabbits results in decreased breaking strength, weight, density, length, and ash content of humeri (Smith, Medlicott, and Ellis, 1934). Deformed front legs is the most evident gross symptom resulting from enlargement of the head and shoulder of the distal end of the radius and ulna and some decalcification, which frequently results in slipping or displacement of the tendons and extreme crippling of the animal. The manganese requirement for the growth of rabbits appears to be about 1.0 mg. daily.

Iodine in the diet, provided in the form of Merck's iodine suspensoid, appears to increase the resistance to coccidiosis, possibly as a result of a general stimulation of the resistance of the rabbit rather than through its effect on the organism. Chesney et al. (1928) reported goiter in rabbits fed a diet consisting solely of cabbage. These goiters were associated with a lowering of the metabolic rate, and feeding of iodine raised the metabolic rate and prevented thyroid hyperplasia. Bauman et al. (1931) confirmed these findings and concluded that cabbage contains a goiterogenic substance which acts by depleting the thyroxin store of the thyroid gland.

FEEDING DIRECTIONS

Proper feeding of laboratory animals is an essential to maintaining a vigorous breeding stock useful in experimental investigations. In so far as practical in the laboratory it is desirable to maintain as nearly natural conditions for the animals as possible. Rabbits normally are more active and eat more at night than during the day. This is especially noticeable during the warmer season of the year. Where grain and hay are both fed, a good practice is to feed the grain the first thing in the morning (what the rabbits will consume in about ½ hour) and the hay at night. Where complete pelleted rations are used, it is probably best to feed late in the afternoon. Adult animals which are not being bred are best hand fed in this manner. Fat animals are undesirable, and particular attention may be given to the condition of the animal or the needs of a particular experiment. In most laboratories there seems to be a tendency to overfeed experimental rabbits to the extent that they become excessively fat and sluggish. This is particularly true where the cages are small in size and prevent the normal exercise of the animal. To keep adult rabbits in the most healthy condition they should never be fed quite all that they will eat and should always be impatient for the feed at feeding time. The average adult rabbit will consume approximately 2 to 4 oz. of grain daily, depending on size. In this respect much of the success of the colony depends upon the interest, powers of observation, and judgment of the attendant who has the animals in his charge. Young growing stock in which most rapid development is desired should have both grain and hay in front of them at all times, and the methods of self-feeding recommended by Templeton, Ashbrook, and Kellogg (1946) are essential.

Particular attention should be paid to the condition of the feed and water receptacles at every feeding. Water crocks or bottles that have become contaminated by fecal material, hay, or grain should be replaced before new feed is supplied, and the old container should be thoroughly washed, scalded, or otherwise disinfected.

Although many believe that the rabbit, like the guinea pig, can get along without water being added to the diet, this applies only to the situation in which a large portion of the diet consists of fresh succulent vegetables, and animals fed on dry feeds should have an abundant supply of clean pure water in front of them at all times.

REFERENCES

Bauman et al. 1931. *Proc. Soc. Exp. Biol. Med.*, vol. 28, p. 1017.
Bensley, B. A. 1938. *Practical anatomy of the rabbit.* Blakiston, Philadelphia.
Bishoff, E. S., and B. L. Sansum. 1932. *Jour. Nutrition*, vol. 5, p. 403.

Castle, W. E.   1930.   *Genetics of the rabbit.*   Harvard University Press, Cambridge, Mass.
——.   1940.   *Mammalian genetics.*   Harvard University Press, Cambridge, Mass.
Chesney et al.   1928.   *Bull. Johns Hopkins Hosp.,* vol. 43, p. 261.
Du Vigneaud, V., D. Melville, and P. Gyorgi.   1940.   *Science,* vol. 92, p. 62.
Epstein and Margolis.   1941.   *Jour. Nutrition,* vol. 22, pp. 415–424.
Gyorgi et al.   1941.   *Science,* vol. 93, p. 477.
Hammond, J.   1925.   *Reproduction in the rabbit.*   Oliver and Boyd, London.
——.   1928.   *Züchtungskunde,* vol. 3, p. 523.
——.   1934.   *Jour. Exp. Biol.,* vol. 11, pp. 140–161.
Hammond, J., and A. Walton.   1934.   *Jour. Exp. Biol.,* vol. 11, pp. 307–319.
Hartman, C. G.   1945.   *Annals N. Y. Acad. Sciences,* vol. 46, pp. 23–44.
Heape, W.   1905.   Ovulation and degeneration of ova in the rabbit.   *Proc. Roy. Soc.,* ser. B, vol. 76, p. 260.
Hogan, A. G., and W. S. Ritchie.   1934.   *Missouri Agr. Exp. Sta. Res. Bull.* 219.
Kellogg, C. E.   1936.   *U. S. Dept. Agr. Circ.* 410.
Lambert, W. V., and Fred F. McKenzie.   1940.   *U. S. Dept. Agr. Circ.* 567.
Lease, J. G., H. T. Parsons, and E. Kelly.   1937.   *Biochem. Jour.,* vol. 31, pp. 433–437.
Loosli, J. K.   1945.   *Annals N. Y. Acad. Sciences,* vol. 46, pp. 45–75.
Macironi, C., and A. Walton.   1938.   *Jour. Agr. Science,* vol. 28, pp. 122–134.
Marshall, F. H. A.   1922.   *Physiology of reproduction.*   Oliver and Boyd, London.
Marshall, F. H. A., and J. Hammond.   1946.   Biol. 39, Ministry of Agr. and Fisheries.   His Majesty's Stationery Office, Edinburgh.
Pickard, J. N., and F. A. E. Crew.   1931.   *The scientific aspects of rabbit breeding.*   Watmoughs, London.
Pincus, G.   1936.   *The eggs of mammals.*   Macmillan, New York.
——.   1937.   *Jour. Morph.,* vol. 61, pp. 351–376.
Rosahn, P. D., H. S. N. Greene, and C. K. Hu.   1934.   *Science,* vol. 79, pp. 526–527.
Smelser, G. K., A. Walton, and E. O. Whetham.   1934.   *Jour. Exp. Biol.,* vol. 11, pp. 352–363.
Smith, S. E., M. Medlicott, and G. H. Ellis.   1934.   *Arch. Biochem.,* vol. 4, pp. 281–289.
Templeton, G.   1945.   *U. S. Dept. Interior Wildlife Leaflet* 266.
Templeton, G., F. G. Ashbrook, and C. E. Kellogg.   1946.   *U. S. Dept. Agr. Farmers' Bull.* 1730.
Tietz, E. B.   1933.   *Science,* vol. 78, p. 316.
Vail, Edward L.   1944.   Spastic paralysis in domestic rabbits.   *Jour. Am. Vet. Med. Assoc.,* vol. 104, pp. 334–335.
Wing, F.   1945.   Thesis.   Brown University.
Wooley, J. G., and W. H. Sebrell.   1945.   *Jour. Nutrition,* vol. 29, pp. 191–199.

7· THE DOG

LEON F. WHITNEY
Whitney Veterinary Clinic
Orange, Connecticut

INTRODUCTION

The adaptable dog presents no great problems of care under laboratory conditions. He can stand extremes of heat and cold without much conditioning. He can be happy whether maintained in a cage or running over acre tracts of land. He can live on such a variety of foods that the feeding problem, like most of the other problems, is a matter of economics. He can live alone or with a group. All he asks is food, sanitation, freedom from parasites and disease; with these, he makes a good and faithful servant, a willing and eager slave.
In return for his service, he deserves all the comfort we can afford him.

ACCOMMODATIONS

There are many ways of maintaining dogs, ranging from the single cage to the large outdoor kennel. Every institution has its own individual problems. Some must of necessity maintain its dogs in basements under classrooms, laboratories, offices. Others own country property where the dogs may be maintained by a staff and transported back and forth to the laboratories. Some institutions are able to keep part of their dogs conveniently in the laboratories and part in country kennels where the staff goes to do their testing. Such a combination is ideal, especially for studies in the fields of nutrition, psychology, and genetics and long-range studies such as those in pathology and physiology.
Almost every study starts out with a limited budget. The student is desirous of stretching every dollar to its fullest extent. It costs far more for assistance in the single-cage indoor kennel than it does in the outdoor community-pen system. It even costs less for help when every dog in outdoor runs is maintained singly, depending, of course,

on the methods employed. The author took entire care of a kennel of 200 dogs and puppies single handed for a year, kept careful time records of feeding, watering, weighing, grooming, quieting dogs from barking, cleaning runs, and renewing sand occasionally. All these activities for the 200 dogs were handled in five hours a day. They were done with stopwatch efficiency, which the help of today cannot be expected to duplicate.

In the present Whitney Veterinary Clinic, with accommodations for 200 individual dogs, half in the hospital and half outside, it is possible

Fig. 1. Pens used in rabies studies, where positive control is essential. Part of Rockefeller Institute rabies laboratory at Montgomery, Ala.

to compare costs carefully. Aside from medical treatments, two men are required to manage the indoor dogs properly, as compared with one who easily manages the outdoor facilities and whose dogs are healthier.

Indoor kennel work requires constant changing of newspapers from the cage bottoms, everlasting diligence in hunting odors, frequent watering, repairs, and treatments, not to mention scrubbing floors and paint. The outdoor man, half of whose charges are kept in kennels with indoor rooms and outdoor runs and the other half with individual houses for each run, feeds, waters, cleans runs, renews sand, powders dogs against external parasites, and in the afternoon has time for repairs—painting, carpentering, etc.

We find the following types when considering outdoor accommodations:

1. Single pens, isolated, with individual houses.
2. Multiple pens, isolated, with individual houses.
3. Multiple pens together with individual houses.
4. Buildings with individual rooms and runs attached.
5. Buildings with community rooms and runs attached.

Isolated single pens have the advantage of keeping one dog entirely away from the others except for access of insects, birds, and rodents which may transfer disease or parasite eggs.

FIG. 2. Adequate pens attached to long building serve to contain one dog or many in each pen, as requirements demand. There are an equal number of pens at rear of building.

Multiple pens permit somewhat more efficient management.

Houses within or attached to any type of outdoor runs can be of simple construction. A vinegar barrel, having an oval hole in one end and resting on four legs driven into the ground, is inexpensive

FIG. 3. One of the author's isolated pens, built in the shelter of the woods. Dogs thrived and enjoyed the shade in summer.

and thoroughly satisfactory. Dogs lie to one side to escape the sun in hot weather when no other shade is provided, and they soon learn to use the barrels in rainy or snowy weather. Deep straw bedding in them is sufficient to provide warmth.

Kennel buildings with attached runs may be of a variety of designs from octagonal buildings with runs radiating out to plain straight

Fig. 4. A simple, crude pen, yet useful in confining these large crossbred dogs.

alleys with indoor rooms on either side and runs outside. Dogs get exercise running the length of the run or playing in the middle. Therefore long narrow runs are probably preferable to wide square ones.

Fig. 4A. Larger pens are advisable for larger breeds.

One of the first ideas of the neophyte kennel designer is a concrete run. But most kennel managers soon tear them out or wish they had never installed them. Dogs kept on concrete are usually badly infested with internal parasites as compared with those kept on sand. The finest pores of concrete are large compared with eggs of parasites, which lodge in them, secure against the hoe or shovel which scrapes the surface. Constant washing with water provides an excellent environment in the warm months. Many institutions which have laid concrete have great difficulty with whipworms. To properly wash such runs every day requires a great amount of water under pressure, scrubbing with a heavy brush, and ten times the hours which sand-run cleaning takes. The alternative to washing is burning—efficient to kill parasites and their eggs but a slow and time-consuming task.

Fig. 5. Plans for construction of wire-bottomed cages. In such cages the author raised hundreds of small puppies. The wire bottoms do not harm the feet.

Gravel or cinder runs are not advisable. In cleaning sand runs the assistant takes a shovel and hoe and tips the stools onto the shovel, spreading them as little as possible and taking a little sand when the stools are soft. At the end of 6 months—in spring and fall—2 inches of sand can be shoveled away and 4 to 6 inches more added, to replace what was removed during the 6 months, plus the newly removed 2 inches. It soon packs down. In this way, the runs are kept clean and neat except when the dogs attempt to dig holes in hot weather. This inconvenience is more than offset by the greater comfort to the dogs because lying on concrete in hot weather is exhausting. Concrete runs often cause huge calluses to develop on the elbows and other parts of dogs of larger breeds.

Fences seldom need be over 6 feet high. It is less expensive to stretch 16-gauge, 2-inch mesh hexagonal netting over the top than to make runs 7 or 8 feet high, especially when the cost of posts and labor is considered. Dogs which can negotiate a 6-foot fence usually do it by climbing, and these same dogs are able to climb an 8-foot fence.

Link fence wire is excellent, but metal posts are needed and gate frames must be neat. It has few advantages over the wire shown in figure 4A. This has lasted for 20 years only 3 miles from the seacoast, where ordinary hexagonal poultry netting often rusts away in 5 years. The portion buried in the ground lasted 15 years.

It pays to use cedar or cypress posts. Oak seldom lasts over 8 years. Chestnut is still available and is excellent. Impregnating posts in coal-tar preservatives may add a year or two to their lives.

FIG. 6. Puppies happily growing in a wire-bottomed pen.

FIG. 7. A battery of wire-bottomed cages used at Alabama Veterinary College was completely successful.

Very few laboratories have learned the value of wire-bottom outdoor cages. Figures 5, 6, 7 and 8 show one practical type and a group used

FIG. 8. Outdoor wire-bottomed pens may be used in various ways. Here, the covered one contains puppies testing a diet for vitamin-D content.

by a veterinary college. These cages offer excellent ways to keep several companionable small dogs together. The author has raised

FIG. 9. One side of the author's experimental kennel. In this type of pen he has raised dogs successfully for 20 years.

over a thousand puppies in such cages. The wire bottoms strengthen rather than weaken feet. Dogs such as cocker spaniels and beagles can be raised to full growth in them and allowed out for romps in large exercise runs if it is desired.   They are practical, easy to manage, and inexpensive to build.

Some kennel managers favor taking dogs into buildings at night and putting them out in separate runs when the weather is propitious. The system has some advantages but even more drawbacks.

At the Whitney Veterinary Clinic, where we have 4 acres fenced, and our many types of pens within, we have found it especially desir-able to keep some sheep.   Two or three to the acre will crop the grass and eat weeds and poison ivy, which grows luxuriously if not repressed. Contrast the illustrations (figs. 2, 9) showing the lawn about the kennel runs with the appearance of the ground (fig. 4A) before sheep became an institution.   Until they were in-troduced, upkeep was a serious item. Now, the sheep maintain an excel-lent lawn between pens and furnish half a dozen lambs a year for food.

Sheep have two principal draw-backs: they excite some dogs to bark, and if grass is short because of over-crowding or drought, they will eat the bark of certain species of trees and kill them.

The indoor cage question has been settled by various laboratories and hospitals in ingenious fashion.   The body may be of heavy metal with

Fig. 10.   Daytime pens of link-mesh fabric are found to be satisfactory in some institutions.   Dogs can be trained to run out and return, so they need not be carried.   The illus-trations show a walk onto which 38 pens open.

or without a pan.   Cages generally are built in tiers of two or three. Pans are preferable even though slovenly assistants have to be watched to see that they clean on both sides and the space under them.   Dogs sometimes urinate against the side of the cage.   Small flanges are sometimes soldered to the cage insides, standing out a quarter of an inch over the pan edge, to guide urine and liquid feces into the pan and not behind it.

Cages may be on rollers or built in solidly.   Some institutions have steam rooms into which whole cages can be pushed for disinfection. This is seldom necessary if proper disinfection is applied daily.

Fig. 11. Cocker spaniels thrive outdoors in such pens, provided they are well bedded in winter. The sand in such heavily populated runs needs renewal four times a year.

Cage doors and latches may be purchased from several concerns and generally for much less than they can be made. Usually link wire is

Fig. 12. In these and other, similar outdoor pens, the author has raised many thousands of dogs. The netting is 1½-inch 16-gauge fox netting. It has lasted from 1928 to 1948.

laced into frames. Some may prefer iron bars for their greater ease in cleaning. Cage doors should be attached so that the hinge pins may be taken out and the doors taken to a large washing space at regular intervals, where they may be scrubbed clean.

Cages may also be equipped with 1-inch square mesh wire bottoms made of 11-gauge or heavier wire. This can be electric welded to angle iron or pipe frames. Every wire has to be welded or dogs will soon work the edges loose.

Feeding pans may be of a variety of materials. A heavy aluminum pan has been introduced with an extra-heavy edge which dogs do not chew. Enamelware and thin aluminum are toys on which idle dogs exercise their teeth and which they soon ruin. A pan which holds 2 quarts is an excellent average size.

Water may be supplied by faucet connections with float shut-offs, but most dogs interfere with these devices and soil the water so quickly, thickening it on hot days, or slopping it by pawing at it, that water is best given them in pans. Those animals which play with the water cannot have it left in their cages. Pails attached to the sides of outdoor pens are satisfactory, but any type of watering device must be frequently cleaned.

ISOLATION

It is the usual practice to isolate all newly obtained dogs for 2 weeks. This period will generally allow any dog incubating a disease (except rabies) to manifest the symptoms. Where possible, the isolation section can be used to immunize dogs against disease, as well as to deworm them and kill external parasites. Careful centrifuged fecal examinations should be made of every dog used in research. Many misinterpretations of data have been made because of the effects of parasites.

DISEASES

Carré Distemper. In some areas, the principal dog disease is Carré distemper. It first manifests itself, after a 5-day incubation period, by nausea, occasionally fits, and photophobia. Soon the nose and eyes fill with a stringy mucous discharge; the stool is loose, evil smelling, and sometimes bloody. Pustules may form on the skin of the belly (not a constant or a diagnostic feature). Temperature averages around 104.0°. The appetite is gone; dehydration may occur. The disease may drag on, with one secondary bacterial disease after another setting in until death or recovery occurs. Encephalitis symptoms may ensue, and from it, when it takes the form of convulsions, dogs seldom recover. Chorea twitches may be another form of encephalitis.

Carré distemper in some sections is an exceedingly rare disease because of general successful vaccination of dogs against it. There are today a dozen methods by which dogs can be immunized. Most of them depend for their success on the use of potent biologics. The original Laidlaw-Duncan method embodied a dose of formalin-killed fresh vaccine made of a 20 per cent suspension of spleen tissue from a dog sick with the disease. This was followed in 5 to 7 days with a dose of live virus. An American modification of this was two 5-cc. doses of vaccine 2 weeks apart followed in 2 more weeks by a dose of live virus.

Most of the commercial "live" virus has been long since dead. It
has been found that 5-cc. doses of stale vaccine (most commercial
vaccine is very stale when it reaches the user) is insufficient to immunize large dogs. The method used by the author on at least 10,000
dogs with great success is as follows: Only fresh vaccine sent by
refrigerated air express from the producer is used. The dose is graded
to the size of the dog. Small dogs receive 1 vial (5 cc.). Large dogs
of 100 lb. or more receive 6 vials (30 cc.). One dose elicits excellent
antibody response, far more effective than two small doses—usually
too small to give reasonably solid immunity.

The Green method is to administer live virus. This has been passed
through several successive ferrets, which have attenuated it to such
an extent that it causes a feeble disease in dogs but renders them
immune to ordinary strains of distemper. There has been much
argument over the method. In general, the response has been favorable provided fresh virus can be obtained.

Several other methods are in use, but those outlined above are most
general.

Pharyngolaryngotracheitis. Pharyngolaryngotracheitis (housedog
disease) is a new disease, discovered in 1940, which has swept the
world and is most troublesome in laboratory dogs. The true disease
is often unrecognized. It is manifested by gagging, varying in intensity, during which the dog raises some froth from his throat. He
may spit this out or swallow it. The gums, pharynx, tonsils, larynx,
and trachea usually show inflammation and redness. Tonsils may
protrude from their crypts across the throat until they touch. Temperatures in the uncomplicated disease range around 102.6° F.

This throat phase—the real disease—continues for a week to 10
days. The causative agent is probably a virus. A bacterial pneumonia often follows but may be prevented with sulfa drugs or anti-
biotics which have no effect on the real disease. The dogs usually
appear entirely well, and for most of them, the disease is then at an
end. In some (the younger the more likely), encephalitis develops,
when the virus invades the brain. It has been observed in nine forms.
Only about 10 per cent of dogs over a year and 35 per cent of those
under 6 months old develop this later condition. In some breeds, the
mortality in puppies is close to 100 per cent.

As yet we have no defense against housedog disease. If a roomful
of laboratory dogs starts to gag, it is better to start no vital studies
with them until a month or so has passed to avoid all danger of

encephalitis. It is not known whether such dogs are immune for life or only for a year. If a dog develops encephalitis, he often eats hungrily. This alone helps to differentiate housedog disease from Carré distemper.

Dogs immunized against distemper either artificially or by having recovered are not immune to housedog disease and vice versa. Carré distemper biologics do not immunize against housedog disease.

Hardpad Disease. Much has been written lately about a new disease which has appeared in England and America. This disease has symptoms similar to those of Carré distemper but causes the feet pads and the nose to become hard and keratinized. At present investigations indicate the possibility that this disease may be a mutant form of Carré distemper virus. An English study showed that the disease could be cured by injections of small doses of live Carré distemper virus.

Temperatures in hardpad disease are within the range of those of Carré distemper. Encephalitis ensues in many cases. Further research results are awaited.

Rabies. In areas where rabies is known to exist, and where dogs are shipped to laboratories in rabies-free areas from rabies areas, great caution must be exercised. Attendants should always be on the watch for dogs with personality changes, perverted appetites, insensibility to pain, and throat paralysis. Since many dogs with housedog disease which develop encephalitis show similar symptoms to rabies, there is considerable confusion. Rabies generally has a downhill course, while housedog disease encephalitis most often manifests itself in a series of fits with intermittent periods of normalcy and secondarily as chorea jerks. But there have been many cases of housedog disease encephalitis so similar to rabies that only mouse inoculation tests could differentiate them. Distemper produces mostly intracellular inclusion bodies; rabies, when the disease has progressed long enough, produces intranuclear inclusions; housedog disease produces no inclusions.

Dogs in laboratories or in confinement seldom need rabies vaccinations, especially when they are kept in individual cages.

Skin Diseases. Skin diseases are prevalent in laboratory dogs, especially during summer and damp weather.

Manges may be quickly determined by skin scrapings and may be treated with solutions bearing 1 per cent rotenone, DDT, or benzene hexachloride.

Most of the summer itches with scab-forming infections are thought to be of fungous origin. Oily suspensions of fine sulphur powders or other fungicides generally afford prompt cures. Use of insect powders containing efficient fungicides acts as a deterrent or preventive.

Bacterial Diseases. Bacterial diseases occasionally cause losses. Leptospirosis (a disease spread by rats) is one. It appears in two forms. One (icterohemorrhagic) induces jaundice and hemorrhages. After these symptoms appear in intensity, there is usually no hope for the dog. The canicola form produces symptoms very difficult to diagnose until too late for treatment. Penicillin seems quite specific against this spirochetal disease when used in the early stages. Because of a relatively long incubation period, dogs incubating the disease are often bought—a valid argument for an isolation section in every institution using dogs.

Tumors. Tumors occur with considerable frequency. One dog in every fifteen of all ages seen at the Whitney Veterinary Clinic had a growth of some sort. Some were highly malignant; some benign; some mere hematomas; some inflammatory lesions. Older dogs appear to have the greater tendency to neoplasms. It is therefore advisable to obtain younger animals for critical research.

PARASITES

Internal Parasites. Parasites are easily managed. Periodic fecal examinations should be a part of every well-run dog laboratory. Normal psychological responses are not always evoked in anemic animals heavily parasitized with hookworms. Nutrition tests are practically worthless if part of the animals are infested with parasites. One study was about to credit a deficiency of a vitamin with producing baldness in a dog until it was found that the dog had a heavy infestation of tapeworms.

There are several ways of eliminating intestinal parasites, but it is doubtful if more efficient or safer means than the following have been evolved:

ROUNDWORMS AND HOOKWORMS. Starve dogs 24 hours. Then administer in capsules 1 cc. of tetrachlorethylene for each 8 to 10 lb. of weight. Follow in an hour with a dose of arecolene hydrobromide (see under tapeworms).

WHIPWORMS. Starve dogs 24 hours. Then give sodium pentabarbital at a dosage of 1½ grains for each 25 lb. This will act as an antiemetic. In 15 minutes, give normal butyl chloride in capsules, 1 cc. for each 3 lb. of body weight up to 75 lb., and 1 cc. for each 4 lb. above that. Feed 2 or 3 hours later. Give no physic. The drug will be flushed into the caecum, where many whipworms reside.

TAPEWORMS. Starve dogs 24 hours. Then give arecolene hydrobromide in sugar-coated pills, $\frac{1}{8}$ grain for each 15 lb. of body weight. If a dog is easily nauseated, give the drug in three split doses 7 to 10 minutes apart. As mentioned above, arecolene may be used as a physic following tetrachlorethylene, when it will destroy any tapeworms the dogs harbor—often unknown to the attendant.

COCCIDIOSIS. No effective treatment is known. A high-fat diet helps the dog through the disease.

External Parasites. Fleas and lice can be destroyed by the use of any of the better flea powders. Some persons like derris or cube roots with 5 per cent rotenone; some prefer DDT. Casualties have been reported from a too liberal dusting of DDT in closed areas. Benzene hexachloride diluted with clay in powders is effective. The gamma-isomer percentage should be 1 per cent.

Fleas are effectively destroyed by dusting only one spot on the dog, but lice of course must be touched by the powder, and since they do not move like fleas, the entire dog must be covered. When dogs sleep on bedding, a potent powder mixed with the bedding at each change kills the insects on the dog by the dust he stirs up when he lies down.

Ticks can be destroyed with 10 per cent benzene hexachloride dust, provided the gamma isomer is up to 1 per cent, or with a 1 per cent Lindane dust. (Lindane is the name for the gamma isomer.) Chlordane is another effective component of insect powders.

DEBARKING

While not a health measure to the dog, debarking often seems like one to attendants and those trying to work near a roomful of barking dogs. Debarking is a simple operation which is very effective when properly done. Under general anesthesia, the vocal cords are trimmed off, preferably with scissors with serrated edges to cut down on bleeding. After the operation, the dog is left with the head lower than the body to allow any blood to drain out. The dog can have as much fun making the motions of barking without producing an irritating effect. Occasionally tissue grows back, but it may be retrimmed.

DISINFECTANTS

For cleaning there are standards not only for destruction of bacteria but also for elimination of odors. The kind of disinfectants *not* to use are the obvious phenols and creosote which may accomplish the purpose. These are often used by slipshod attendants to mask odors instead of eliminate them.

Solutions such as sodium hypochlorite which liberate chlorine are eminently satisfactory. The slight temporary chlorine odor soon passes. If dog odors develop, the attendant must locate the source. He should be made to understand that no odors are permissible for long periods. Some laboratories use disinfectants which permeate whole buildings. There are phenol derivatives which have no unpleasant odor whatever. Soaps, too, are excellent. If cages have common drain pipes under them, as are utilized in many dog rooms, a large amount of water should be flushed through the pipes or the urine odor will soak into the clothes of everyone who enters the room; this odor dissipates only after several hours.

VENTILATION

Ventilation is imperative but need not be of so great an intensity as to cause drafts. Suction fans which pull the stale air out of vents need not be strong enough to waste valuable heat in the winter. Dogs rooms may be equipped with fans to go on automatically and change the air at regular intervals.

IDENTIFICATION

Dogs need identification. Many methods are used from metal tags in chains to leather straps. There is no need for any of these appurtenances if a small tattoo outfit is obtained. Those operated electrically can be bought, but these are no better than letters made up of many small chisels. The ear is shaved; an area to be tattooed is painted with the sterile antiseptic ink. The holder containing the letters is placed over the inked area, and the handles are squeezed together. The chisels push the ink through the skin. Then the device is removed, and the ink is scrubbed over the area again and allowed to dry or the surplus may be at once wiped off. Three places for letters or numbers are afforded so that an immense number of combinations is possible. The dog can never lose such identification as he can a tag or collar.

FEEDING

Dogs are as adaptable to foods as to environments. What to feed may be a matter of what is available. There may be table scraps from the kitchen of a large hospital. A budget can be stretched greatly when foods cost nothing. One might use all this book on the question of feeding.

It is not necessary to buy meat. Probably the best buys in complete rations consist of the better types of dog meals. There are at least a dozen available which have been built on careful research. One of these has grown fifteen generations of dogs; others, eight and ten.

Probably no simpler, more complete, and economical ration has yet been devised than a high-class meal-type food to which has been added 20 per cent of fat. This may be any edible fat of vegetable or animal origin. It may be ground or melted and mixed. Institutions have large quantities of old fat used in cooking which they dispose of at a few cents a pound. Butchers sell their fat trimmings to fat-rendering companies for considerably less than dog food costs the laboratory. This fat runs 3500 to 4000 calories per pound, whereas the dog food runs about 1500. Since there are over twice as many calories in a pound of fat as there are in a pound of a good meal-type food, fat costing as much as meal per pound gives twice as much for the money. Besides the money-saving feature, fat has many other virtues, such as increasing palatability and conserving B vitamins.

Some laboratories attempt to furnish variety in food to their dogs, but this is unnecessary since dogs can smell each of the ingredients of food and obtain variety at every meal.

Besides the meals, there are pellets consisting of meal compressed into various shapes and held together with a binder. Dogs generally relish them more than the meal types, but they have the disadvantages that they are more expensive and, because of the large lumps, other foods are not readily mixed with them.

Dog biscuits, most of which are made of 75 to 85 per cent second-grade wheat flour, are poor dog food indeed. Canned foods, while more palatable, are far too expensive for laboratory dogs and no better biologically than meals. Each one-pound can contains about 450 calories vs. 1500 in a pound of meal, and the can usually costs considerably more than meals when bought in the quantities in which laboratories buy their foods. With 75 per cent water, canned foods have to be fed in bulk; the empty cans have to be disposed of and greater storage space is needed. The cost is about four times the cost of feeding meal-type foods.

The question of whether money can be saved by individual laboratories mixing their own foods often arises. If the time of assistants is of no consideration, and if there is plenty of storage space for the unmixed ingredients, a little money can be saved. But the manufacturer, by buying his ingredients in carload lots, obtains a much lower price than can an individual purchasing in hundred-pound or even one-ton lots. Most manufacturers are happy to supply foods at a cost of one dollar a hundred pounds above the cost of materials and the labor of mixing, and their efficient mixers can provide far more uniform foods than manual mixing can accomplish.

BREEDING

Shall the laboratory breed its own stock? Many have tried and discontinued it. This is not alone because of epizootics and the difficulties involved in raising puppies but also because breeding costs too much, and would cost too much even if there were no puppy mortalities. Students have come to expect laboratory dogs to cost no more than $5 or $10. If all the costs of raising puppies to maturity are considered, dogs will cost at least $25 each for the small breeds and more for the larger breeds.

A case in point is this author's experience in raising a strain of uniform pups. For five years and six generations, a strain of beagles was developed by inbreeding. The offspring were uncommonly uniform. The work was encouraged by students who needed dogs and had frequently urged the creation of such a type. The usual advantages to laboratories were cited. After several hundred dogs had been bred and a nice lot was ready, these men and anyone else who might need them were notified. The price of $25 each was too high. If the dogs could have been sold for $5 or even $8, they would doubtless have found wide usage. Budgets would not permit the extra costs. These beagles were finally sold to rabbit hunters for $20 each.

If anyone wants to breed uniform dogs, beagles offer an excellent opportunity because they are uniform already even though not inbred to any extent. But the student must expect to pay well for the uniformity. He must decide whether or not the advantages are worth the cost.

The Mating Cycle. For those who are desirous of raising puppies, a few points based upon considerable study, some of which upset dog-breeding tradition, may be worth consideration.

The mating cycle of the average bitch lasts about 22 days. The vulva swells, the appetite increases, a bloody discharge of variable amount appears in the vulva; the bitch is inclined to urinate frequently and in small amount. If allowed her freedom she likes to ramble for long distances. She becomes more playful in her enclosure but will not copulate. The period is variable as to length but lasts about 10 days. In some bitches the duration is no more than 5, and in others over 14, days.

The acceptance period is that part of the mating cycle which begins with the first time the bitch will voluntarily copulate, and ends with the last day, if that day is judged by the refusal to copulate despite the amorous teasing of a vigorous male. Both these days—first and last—might be judged differently by different investigators. Also, the dog used has much effect on the result, as has the tenderness of the vulva and clitoris. Many bitches will appear to be willing to copulate but will jerk away when the dog's penis touches the clitoris. Some dogs seem unable to evoke a willingness to copulate on the bitches' part, while others will be teased by the bitches which a moment before refused copulation with previous studs. Determining the first day of the acceptance period is not always easy.

The same must be said for the final day. At this time the luteal hormone has shut off desire for copulation, but how potent its effect is cannot be judged accurately. Many a bitch has shown no response to a certain stud and been judged over her period. When allowed to run with a different stud, she may suddenly allow copulation.

What is of most import in breeding for fertile matings is the day of ovulation. This has been placed by two investigators as about the first day of acceptance. My studies place it well along in the acceptance period—about the fifth day for the average bitch. With some it undoubtedly is the first day, while in others it may be the twentieth day of a long acceptance period, but in the greater part it falls at the fifteenth or sixteenth day of the mating cycle, counting from the first day. These times were arrived at by correlating birthdays with known copulation days.

Because the average bitch is mated several days before ovulation, the gestation period is generally calculated as being 63 days. It is probably 61 days from ovulation.

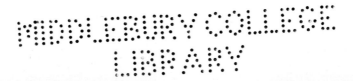
MIDDLEBURY COLLEGE LIBRARY

The optimum time for mating a pair of dogs to obtain the largest possible litter is any time from the middle of the acceptance period to the end. Fertile matings have resulted from copulations which occurred a day after the bitch refused the male, when she had to be restrained to permit the mating. Matings made at the first day of acceptance tend more often to result in "misses."

Fɪɢ. 13. A useful adjunct to any research kennel, these wire-bottomed cages are helpful in deworming, skin-treating, or drying dogs in summer after baths.

ACCESSORIES

Every dog laboratory needs certain accessories. These are combs, brushes, and a clipper, besides some minor gadgets to fit individual needs.

Combs are of all types. Most of them are of little value because the teeth bend too easily. The best and strongest combs can be had in drugstores in Negro sections of any city. They are used in hair straightening and are brass with wooden handles. Sometimes the teeth are too blunt and need pointing with a file, but most are excellent as they are sold. The teeth cannot bend.

Hard scrubbing brushes are enjoyed by the dogs and are inexpensive and efficient.

Clippers are made by only two companies in America, and one is superior to the other in our tests. Blades can be had in several sizes, each size cutting the hair a different length. One blade shaves, one cuts hair about $\frac{1}{8}$ inch, another $\frac{1}{4}$ inch, and another $\frac{3}{4}$ inch.

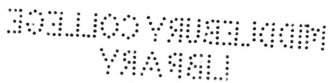

If long-haired dogs are kept clipped, they are easier to care for and are more comfortable in indoor quarters. For outdoor accommodations, the hair may be kept long, especially when fall and winter come. A dog trimmed as late as June will have grown a luxuriant coat by fall.

Bathing may be resorted to if dogs develop too strong a "doggy odor." It does not harm them. But most laboratory and kennel dogs are not bathed all their lives. The old soiled hair falls out in early summer; shedding can be assisted by combing, which also adds much to the dogs' comfort.

Some dogs in laboratories, like those in homes, shed the year round, whereas outdoor dogs have a tendency to shed during a definite period. This variation is believed to be due to the light, which lasts longer to the outdoor dog for part of the year and shorter for the rest. When the days are longest, shedding occurs and the new strong coat grows in as the days slowly shorten. The laboratory dog and the housedog are subjected to the same length day the year round because electric lights are turned on when the natural day grows shorter. This may account for the phenomenon of constant shedding, which, if it is a nuisance, may be corrected by adjusting the hours of light.

It has been found a distinct asset to provide special cages for drying dogs and deworming them, wherever large numbers of dogs are kept. These can be supplied with wire bottoms so breezes can dry dogs from underneath. In the deworming procedure, the liquid part of the stools drops through. Figure 13 shows an efficient battery of such cages.

8· THE DOMESTIC CAT

Felis catus Linnaeus, 1758;

Felis domestica Gemlin, 1788 *

ALDEN B. DAWSON
The Biological Laboratories
Harvard University
Cambridge, Massachusetts

INTRODUCTION

Most cat colonies are holding colonies only, where cats are held temporarily for special purposes and usually no attempt is made to maintain or expand a colony by breeding. Even in cases where animals are mated for the study of reproduction, the young are seldom kept beyond weaning. Ordinarily it is more economical to purchase animals as they are needed rather than rear them to sexual maturity. Although pure-bred animals are available for the establishment of special breeding stocks, no one other than commercial breeders appears to have felt the need, or have had the necessary financial support, to develop genetically homogeneous material. Consequently, except for genetic studies, the cats on which scientific observations are made represent for the most part a heterogeneous stock derived from random matings of various breeds and crosses. The divergent statements in the literature regarding certain phases of reproductive activity in the cat may be attributed in part to genetic differences, but it seems more probable that they are environmental in origin. The data are based on observations made in widely scattered geographic areas, including Warsaw, Algiers, England, and in the states of Illinois, Massachusetts, and Maryland, where climatic conditions are quite diverse. In addition age, nutritional level, and general health may greatly modify the breeding behavior of cats.

Everyone seems to know how to care for a cat about the home, and no one apparently has difficulty in maintaining conditions favorable

* See Mellen (1946, page 2) and Asdell (1946, page 166).

to reproductive activity. However, many investigators comment on the difficulty of breeding cats in confinement, but few record the precautions and positive measures that should be taken in the laboratory to insure success. Because of the paucity of this kind of data in the literature it seems best to describe the methods that were evolved in our laboratory which enabled us to maintain an actively breeding colony for a period of six years, without any serious difficulties, outside of one epidemic of distemper which was almost disastrous. Emphasis was placed on simplicity of procedure compatible with economy and efficiency.

CARE AND HOUSING

Our colony was recruited annually by purchase, and the bulk of the females was bought in early December before mating activity had begun, although additional animals were introduced into the colony during the breeding season if there was too much depletion as a result of killing animals for study. Females in late pregnancy should never be transported, as abortion usually follows shipment by train or truck.

Although many kittens were born in our colony, they were seldom kept beyond weaning, since our interest did not extend beyond this period.

The prime factor in the successful breeding of cats is the securing of tractable and cooperative males. These are rare and are found only after repeated trials. The underlying cause of the continued refusal of normal healthy males to give service to receptive females under laboratory conditions is not known, but it is a familiar problem to those who attempt to breed cats in the laboratory. Females on the whole behave much more normally in the laboratory but do not readily become sexually receptive when confined to cages. They do much better when given some freedom to roam about. Incidentally, the male also should be given considerable freedom of movement.

Consequently, the females were kept unrestrained in individual rooms (dimensions, 14 by 22 feet). Such a room will carry from 20 to 25 animals without undue crowding. Air conditioning solved the problems of temperature, humidity, and ventilation. The floor was of smooth cement, gently sloped to a center drain. Both floor and walls were painted. Sleeping quarters consisted of two tiers of separate bunks, situated about the periphery of the room. The lower tier was raised 8 inches from the floor. It is essential to provide a slight excess of sleeping quarters to avoid quarreling. A broad board on top of the bunks provided "lounging" space. The walls should be kept bare, without shelving or racks, so that frightened or excitable animals have no opportunity to climb about or crawl into inaccessible places.

A bottomless, wooden frame, 4 feet square and 6 inches high, filled with sawdust or shavings, was placed on the floor to receive urine and feces. This accommodation was used rather conscientiously unless it became too greatly contaminated. Ordinarily, the frame was lifted morning and evening and the contents shoveled up. The floor was washed down with a hose each morning, and at this time the cats retreated to the bunks or lounging space until the floor had dried.

The animals should never be startled and should be kindly treated at all times. The caretaker should take a few minutes each day to pet the animals, as it makes subsequent handling involved in the determination of estrus and the introduction of the female to the mating enclosure a much less precarious task.

The mating pen was a temporary structure set up by assembling five wooden-framed, chicken-wire panels which were simply hooked together to provide a covered enclosure 4 feet wide, 6 feet long, and 30 inches high. Frequently, no restraint of either of the participating animals was necessary.

Pregnant females should be removed from the colony at least one week before parturition. A mating pen, when supplied with a corrugated-paper box with a simple opening, and bedding, provides inexpensive but adequate maternity quarters. A female with kittens should be given a considerable measure of seclusion if she is going to rear her young successfully.

Several authors emphasize the importance of animals having access to outside runs where they will receive ample sunshine and fresh air. Such quarters are unquestionably desirable but do not seem essential to successful breeding. This type of accommodation should be provided if plans are being developed for a permanent building to house cats.

Some object to cement flooring and recommend wood or even cork. The concrete does not appear objectionable if the quarters are kept warm at 70 to 72° F. and the cats have resting and sleeping quarters off the floor. Painted cement has the advantage of being readily cleaned and disinfected, since there are no cracks in which debris can accumulate.

SANITATION

The maintenance of strictly clean quarters is about the only precaution that is needed to protect the health of a colony. Quarantine of animals before their addition to an established colony is desirable, but most of the cats, acquired by purchase, appear to have developed a high degree of immunity to most diseases that normally affect these animals. A well-nourished adult animal seldom requires "worming," and fleas are readily exterminated. Protection against distemper is very desirable, since if it appears in the breeding season it will suppress practically all sexual activity in the colony for the year, i.e., of the animals that survive.

The term feline distemper has been loosely used by animal breeders, and in the interests of proper identification of pathological conditions in the cat it should be discarded. Present-day usage indicates that the most specific designation of this disease is feline infectious, or acute feline, panleukopenia, although it is also referred to under various other names, such as feline infectious gastroenteritis, hemorrhagic enteritis, and infectious enteritis (Brumley, 1943; Bentinck-Smith, 1949). This disease is also known as cat plague but the terms show fever and influenza are apparently not applicable. It should also be distinguished from feline infectious coryza (feline pneumonitis).

Weaned kittens are most susceptible, but adult animals may also readily contract this disease. Active immunity may be conferred by using a specific vaccine, an inactivated virus suspension prepared from the spleen and liver tissues of artificially infected kittens, but there does not appear to be complete agreement regarding the degree or duration of such protection. A temporary passive immunity, lasting two to three weeks, may be produced by use of an homologous serum. This serum may also be used in treatment of the disease, especially in its very early stages.* A brief discussion of methods of protection is also given by Mellen (1946, page 167).

Feline pneumonitis, involving lungs, pharynx, and sinuses, and apparently of rickettsial origin, is also a troublesome infectious disease in cat colonies. Aureomycin has been found to be effective in

---

* The above statement regarding panleukopenia in cats represents my personal conclusions after consultation with various persons experienced in the handling of cats. I am especially indebted to Dr. C. R. Schroeder, Lederle Laboratories Division of American Cyanamid Company, and Dr. Gerry B. Schnelle, Angell Animal Hospital, Boston, for information and references to the literature in this field. A product circular of Lederle Laboratories was very helpful.

treatment of this disease. Dr. C. R. Schroeder recommends the following dosages of aureomycin for pneumonitis when given orally: 25.0 mg. per pound body weight initially, and 12.5 mg. per pound body weight every twelve hours thereafter for several days.

It is advisable to have some treatise on the diseases of cats, such as Brumley (1943) or Kirk (1925, 1947) available for consultation. There is also an excellent chapter on the diseases and parasites of the cat in a recent book by Mellen (1946), written primarily for amateur and professional breeders. In many ways it is more useful than the technical textbooks written by and for veterinarians.

FEEDING

Feeding cats is an economic rather than a nutritional problem, and much ingenuity is needed to provide a cheap but adequate diet. The simplest but not the least expensive method is to buy prepared cat foods. An adequate basic ration can be provided with cheap grades of canned salmon, fresh or canned horse meat, and freshly boiled cod and haddock. Boiled white potatoes, raw or stewed tripe, raw liver, beef heart, and kidney may also be used to provide variety in the diet. This should be supplemented by whole milk—as much as the animals will take—and it was our custom to keep cod-liver oil and water available at all times. This diet was sometimes spiced by discarded animals from the rat and mouse colonies. Eggs were seldom used. The animals were fed meat once daily, in the morning, and after the feeding period all uneaten food was removed and the containers sterilized. Milk may be given in the late afternoon and contributes to the comfort and quiet of the colony. Feeding containers, each adequate for the accommodation of five cats, were used. Some groups were congenial and fed simultaneously, while in other groups an order of dominance was developed and feeding took place in a rather rigidly determined sequence. However, with adequate amounts of food available no undernourishment of the more timid animals occurred.

Hall and Pierce (1934), in their study of the growth rate of kittens, kept a small group of animals on a special diet designed to meet all known dietary requirements. Salt and mineral content (iron, copper, magnesium, manganese, etc.) were carefully controlled, and adequate vitamin intake was assured by including lettuce, dried yeast, cod-liver oil, and corn oil. The growth rate of kittens in this group did not differ significantly from others in which the food of the mothers consisted of beef heart and milk with occasional supplements of liver and cod-liver oil.

REPRODUCTION

According to Marshall (1922), the cat will breed when 10 months old or even earlier, but Asdell (1946) states that the female cat reaches puberty at 15 months of age. Cats rarely produce young after they reach an age of 14 years but may cease breeding activity when considerably younger (Marshall, 1922). According to Mellen (1946), females are considered to be in their sexual prime when from 2 to 8 years old; males are best for breeding from 3 to 14 years of age; but both sexes may live to be much older, 21 or even 23 years (Marshall, 1922; Todd, 1939). Higher records, up to 31 years, have been reported.

Breeding Season. The breeding season apparently varies considerably in different geographical locations. However, in some instances the term breeding period or periods appears to refer only to the time of occurrence and number of pregnancies in a year and not to indicate the total period of time in which sexual receptivity may occur at rather close periodic intervals. The continued recurrence of heat periods of course can be interrupted at any time by pseudopregnancy in an infertile mating, or by pregnancy, with or without subsequent lactation. When estrus reappears after these interruptions, to be followed again by pregnancy, the statement that the cat exhibits two or three breeding periods a year, as the case may be, is a description of the facts, but it must be borne in mind that without mating the female might have exhibited periodic sexual receptivity throughout the entire time of the so-called breeding periods. Accordingly, I should like to define the breeding season as the total period of the year in which females may exhibit sexual receptivity. This is admittedly a definition in terms of potential breeding but avoids the complication of having to take into consideration the usual disruptions of gestation and lactation.

According to Longley (1911), the sexual season of the cat, in the region of New Haven, Conn., is at its height during the latter half of February and March. Marshall (1922) states that in England cats generally breed two or three times a year but also reports one case of an unmated female in which estrus regularly recurred at intervals of about a fortnight from December until the following August. The data of Foster and Hisaw (1935) indicate that in the vicinity of

Madison, Wis., cats are usually anestrous from September to January. Windle (1939) regarded the period from February to June, inclusive, as the normal estrous season in Chicago, Ill., while Dawson (1941b) found that in Cambridge, Mass., cats had a normal breeding period from January until July. In northern Europe (Warsaw) heat periods appear twice a year, in the spring and early autumn (Liche, 1939), but in Algiers the estrous seasons are from December 15th to the end of January and from July 15th to the end of August (Gros, 1936). In the northern United States it appears evident that cats may be found in heat at any time from January to July inclusive, with varying diestrous intervals from 2 to several weeks, and that the breeding cycle is characterized by a single definite, anestrous period lasting about 5 months. Doubtlessly, late gestation and lactation resulting from a July mating might delay the time of onset of the following estrous season. This period appears to be determined in part by the length of day. The anestrous period may be shortened by increasing the period of daily illumination. Cats exposed to longer periods of lighting, beginning in October, were brought into heat in late November after about 50 days of treatment, much earlier than the usual time (Dawson, 1941b).

Estrus. In the laboratory, heat periods may not recur at regular intervals, but occasional females may maintain a regular two-week cycle for several months (personal observation). The duration of estrus varies greatly from individual to individual and even in the same individual, from time to time. Van der Stricht (1911) states that it will last 2 or 3 days, presumably after the first copulation, while in the absence of the male, or in the presence of a restrained male, estrus will often continue 9 or 10 days (Longley, 1911; Liche, 1939), although there are reports of estrus persisting for 2 or 3 weeks (Mellen, 1946).

The literature on the behavior of the estrous cat has been recently reviewed by Young (1941). Good descriptions are also given by Van der Stricht (1911), Bard (1934, 1936, 1939), Greulich (1934), Gros (1936), and Maes (1939, 1940a, 1940b). Full details of this rather elaborate pattern cannot be given here. Estrous behavior may be divided into three phases, the courtship activities, the presentation and reception responses, and the postcopulatory responses. An early proestrous activity, consisting of playful rolling and excessive rubbing, may frequently be recognized 2 or 3 days before sexual receptivity actually appears. The estrous crouch, and the treading response with the hind legs to mild rubbing or tapping on the genital region, are characteristic of estrus. Successful intromission is marked by a loud cry or growl. The after-reaction consists of vigorous, almost frantic, rolling, sliding, rubbing, squirming, and licking. This is followed by a period of contented inactivity. Mating may be repeated within intervals of 15 minutes to a half-hour. The full mating response may also be elicited by mechanical stimulation of the uterine cervix (Greulich, 1934). Estrus has also been induced in a variety of ways. Estrogens alone are sufficient to induce heat in diestrous, anestrous, or spayed animals (Bard, 1936, 1939); Maes, 1939, 1940a, 1940b). Treatment with other hormones such as the pituitary gonadotropins, and gonadotropic hormones extracted from human menopausal or pregnancy urine, may in addition induce ovulation with ensuing pseudopregnancy (Foster and Hisaw, 1935; Friedgood, 1939a; Snyder and Wislocki, 1931; Windle, 1939). During anestrus there may be some seasonal refractoriness to hormonal stimulation. The neural factors, central and sympathetic, involved in estrous and mating behavior of the female cat have been studied experimentally by several investigators (Bard, 1934, 1935, 1936, 1939, 1940; Bard and Rioch, 1937; Dempsey and Rioch, 1939; Maes, 1939, 1940a, 1940b; Ranson, 1934; Simeone and Ross, 1938). Since the results of these experiments are not too pertinent to the present discussion and have been critically reviewed by Young (1941), no more than this passing reference will be made to them.

Vaginal Smears. The vaginal smear during anestrus consists predominantly of small, nucleated epithelial cells of varying size, shape, and affinity for stains. In proestrus the cells become more numerous and greatly flattened until at the onset of estrus only a few small epithelial cells remain. In estrus most of the cells are usually large and non-nucleated. Progressive cornification continues until about three days after copulation, when leucocytes invade the vaginal epithelium and masses of leucocytes appear in the smears. This is characteristic of metestrus (Asdell, 1946; Foster and Hisaw, 1935; Gros, 1936; Liche and Wodzicki, 1939). Nucleated, epithelial cells with large vacuoles are diagnostic of early pregnancy, and after implantation the smear gradually comes to resemble the proestrous condition with small and large epithelial cells plus a few leucocytes (Foster and Hisaw, 1935).

Ovulation. It is generally conceded that ovulation in the cat is not spontaneous but is induced by coitus or a comparable artificial stimulus (Greulich, 1934; Dawson and Friedgood, 1940). A series of observations (Van der Stricht, 1911; Longley, 1911; Manwell and

Fig. 1. A follicle from an unmated cat on the third day of estrus, showing extensive lacunar spaces in the cumulus. ×90. (Dawson and Friedgood, 1940.)

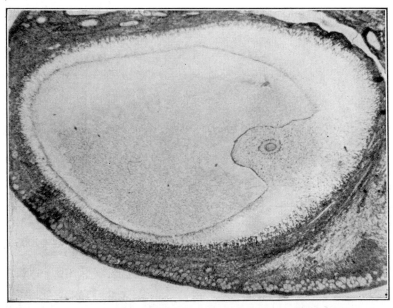

Fig. 2. Section of an entire follicle from a cat 22½ hours after mating, showing the secondary follicular fluid, the modification of the granulosa and cumulus, and the development of the corona radiata. ×30.

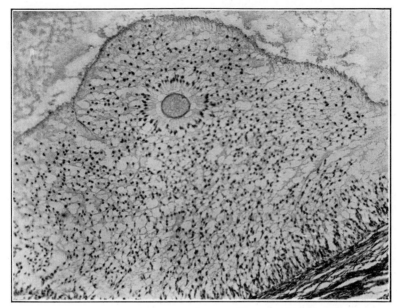

FIG. 3. Follicle from a cat 24 hours after the first mating, showing a maximum reaction in the cumulus and a well-developed corona. The ovum has a second maturation spindle and a polar body. ×90. (Dawson and Friedgood, 1940.)

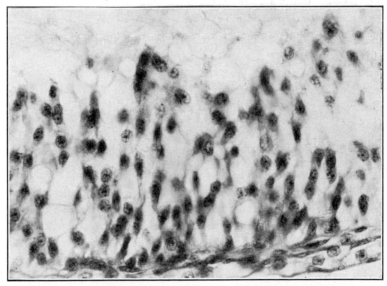

FIG. 4. The modified granulosa of a follicle 24 hours after mating. The granulosa cells are irregular in form and widely separated. They are slightly hypertrophied with vesicular nuclei, indicating an initial stage of the transformation into luteal cells. ×525. (Dawson and Friedgood, 1940.)

Wickens, 1928; Courrier and Gros, 1933; Gros, 1936) indicates that coitus is essential, although Windle (1939) appears to feel that spontaneous ovulation has not been completely ruled out and reviews some of the evidence which at best is inconclusive. Ovulation can also be induced experimentally by appropriate treatment with pituitary and other gonadotropic hormones (Snyder and Wislocki, 1931; Bourg, 1933; Foster and Hisaw, 1935; Friedgood, 1939a; Windle, 1939).

There appears to be some variation in the interval between copulation and ovulation. The observations of Longley (1911) place the time of ovulation at about the end of the second day. According to Van der Stricht (1911), ovulation occurs early in the second day. Manwell and Wickens (1928) found tubal eggs in one animal 21 hours after a timed mating. More extensive observations of Courrier and Gros (1933) and Gros (1936) indicate that ovulation usually takes place 26 to 27 hours postcoitum. Greulich (1934) found that the minimal time after stimulation was approximately 25 hours, but Liche (1939, Warsaw) gives the interval as 40 to 54 hours. Dawson and Friedgood (1940) conclude that the normal expectancy for ovulation appears to be between 24 to 30 hours after the first mating. The changes in the follicle and ovum which occur during preovulatory swelling have been described by Van der Stricht (1911), Longley (1911), and Dawson and Friedgood (1940).

There are no obvious changes in the mature follicle (fig. 1) during the first 6 hours after mating. After this period the cumulus is enlarged because of the accumulation of secretion between the cells, and the reaction gradually spreads around the wall of the follicle to modify the entire granulosa (figs. 2, 3, 4). This is accompanied by the production of secondary follicular fluid which distends the follicle. The membrane of the germinal vesicle disappears between 10 and 12 hours, and the first maturation spindle is present at 18 hours. The first polar body is given off and the second maturation division is formed at $22\frac{1}{2}$ hours. Further maturation changes do not occur within the follicle. The second polar body is formed soon after the entry of the sperm. The entire cumulus, with enclosed ovum and corona radiata, is extruded as a compact mass at ovulation. There is no bleeding from the ruptured follicle.

Early laparotomy, performed 15 to 55 minutes postcoitum, will inhibit the ovulatory response, but if the operation is delayed for 6 hours after mating, ovulation will occur normally (Friedgood, 1938,

1939b; Friedgood and Foster, 1938; Friedgood and Dawson, 1942). Anesthesia alone does not seem to block the ovulatory reaction.

Fertilization and Early Cleavage. Fertilization apparently takes place sometime during the first half of the second day after mating (Van der Stricht, 1911). At 41½ hours, four eggs were recovered; one in the infundibulum, two in the middle third, and one in the uterine third of the oviduct. Three of these ova possessed pronuclei. The rate of the ascent of the sperm is not known. Sperm were found in the lumen and glands of the uterus as early as 1 hour and 40 minutes after the first mating (personal observation).

Early cleavage takes place as the fertilized ova pass down the oviduct. The ovum enters the uterus about 7 days after mating (Andersen, 1927; Manwell and Wickens, 1928) and is in a late morula or early blastodermic vesicle stage (Hill and Tribe, 1924). The time of implantation is about 2 weeks postcoitum (Foster and Hisaw, 1935; Courrier and Gros, 1932a, 1932c, 1933; Gros, 1936). At the fourteenth day the embryonal area is 4 mm. long, and six to eight pairs of somites are present. Internal migration of the ova from one cornu to the other apparently may take place, so that the number of corpora in one ovary may not correspond to the number of fetuses found on the same side (Hill and Tribe, 1924; Markee and Hinsey, 1933).

Embryonic and Fetal Growth. Considerable data on the rate of embryonic and fetal growth have been compiled. Crown-rump measurements have been made by Windle and Griffin (1931), Windle and Fish (1932), and Windle, O'Donnell and Glasshagle (1933) on a series of litters from the seventeenth day until term. These data, supplemented by a few measurements made in this laboratory, are given in table 1. Such records are of value in determining the period of pregnancy when the time of mating is not known. More complete analyses of later stages of prenatal growth, including increase in dimensions and weight of the head, trunk, limbs, digestive tract (including liver and pancreas), brain, spinal cord, and eyeballs, have been made by Latimer and his collaborators (1931–1938).

Gestation. The records indicate that the duration of gestation may vary from 56 to 65 days after coitus, but the normal period is usually 62 to 63 days.

Parturition. Parturition usually occurs in a normal manner with few complications. Some cats have their kittens in half an hour, but labor may be prolonged for several hours. The mother bites off the umbilical cords and eats the placentae.

TABLE 1

CROWN-RUMP MEASUREMENTS OF EMBRYONIC AND FETAL GROWTH

Time since Mating (days)	Size (mm.) (C.R. Length)	Time since Mating (days)	Size (mm.) (C.R. Length)
17–18	5.5– 6.0	29	26.0– 28.0
19	7.0	30	27.0– 30.0
20	8.0– 9.0	31	36.0
21	10.0	32	37.0– 38.0
22	11.5–12.0	33	37.0– 38.0
23	13.5–14.5	35	45.0
23.5	15.0–16.0	36	50.0
24	14.0–15.5	40–41	68.0
24+	16.0	42	80.0
24.5	17.0	45?	90.0
?	17.5	47	90.0
25	19.5–20.0	49	100.0–102.0
26	21.0–22.0	50	102.0–103.0
27	22.5	54	112.0–115.0
28	26.0	60–63 *	135.0–150.0

* Newborn.

Litter Size. The size of litters varies, tending to be smaller in very young and old females. The average number of ovulations is slightly over four (Gros, 1936, Algiers), and the mean litter size is 3.88 kittens (Hall and Pierce, 1934, California), but the range is from one to eight or occasionally more kittens. One factor influencing litter size is the weight of the mother. There appears to be a tendency for small mothers to have large kittens. The birth weight of kittens varies from 70 to 144 gm., the mean being 106.4 gm. (Hall and Pierce, 1934).

Lactation. The female cat possesses five pairs of mammary glands, two of which are on the thorax and the remaining three on the abdomen (Reighard and Jennings, 1935), but Turner and De Moss (1934) state there are normally four pairs. A litter larger than eight is seldom successfully nursed. Suckling is almost continuous during the first 24 hours, but beginning on the second day the mother takes periodic leaves from the nest box.

There is no postpartum estrus in the cat, and lactating females remain anestrous. The lactation period may extend to 60 days, but at from 4 to 6 weeks, the kittens, by way of gradual weaning, may be given supplementary feeding. It is usually best to leave the kittens with the mother for 2 months, although some mothers will wean their kittens by 50 days.

Lactation is quickly inhibited by weaning. There is an accumulation of milk in the glands for about 24 to 48 hours, but by the fifth day practically all milk has disappeared. The lobule-aveolar system will undergo complete regression if there is a period of 80 days or more between one lactation and the initiation of another pregnancy (Turner and De Moss, 1934). Lactation is suppressed after hypophysectomy in late pregnancy even in animals which deliver living kittens (Allen and Wiles, 1932; McPhail, 1935).

Estrus may occur as early as 2 to 3 weeks after weaning. Bourg (1935) cites a case of impregnation 6 weeks after a parturition which was followed by lactation. The anestrous period, following pregnancy without lactation, lasts, according to Evans and Swezy (1931), from 2 to 4 weeks; but Longley (1911) gives it as from 3 to 6 weeks, in agreement with von Winiwarter and Sainmont (1907). Gros (1935) states that a new follicular phase may begin as early as the fifteenth day after parturition without lactation.

Superfetation. Several cases of probable superfetation in the cat have been recorded (Harman, 1917; Hunt, 1919; Marshall, 1922; Markee and Hinsey, 1935a). In Harman's case almost full-term fetuses (90 mm. crown-rump length) were present in both uterine horns. A smaller embryo, 10 mm. in length, was also found in one. In Hunt's case a cat gave birth to an apparently normal embryo 14.1 mm. in length, and 3 hours later a full-term living kitten was delivered. In the case described by Markee and Hinsey two normal kittens were born from the right horn, followed 13 days later by two more normal ones from the left horn. It is difficult to relate these phenomena to the facts known about the reproductive activity in the cat. The cases described by Harman and Hunt suggest a fertile mating followed by ovulation and implantation relatively late in pregnancy, about 30 to 40 days after the first mating. The third case implies a second successful mating, either in a prolonged original heat period or in a normally succeeding one, i.e., 2 weeks later. In our colony occasional females gave external evidence of heat after the fourth week of pregnancy and responded typically to artificial stimulation of the cervix (Dawson, 1946a). In the case recorded by Marshall (1922) the cat experienced heat and mated, after being pregnant for 6 weeks. Three weeks later she delivered five kittens, four of normal size, and one small one, apparently a three-week-old embryo. In midgestation, heat may be induced by injections of estrogen without causing abortion (Courrier and Gros, 1935b).

Synchorial Litter Mates. These cases are of considerable interest, since they provide a possible explanation for the atypical sexual conditions frequently found in male tortoise-shell cats. In fused placenta, when the litter mates are of opposite sex, the male partner might exercise a free-martin effect upon the female sufficient to cause complete sex inversion, an alternative hypothesis to the view that the condition in the male tortoise-shell cat is of genetic origin (Bamber, 1922; Bamber and Herdman, 1927; Bissonnette, 1928b, Bonnevie, 1925; Doncaster, 1904, 1920; Hayes, 1923; Little, 1920). The hormonal hypothesis was strongly urged by Bissonnette (1928a), but the morphological evidence was inconclusive, since the embryos were at a stage in which the gonads were not sexually differentiated. In a later case described by Wislocki and Hamlett (1934) the male and female partners were of sufficient age (male 50 mm., female 61 mm., crown-rump length) that the reproductive tracts were well differentiated. There was no evidence of hormonal modification, and the authors concluded that "the present findings do not lend weight to the hypothesis that sterile male tortoise-shell cats may be sex-inverted females, transformed by a hormonal influence resulting from the conjoined condition of two litter mates" (page 102).

Pseudopregnancy. Pseudopregnancy occurs after any non-fertile ovulation, whether brought about by mating with a vasectomized male, by mechanical stimulation of the cervix, or by hormonal treatment. Luteinization in situ may also be effective. The preimplantation transformation of the uterine endometrium is not dependent on the presence or development of fertilized ova, as has been shown by studies on non-gravid cornua and pseudopregnant uteri. Such uterine changes may also be simulated by proper hormonal substitution even in spayed animals. Normal pseudopregnancy, i.e., following sterile mating, apparently lasts for 30 to 40 days after the application of the inducing stimulus. According to Gros (1935), the uterus is in repose on the thirty-third day, and the female is in estrus by the fortieth day, while Liche (1939) states that pseudopregnancy lasts 36 days. The data on hormonally induced pseudopregnancy are much more variable.

Postnatal Development. Kittens are born temporarily blind and deaf. The time of opening of the eyes is variable, but they are usually open by the tenth day. The eyes may begin to open during the first or second week and may not be fully open until the end of the third week (Mellen, 1946).

Milk teeth are cut during the period from 11 days to 4 or 5 weeks after birth and are displaced by permanent teeth during the fourth to seventh months (Mellen, 1946). The incisors and canines appear first; the second upper deciduous premolars appear last (Stromsten, 1947, page 152).

There is little evidence for a postnatal decrease in weight (Latimer and Ibsen, 1932; Hall and Pierce, 1934). The data of Latimer and Ibsen on the growth of twelve kittens (six males, six females) cover a period from the first to the thirteenth week after birth. The average weight of female kittens at birth was 103.92 gm., that of males 97.75 gm. At the thirteenth week the male kittens had reached a little over 40 per cent of the weight of the adult male (Latimer, 1936). The female average was 320 gm. less. Both sexes grew at the same rate until the eighth week; after that the males grew more rapidly. Hall and Pierce followed the rate of growth of kittens from birth until the fiftieth day. The influence of sex was not studied, and their data are in fair agreement with those of Latimer and Ibsen.

Ovary. The ovary of the cat has been a favorite subject of study. The early and comprehensive morphological studies of Sainmont (1905), von Winiwarter and Sainmont (1907, 1909), and earlier investigators have been reviewed by Kingsbury (1913 to 1939) in several papers on the ovary of the cat. Differentiation between ovary and testis is possible on the twenty-ninth day (Sainmont, 1905). In the forty-day fetus the morphology of the testis is well established, and the primitive medulla and cortex of the ovary are evident (Kingsbury, 1913). An account of the histogenetic changes from this period until sexual maturity will be omitted. In early stages the interstitial cells are formed from modified stromal cells. After the appearance of well-defined follicles the interstitial cells form conspicuous bodies and are produced by the hypertrophy of thecal cells during follicular atresia (Kingsbury, 1914, 1939). Kingsbury (1938) found no evidence of a new formation of egg cells from the surface epithelium either before or after sexual maturity. He concluded that the stock of primary follicles established in the first few weeks after birth was adequate for the entire reproductive period. Interesting abnormal nuclear phenomena of the oöcytes of three-week-old kittens, including multipolar mitosis and multinucleate cells, have been described by Sneider (1937). Polyovular follicles are relatively common in older animals (Dederer, 1934).

Corpus Luteum. CORPUS LUTEUM OF PREGNANCY. The development and morphology of the corpus luteum during pregnancy have been recently described by Dawson (1941a). After ovulation the follicle is partially collapsed and filled with a clear, viscous—almost gelatinous

—matrix (fig. 5). There is no hemorrhage or leucocytic invasion. The lutein cells are formed from the granulosa; the interstitial supportive tissue is derived from the theca interna. Preluteal changes in the granulosa cells (fig. 4) occur even before ovulation (Dawson and Friedgood, 1940), and by the seventh day the transformation into typical luteal cells is almost complete. Corpora lutea reach maximum

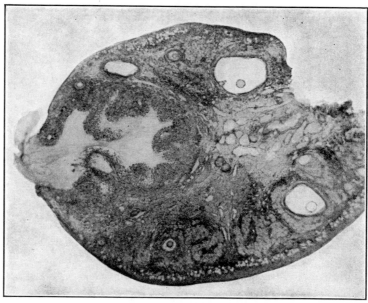

FIG. 5. Section through a ruptured follicle 36 hours after the first mating, showing the plications of the wall which involve both the mural granulosa and theca interna. The lumen is filled with a viscous matrix which protrudes through the point of rupture. ×18. (Dawson, 1941a.)

size from ten to sixteen days after coitus (fig. 6). After the sixteenth day the fine peripheral vacuolation of the cells disappears, and the cytoplasm of the luteal cells gradually assumes a fibrous appearance which is complete at twenty-seven days. This change in the histological appearance of the luteal cells may be correlated with a transfer of control from the pituitary gland to the placenta (Dawson, 1946a). Later the luteal cells become filled with numerous large vacuoles. This condition is highly developed by the fiftieth day and persists during the remainder of pregnancy. The fatty infiltration is regarded as evidence of regression and loss of functional activity. This conclusion, based on morphological evidence, is in agreement with the experimental evidence of Courrier and Gros (1935a, 1936), who found that from the eighteenth to the forty-fifth day, ovariectomy was invariably followed by abortion, but after the forty-ninth day gestation was not interrupted by this operation.

CORPUS LUTEUM OF PSEUDOPREGNANCY. Pseudopregnancy induced by service with a vasectomized male lasts for about 36 days. The corpora

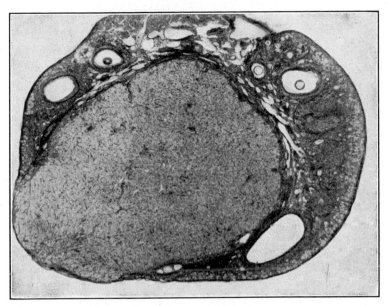

FIG. 6. A corpus of approximately maximum size, *ca.* 14 days after mating: the embryos recovered had seven pairs of somites, a level of differentiation attained at the time of implantation. ×18. (Dawson, 1941a.)

are fully developed by 13 days, and 28 days after copulation regression begins, simultaneously with the appearance of growing follicles which will initiate a new cycle (Liche, 1939). Our observations on corpora lutea of pseudopregnancy, induced by cervical stimulation, do not extend beyond the twenty-seventh day. At this time coarse vacuolation, characteristic of regression in late pregnancy, was becoming apparent. Other criteria such as the time of onset of heat after pseudopregnancy and the histological appearance of the uterus also serve to delimit the period of pseudopregnancy, although they do not yield direct evidence on the functional life span of the corpora. Thus Gros (1935) found that after sterile copulation the uterine changes were maintained for 23 days, with regression to the resting level on the thirty-third day.

The data on the length of activity of corpora formed, after ovulation induced by hormonal treatments, are more variable. In most instances the physiological response of uterine muscle to adrenalin was used to test for the presence of progesterone. In non-pregnant uteri adrenalin causes relaxation of the smooth muscle, while in pregnant uteri it causes contraction—the so-called reversal-response (Van Dyke and Gustavson, 1929). Foster and Hisaw (1935) found that the adrenalin reversal-response was maintained in experimental pseudopregnancy until the forty-fourth day, but Van Dyke and Li (1938) failed to secure this response after the twentieth day. The significance of this response as a test for the presence of progesterone is further discussed by Van Dyke and Chen (1939).

POSTPARTUM CORPUS LUTEUM. The corpora lutea of lactating animals give some evidence of a mild rejuvenation by the end of the first week after parturition. There is increased vascularity and loss of coarse vacuolation which is characteristic of corpora of late pregnancy. This response may be due to the rapid release of pituitary prolactin (luteotropin). These changes were not observed in animals from whom the kittens had been removed at birth. It is estimated that the total life span of the corpora may be as long as 6 to 8 months, and it is suspected that lactation tends to prolong the life of the corpora (Dawson, 1946b).

Oviduct. During anestrus the total epithelium is low and ciliation is inconspicuous. There is no mitosis. With the onset of estrus, mitoses are found and the epithelium becomes much taller, glandular, and well ciliated (Foster and Hisaw, 1935; Gros, 1936).

The oviduct joins the uterus at its extreme anterior end and on the mesenteric side; then turns to terminate in a low papilla. There is no sphincter-like thickening of the muscle, but the tubo-uterine junction has a valve-like action which prevents the passage of fluids from the uterus to the tube (Andersen, 1928).

Uterus. UP TO IMPLANTATION. The morphological changes in the endometrium of the cat uterus from the anestrous condition until the time of implantation have been described by Foster and Hisaw (1935), and by Courrier and Gros, and by Gros in a series of short papers which are summarized and extended in the doctoral thesis of Gros (1936). A comparable study has been made by Dawson and Kosters (1944) on carefully timed material in which the condition of the uterus was related to the stage of development of the corpora lutea.

Three more or less distinct phases are recognized: an anestrous, a proestrous and estrous, and a luteal phase. The luteal phase may be divided into two stages coinciding with the first and second week of the preimplantation period. In the anestrous uterus the glands are straight, narrow, and relatively short. Their epithelium and that of the surface are low and flattened and no mitoses are present. During proestrus and estrus, the glands dilate but remain straight; their epithelium as well as the surface epithelium increases in height, and an appreciable number of mitotic figures appear (fig. 7).

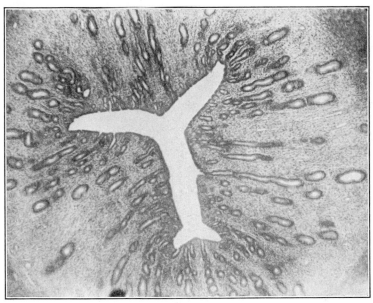

Fig. 7. Uterus of a cat in the second day of estrus. The glands are still almost straight and extend to the base of the endometrium. They show some dilatation, and the epithelium is tall, columnar. ×40. (Dawson and Kosters, 1944.)

In the early luteal phase the glands enlarge rapidly and become tortuous, and the intervening stromal tissue is greatly compressed. Large deposits of glycogen accumulate in both the surface and glandular epithelium, chiefly in an infranuclear position, displacing the nuclei apically (figs. 8, 10, 11). After the beginning of the second week the glycogen deposits are rapidly reduced and the superficial mucosa is thrown up into narrow folds so that the endometrium is divided into a superficial dentate or fringed zone and a deep spongy (glandular) zone (fig. 9). The glycogen stores are depleted at the time of implantation.

The location of the sites of future embryonic attachment may be recognized as early as the beginning of the thirteenth day and are marked by uterine dilatations from 9 to 10 mm. in diameter (Courrier and Gros, 1932c). As each of the free, oval blastocysts enlarges in the cornua, a circular, equatorial zone of contact is established with the endometrium and implantation begins. The placenta is accordingly

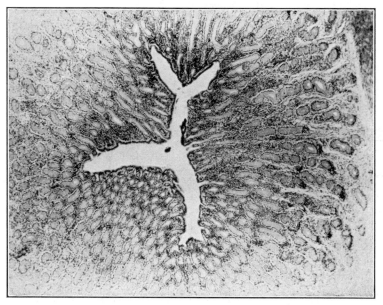

FIG. 8. Uterus of a cat 6 days after cervical stimulation on the first day of estrus, showing the hypertrophied uterine glands and the apical displacement of the nuclei of the surface epithelium. ×40. (Dawson and Kosters, 1944.)

zonary in form. Histologically it is classified as a placenta endotheliochorialis. The paraplacental chorion derived from the two polar regions of the blastocyst remains smooth but closely associated with the uterine epithelium. The junction between the unmodified chorion and the placenta is marked by "the brown border," produced by hemorrhages of maternal blood into a restricted zone between the membranous chorion and the uterine surface (Wislocki and Dempsey, 1946).

IMPLANTATION AND PLACENTATION. Further details regarding the development and organization of the placental labyrinth may be found in some very old and some recent studies (Duval, 1893 to 1895; Jenkinson, 1913; Heinricius, 1914; Grosser, 1927; Mossman, 1937; Wislocki and Dempsey, 1945, 1946). Certain of the factors influencing uterine

growth during pregnancy have been studied by Hinsey and Markee
(1935) and Markee and Hinsey (1935b). Interruption of the thoracolumbar sympathetic innervation of one horn had no observable effect
on the growth of the uterine horn or on the growth and distribution
of fetuses. A local factor, distention, seems to exert its influence on

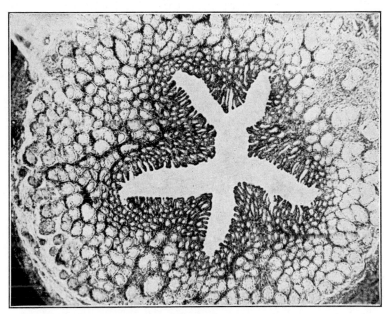

FIG. 9. Uterus of a cat 10 days after a successful mating. The surface of the
endometrium is highly modified, i.e. is thrown up into narrow tooth-like folds so
that the uterine glands now open into crypts lined with typical superficial epithelium. The zonation of the mucosa into an outer compact and inner spongy
zone is pronounced. ×40. (Dawson and Kosters, 1944.)

all components of the uterine horn, including the endometrium. Total
sympathectomy (Cannon and Bright, 1931; Simeone and Ross, 1938)
may impair the processes of reproduction, resulting in frequent abortions and stillbirths, and interfere with lactation even when parturition
is normal.

NORMAL PSEUDOPREGNANCY. The uterine changes during normal
pseudopregnancy after sterile coitus or cervical stimulation parallel
closely those of pregnancy, especially in the preimplantation period,
although they may not always occur at a comparable rate. With the
loss of luteal activity there is a leucocytic infiltration of the endometrial fringe, which is gradually reduced by dissolution and sloughing.

ARTIFICIAL ACTIVATION OF THE UTERUS. Artificial pseudopregnancy may be induced by the substitution of injections of appropriate extracts for the normal secretions of the pituitary gland, thereby evoking the normal activity of the ovary, follicular maturation, ovulation, and corpus luteum formation with results quite comparable to normal pseudopregnancy (Courrier and Kehl, 1929; Foster and Hisaw, 1935).

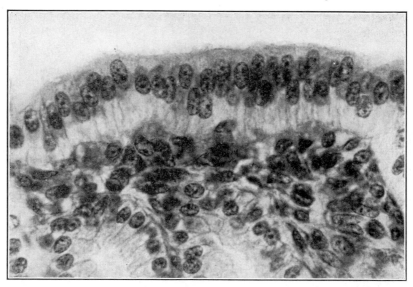

FIG. 10. Epithelium of uterine surface at the end of the sixth day after mating. The nuclei are displaced apically because of basal accumulation of glycogen. ×720. (Dawson and Kosters, 1944.)

Usually, less complete activation of the endometrium is obtained by making direct substitutions for either or both of the normal ovarian hormones, and in many instances the degree of success has been measured in terms of the uterine response (Rowlands and McPhail, 1936; Gros, 1939). Many different extracts and hormones have been used on cats, under widely varying physiological and experimental conditions, including immature, anestrous, spayed, adrenalectomized, hypophysectomized, and pregnant animals. The results cannot be analyzed here, but a summary citation of the more pertinent literature will be given. The effects of pregnancy urine extract on immature and mature female cats have been studied by Bourg (1931 to 1935). Windle (1939) induced mating and ovulation in anestrous females after treatment with pregnancy urine extract and a pregnancy serum. Starkey and Leathem (1939 to 1943) have secured variable degrees of uterine activation by treating immature females with menopause urine extract, estrone, normal male urine extract, and pregnant mare serum. The age of the animals was a significant factor in conditioning the degree of response, although the early observations of Courrier and Gros (1932b) show activation of the uterine gland in newborn kittens after

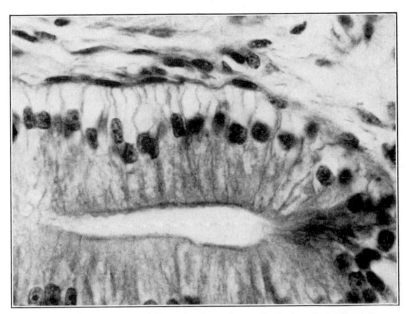

Fig. 11. Epithelium of a uterine gland at the end of the sixth day after successful mating, showing an advanced stage of glycogen storage with basal vacuolation of the cytoplasm. ×720. (Dawson and Kosters, 1944.)

injections of folliculin. The supplementary action of maternal hormones is not precluded. Testosterone propionate may partially substitute for progesterone (Courrier and Gros, 1938a, 1938b), but desoxycorticosterone acetate can produce a better uterine response (Leathem and Crafts, 1940).

POSTPARTUM INVOLUTION. During the first two weeks of postpartum the involution of the myometrium and reconstitution of the endometrium proceed at the same rate in lactating and non-lactating females. Thereafter the differential effects of lactation become obvious. In non-lactating animals retrogressive changes in the myometrium cease, and by the fourth week or earlier the endometrium may be in a resting or proestrous condition. With continuing lactation both the myometrium and endometrium exhibit progressive involution which continues for the duration of lactation (fig. 12). The degree of regression is comparable to that seen in females ovariectomized for a similar period (Dawson, 1946c).

Vagina. The anestrous vaginal epithelium consists of few layers of cells. In proestrus there is a gradual increase in the number of cell

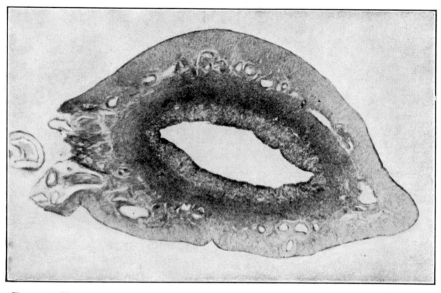

Fig. 12. Uterus of a cat which had delivered three kittens and had nursed them for 63 days. There is great involution of both myometrium and endometrium. ×20. (Dawson, 1946c.)

layers which reaches a maximum in estrus with several layers of superficial cornified cells. There is a continuous sloughing during heat, which becomes accentuated at the close of this period, accompanied by leucocytic infiltration. The vagina is relatively quiescent during the preimplantation period and after implantation (Foster and Hisaw, 1935; Gros, 1936).

Hypophysis in Reproduction. The anterior pituitary gland is essential to the maintenance of pregnancy until the last 8 to 10 days of gestation (Allen and Wiles, 1932; McPhail, 1935). Cytological evidence suggests that a specially modified acidophile may be the source of the luteinizing and luteotropic hormones (Dawson, 1946a). The gonadotropic potency of the anterior lobe, tested by the ability to induce ovulation in the rabbit (Hill, 1934), is the highest in the normal male (3120 units per gram of tissue), reduced in the castrated male (1230 units), and lowest in the female (800 units). The stage of the female reproductive cycle was not indicated. However, Magistris (1932) found that the female hypophysis was twice as potent gonadotropically as that of the male (immature mouse test). The lactogenic content is low in the anestrous female (65 B.U. per gram of tissue), high in the breeding season (224 B.U.), and lowest in the male (37 B.U.) in the non-breeding season (Reece and Turner, 1937).

Genetics. Although information in inheritance is important to animal breeders, no attempt will be made to review the extensive literature on the genetics of the cat. Reference has already been made to the sex-linked, yellow mutation in discussing the sexual condition in tortoise-shell males, which are nearly always sterile. The early literature on inheritance of cats has been reviewed by Bamber (1927), and Castle (1930, 1940) has summarized the information on the more common mutations and their mode of inheritance (table 2).

TABLE 2

SOME GENE MUTATIONS OF THE CAT *

(The dominant character appears first in the column.)

1. Tabby, black.
2. Black (of Siamese), tabby.
3. Black, yellow (sex-linked). Heterozygotes, tortoise shell.
4. Intense, dilute (blue, cream).
5. Full color, silver, Burmese, Siamese (an albino series).
6. Coat white (eyes colored), coat colored, dominant sublethal, deaf when blue-eyed.
7. Unspotted, piebald (Dutch type).
8. Short hair, long hair (Angora).
9. Short tail (Manx), normal tail. Homozygotes, tailless.
10. Polydactyl, normal toes.

* Castle, 1940, p. 159.

REFERENCES

Allen, H., and R. Wiles. 1932. The role of the pituitary gland in pregnancy and parturition. I. Hypophysectomy. *Jour. Physiol.*, vol. 75, pp. 23–28.

Andersen, Dorothy. 1927. The rate of passage of the mammalian ovum through various portions of the fallopian tube. *Amer. Jour. Physiol.*, vol. 82, pp. 557–569.

——. 1928. Comparative anatomy of the tubo-uterine junction. Histology and physiology in the sow. *Amer. Jour. Anat.*, vol. 42, pp. 255–305.

Asdell, S. A. 1946. Pp. 166–171 in *Patterns of mammalian reproduction*. Comstock Publ. Co., Ithaca, N. Y.

Bamber, R. C. 1922. The male tortoiseshell cat. *Jour. Genetics,* vol. 12, p. 209.

——. 1927. Genetics of domestic cats. *Bibliographica Genetica,* vol. 3, pp. 1–86.

Bamber, R. C., and E. C. Herdman. 1927. Dominant black in cats and its bearing on the question of the tortoiseshell males. *Jour. Genetics,* vol. 18, pp. 219–221.

Bard, P. 1934. On emotional expression after decortication with some remarks on certain theoretical views. II. Modes of emotional expression evocable in surviving decorticate animals. *Psychol. Rev.,* vol. 41, pp. 424–449.

——. 1935. The effects of denervation of the genitalia on the oestrual behavior of cats. *Amer. Jour. Physiol.,* vol. 113, p. 5.

——. 1936. Oestrous behavior in surviving decorticate cats. *Amer. Jour. Physiol.,* vol. 116, pp. 4–5.

——. 1939. Central nervous mechanisms for emotional behavior patterns in animals. *Res. Publ. Assoc. Nerv. Ment. Dis.,* vol. 19, pp. 190–218.

——. 1940. The hypothalamus and sexual behavior. *Res. Publ. Assoc. Nerv. Ment. Dis.,* vol. 20, pp. 551–579.

Bard, P., and D. McK. Rioch. 1937. A study of four cats deprived of neocortex and additional portions of the forebrain. *Johns Hopkins Hosp. Bull.,* vol. 60, pp. 73–147.

Bentinck-Smith, John. 1949. Feline panleukopenia (feline infectious enteritis). A review of 574 cases. *No. Amer. Veterinarian,* vol. 30, pp. 379–384.

Bissonnette, T. H. 1928a. A case of potential freemartins in cats. *Anat. Rec.,* vol. 40, pp. 339–349.

——. 1928b. A note on the occurrence of fused placentae in cats. *Jour. Heredity,* vol. 19, pp. 87–89.

Bonnevie, K. 1925. Intersexualität bei schildpattfarbigen Katzen. *Roux's Arch. f. Entwick. der Organismen,* Bd. 106, S. 611–641.

Bourg, R. 1931a. Les modifications prevoquées par la gravidine au niveau de l'ovaire et du tractus génital de la chatte impubère. *C. r. soc. biol.,* T. 106, pp. 926–928.

——. 1931b. Les modifications provoquées par la gravidine chez chatte adulte en dehors de la gestation et durant cette period. *C. r. soc. biol.,* T. 108, pp. 216–217.

——. 1932a. Etude de l'évolution de la phase lutéinique provoquée par la gravidine chez la chatte impubère et adulte. *C. r. soc. biol.,* T. 111, pp. 235–238.

——. 1932b. Etude des rapports entre les modifications provoquées au niveau de la première et de la seconde poussées germinatives de l'ovaire de la chatte et celles des tractus correspondants. *C. r. soc. biol.,* T. 111, pp. 148–150.

——. 1933. La phase post-luteinique provoquée dans la tractus genital de la chatte adulte. *C. r. soc. biol.,* T. 114, p. 562.

——. 1935. Etudes des modifications provoquées par la gravidine au niveau de l'ovaire et du tractus génital chez la chatte. I. Modifications chez la chatte impubère. *Arch. biol.,* T. 46, pp. 47–146.

Bromiley, R. B., and P. Bard. 1940. A study of the effect of estrin on the responses to genital stimulation shown by decapitate and decerebrate female cats. *Amer. Jour. Physiol.,* vol. 129, p. 138.

Brumley, O. V. 1943. *A text-book of the diseases of the small domestic animals.* Lea and Febiger, Philadelphia.

Cannon, W. B., and W. M. Bright. 1931. A belated effect of sympathectomy on lactation. *Amer. Jour. Physiol.,* vol. 97, pp. 319–321.

Castle, W. E. 1930. Cats. Pp. 198–199 in *Genetics and eugenics*. 4th rev. ed. Harvard University Press, Cambridge, Mass.

——. 1940. Cats. Pp. 154–159 and 164 in *Mammalian genetics*. Harvard University Press, Cambridge, Mass.

Courrier, R., and G. Gros. 1932a. Contribution a l'étude du cycle génital chez la chatte. *C. r. soc. biol.*, T. 110, pp. 51–53.

Courrier, R., and G. Gros. 1932b. Crise génitale experimentale chez la chatte nouveau-née. *C. r. soc. biol.*, T. 110, pp. 1021–1023.

Courrier, R., and G. Gros. 1932c. Remarques sur la nidation de l'oeuf chez la chatte. *C. r. soc. biol.*, T. 111, pp. 787–789.

Courrier, R., and G. Gros. 1933. Données complementaires sur le cycle génital de la chatte. *C. r. soc. biol.*, T. 114, pp. 275–277.

Courrier, R., and G. Gros. 1935a. Contribution à l'endocrinologie de la grossesse chez la chatte. *C. r. soc. biol.*, T. 120, pp. 5–7.

Courrier, R., and G. Gros. 1935b. Action de la folliculine chez la chatte gestante. *C. r. soc. biol.*, T. 120, pp. 8–9.

Courrier, R., and G. Gros. 1936. Dissociation foetoplacentaire realisée par la castration chez la chatte. *C. r. soc. biol.*, T. 121, pp. 1517–1520.

Courrier, R., and G. Gros. 1938a. Influence du propionate de testostérone sur l'utérus. *C. r. soc. biol.*, T. 127, pp. 921–923.

Courrier, R., and G. Gros. 1938b. Nouvelles recherches sur le action du propionate de testostérone chez la femelle. *C. r. soc. biol.*, T. 128, pp. 194–196.

Courrier, R., and R. Kehl. 1929. Sur le mode d'action des extraits hypophysaires anterieurs. *C. r. soc. biol.*, T. 100, pp. 711–712.

Dawson, A. B. 1941a. The development and morphology of the corpus luteum of the cat. *Anat. Rec.*, vol. 79, pp. 155–169.

——. 1941b. Early estrus in the cat following increased illumination. *Endocrinology*, vol. 28, pp. 907–910.

——. 1946a. Some evidences of specific secretory activity of the anterior pituitary gland of the cat. *Amer. Jour. Anat.*, vol. 78, pp. 347–409.

——. 1946b. The postpartum history of the corpus luteum of the cat. *Anat. Rec.*, vol. 95, pp. 29–51.

——. 1946c. The effects of lactation on the postpartum involution of the uterus of the cat. *Amer. Jour. Anat.*, vol. 79, pp. 241–266.

Dawson, A. B., and H. B. Friedgood. 1940. The time and sequence of preovulatory changes in the cat ovary after mating or mechanical stimulation of the cervix uteri. *Anat. Rec.*, vol. 76, pp. 411–429.

Dawson, A. B., and B. A. Kosters. 1944. Preimplantation changes in the uterine mucosa of the cat. *Amer. Jour. Anat.*, vol. 75, pp. 1–27.

Dederer, P. H. 1934. Polyovular follicles in the cat. *Anat. Rec.*, vol. 60, pp. 391–403.

Dempsey, E. W., and D. McK. Rioch. 1939. The localization in the brain stem of the oestrous responses of the female guinea pig. *Jour. Neurophysiology*, vol. 2, pp. 9–18.

Doncaster, L. 1904. On the inheritance of tortoise shell and related colours in cats. *Proc. Cambr. Phil. Soc.*, vol. 13, p. 35.

——. 1920. The tortoiseshell tomcat. A suggestion. *Jour. Genetics*, vol. 9, pp. 335–338.

Duval, M. 1893–1895. Le placenta des carnassiers. *Jour. anat. physiol.,* T. 29, pp. 249–340; T. 30, pp. 649–715; T. 31, pp. 38–80.

Evans, H. McL., and O. Swezy. 1931. Ovogenesis and the normal follicular cycle in adult mammalia. *Mem. Univ. Calif.,* vol. 9, pp. 119–188.

Foster, M. A., and F. L. Hisaw. 1935. Experimental ovulation and resulting pseudopregnancy in anoestrous cats. *Anat. Rec.,* vol. 62, pp. 75–93.

Friedgood, H. B. 1938. Cortico-adrenal and neural effects on gonadotropic activity of the pituitary. *Science,* vol. 86, pp. 84–95.

——. 1939a. Induction of estrous behavior in anestrous cats with the follicle-stimulation and luteinizing hormones of the anterior pituitary gland. *Amer. Jour. Physiol.,* vol. 126, pp. 229–233.

——. 1939b. Adrenalectomy and coitus induced ovulation in the rabbit and cat. *Endocrinology,* vol. 25, pp. 296–301.

Friedgood, H. B., and A. B. Dawson. 1940. Physiological significance and morphology of the carmine cell of the cat's anterior pituitary. *Endocrinology,* vol. 26, pp. 1022–1031.

Friedgood, H. B., and A. B. Dawson. 1942. Inhibition of the carmine-cell reaction in the pituitaries of cats which mate but do not ovulate. *Endocrinology,* vol. 30, pp. 252–257.

Friedgood, H. B., and M. A. Foster. 1938. Experimental production of ovulation, luteinization and cysts of corpus luteum in adrenalectomized anestrous cats. *Amer. Jour. Physiol.,* vol. 123, pp. 237–242.

Greulich, W. W. 1934. Artificially induced ovulation in the cat (*Felis domestica*). *Anat. Rec.,* vol. 58, pp. 217–224.

Gros, G. 1933. Recherches préliminaries sur le cycle génital chez la chatte. *Bull. hist. appl.,* T. 10, pp. 5–11.

——. 1935. Evolution de la muqueuse uterine chez la chatte. *C. r. soc. biol.,* T. 118, pp. 1575–1578.

——. 1936. Le cycle génital de la chatte. Thèse de Medecine. Alger.

——. 1939. Sur l'action du diethyl-stilboéstrol chez la chatte. *C. r. soc. biol.,* T. 131, pp. 172–174.

Grosser, O. 1927. *Früentwicklung, Eihautbildung, und Placentation des Menschen und der Saügetiere.* Bergmann, Münich.

Hall, W. E., and G. N. Pierce. 1934. Litter size, birth weight, and growth to weaning in the cat. *Anat. Rec.,* vol. 60, pp. 111–124.

Harman, Mary T. 1917. A case of superfetation in the cat. *Anat. Rec.,* vol. 13, pp. 145–153.

Hayes, F. A. 1923. The tortoise shell cat. *Jour. Heredity,* vol. 14, p. 369.

Heinricius, G. 1914. Über die Embryotrophe der Raubtiere in morphologischer Hinsicht. *Anat. Hefte.,* Bd. 50, S. 115–192.

Hill, J. P., and Margaret Tribe. 1924. The early development of the cat (*Felis domestica*). *Quart. Jour. Micro. Science,* vol. 68, pp. 513–602.

Hill, R. T. 1934. Species variation in the gonadotropic activity of the hypophysis. *Jour. Physiol.,* vol. 83, pp. 137–144.

Hinsey, J. C., and J. E. Markee. 1935. Studies on uterine growth. I. Does thoracolumbar sympathectomy affect the growth of the pregnant cat uterus? *Anat. Rec.,* vol. 61, pp. 253–260.

Hunt, H. R. 1919. Birth of two unequally developed cat fetuses (*Felis domestica*). *Anat. Rec.,* vol. 16, pp. 371–377.

Jenkinson, J. W. 1913. The placenta. Chapter IX, pp. 215–258, in *Vertebrate embryology, comprising the early history of the embryo and its foetal membranes.* The Clarendon Press, Oxford.

Kingsbury, B. F. 1913. The morphogenesis of the mammalian ovary: *Felis domestica. Amer. Jour. Anat.,* vol. 15, pp. 345–387.

——. 1914. The interstitial cells of the mammalian ovary: *Felis domestica. Amer. Jour. Anat.,* vol. 16, pp. 59–95.

——. 1938. The post-partum formation of egg cells in the cat. *Jour. Morph.,* vol. 63, pp. 397–419.

——. 1939. Atresia and interstitial cells of the ovary. *Amer. Jour. Anat.,* vol. 65, pp. 309–331.

Kirk, Hamilton. 1925. *The diseases of the cat and its general management.* Baillière, Tindall, and Cox, London.

——. 1947. *Index of diagnosis, canine and feline.* 3rd ed. Williams and Wilkins, Baltimore.

Latimer, H. B. 1931. The prenatal growth of the cat. II. The growth of the dimensions of the head and trunk. *Anat. Rec.,* vol. 50, pp. 311–332.

——. 1933. The prenatal growth of the cat. III. The growth in length of the two extremities and of their parts. *Anat. Rec.,* vol. 55, pp. 377–394.

——. 1934a. Prenatal growth of the cat. IV. Growth in length and weight of the digestive tube. *Anat. Rec.,* vol. 60, pp. 23–41.

——. 1934b. The prenatal growth of the cat. V. The ponderal growth of the liver and pancreas. *Scritti biologici,* vol. 9, pp. 313–322.

——. 1930. Weights and linear measurements of the adult cat. *Amer. Jour. Anat.,* vol. 58, pp. 329–347.

——. 1938a. The prenatal growth of the cat. VII. The growth of the brain and of its parts, of the spinal cord and of the eyeballs. *Jour. Comp. Neur.,* vol. 68, pp. 381–394.

——. 1938b. The weights of the brain and of its parts, of the spinal cord and of the eyeballs in the adult cat. *Jour. Comp. Neur.,* vol. 68, pp. 395–404.

Latimer, H. B., and John M. Aikman. 1931. The prenatal growth of the cat. I. The growth in weight of the head, trunk, fore limbs, and hind limbs. *Anat. Rec.,* vol. 48, pp. 1–26.

Latimer, H. B., and Herman L. Ibsen. 1932. The postnatal growth in body weight of the cat. *Anat. Rec.,* vol. 52, pp. 1–5.

Leathem, J. H., and R. C. Crafts. 1940. Progestational action of desoxycorticosterone acetate in spayed-adrenalized cats. *Endocrinology,* vol. 27, pp. 283–286.

Liche, H. 1939. Oestrous cycle in the cat. *Nature,* vol. 143, p. 100.

Liche, H., and K. Wodzicki. 1939. Vaginal smears and the oestrous cycle of the cat and lioness. *Nature,* vol. 144, pp. 245–246.

Little, C. C. 1920. Is the tortoiseshell tomcat a modified female? *Jour. Genetics,* vol. 10, p. 301.

Longley, W. H. 1911. The maturation of the egg and ovulation in the domestic cat. *Amer. Jour. Anat.,* vol. 12, pp. 139–172.

McPhail, M. K. 1935. Hypophysectomy of the cat. *Proc. Roy. Soc.,* ser. B, vol. 117, pp. 45–63.

Maes, J. 1939. Neural mechanism of sexual behavior in the female cat. *Nature,* vol. 144, pp. 598–599.

Maes, J. 1940a. Hypophysectomie et comportement sexuel de la chatte. *C. r. soc. biol.,* vol. 133, pp. 92–94.

——. 1940b. Le mecanisme nerveux la comportement sexuel de la chatte. *C. r. soc. biol.,* T. 133, pp. 95–97.

Magistris, H. 1932. Quantitative untersuchungen über den Gehalt des Hypophysenvorderlappens an Follikelreifungs—und Luteinisierungshormon bei verscheidenen Tieren. *Arch. f. ges. Physiol.,* Bd. 230, S. 835.

Manwell, E. J., and P. G. Wickens. 1928. The mechanism of ovulation and implantation in the domestic cat. *Anat. Rec.,* vol. 38, p. 54.

Markee, J. E., and J. C. Hinsey. 1933. Internal migration of ova in the cat. *Proc. Soc. Exp. Biol. Med.,* vol. 31, pp. 267–270.

Markee, J. E., and J. C. Hinsey. 1935a. A case of probable superfetation in the cat. *Anat. Rec.,* vol. 61, pp. 241–251.

Markee, J. E., and J. C. Hinsey. 1935b. Studies on uterine growth. II. A local factor in the pregnant uterus of the cat. *Anat. Rec.,* vol. 61, pp. 311–319.

Marshall, F. H. A. 1922. *The physiology of reproduction.* Longmans Green, London.

Mellen, I. M. 1946. *A practical cat book for amateurs and professionals.* Charles Scribners Sons, New York.

Mossman, H. W. 1937. Comparative morphology of the fertilization membranes and accessory uterine structures. *Contrib. Embryol. Carnegie Inst. Wash.,* vol. 26, pp. 129–246.

Ranson, S. W. 1934. The hypothalamus; its significance for visceral innervation and emotional expression. The Weir Mitchell Oration. *Trans. Coll. Phys., Philadelphia,* vol. 2, pp. 222–242.

Reece, R. P., and C. W. Turner. 1937. The lactogenic and thyrotropic hormone content of the anterior lobe of the pituitary gland. *Univ. Missouri Agr. Exp. Sta. Res. Bull.* 266, pp. 5–104.

Reighard, J., and H. S. Jennings. 1935. *Anatomy of the cat.* 3rd ed. Henry Holt, New York.

Rowlands, I. W., and M. K. McPhail. 1936. The action of progestin on the uterus of the cat. *Quart. Jour. Exp. Physiol.,* vol. 26, pp. 109–118.

Sainmont, G. 1905. Recherches relatives à d'organogenèse du testicle et de l'ovaire chez le chat. *Arch. biol.,* T. 22, pp. 71–162.

Simeone, F. A., and J. F. Ross. 1938. The effect of sympathectomy on gestation and lactation in the cat. *Amer. Jour. Physiol.,* vol. 122, pp. 659–667.

Sneider, M. E. 1937. Cytological abnormalities in the oocytes of the 3-week kitten's ovary. *Anat. Rec.,* vol. 70, pp. 13–27.

Snyder, F. F., and G. B. Wislocki. 1931. Further observations upon the experimental production of ovulation in rabbits. *Bull. Johns Hopkins Hosp.,* vol. 49, p. 106.

Starkey, W. F., and J. H. Leathem. 1939a. Some effects of menopause urine extract on sexual organs of immature female cats. *Proc. Soc. Exp. Biol. Med.,* vol. 41, pp. 503–507.

Starkey, W. F., and J. H. Leathem. 1939b. Action of estrone on sexual organs of immature male cats. *Anat. Rec.,* vol. 75, pp. 85–89.

Starkey, W. F., and J. H. Leathem. 1940a. Response of the immature female cat uterus to estradiol benzoate (estrone). *Proc. Penn. Acad. Science,* vol. XIV, pp. 87–92.

Starkey, W. F., and J. H. Leathem. 1940b. Effect of normal male urine extract on immature female cats. *Endocrinology,* vol. 26, pp. 499–502.

Starkey, W. F., and J. H. Leathem. 1943. Ovarian cysts in immature female cats following pregnant mare serum hormone administration. *Anat. Rec.,* vol. 86, pp. 401–407.

Stromsten, F. A. 1947. *Davison's mammalian anatomy with special reference to the cat.* 7th rev. ed. Blakiston, Philadelphia.

Todd, T. W. 1939. Ageing of vertebrates. Pp. 71–82 in *Ageing: biological and medical aspects,* ed. by E. V. Cowdry. Williams and Wilkins, Baltimore.

Turner, C. W., and W. R. de Moss. 1934. The normal and experimental development of the mammary gland. I. The male and female cat. *Univ. Missouri Agr. Exp. Sta. Res. Bull.* 207, pp. 5–15.

Van der Stricht, R. 1911. Vitellogenèse dans l'ovule de chatte. *Arch. biol.,* T. 26, pp. 365–481.

Van Dyke, H. B., and J. S. Chen. 1939. The assay of progesterone by the production of an artificial pregnancy response of the feline uterus. *Endocrinology,* vol. 25, pp. 337–346.

Van Dyke, H. B., and R. G. Gustavson. 1929. On the pregnancy response of the uterus of the cat. *Jour. Pharm. Exp. Therap.,* vol. 37, p. 379.

Van Dyke, H. B., and R. C. Li. 1938. The secretion of progesterone by the cat's ovary following the formation of corpora lutea due to the injection of anterior pituitary extract or prolan. *Chin. Jour. Physiol.,* vol. 13, pp. 213–228.

von Winiwarter, H., and G. Sainmont. 1907. Über die ausschliesslich post-fötale Bildung der definitiven Eier bei der Katze. *Anat. Anz.,* Bd. 32, S. 613 616.

von Winiwarter, H., and G. Saintmont. 1909. Nouvelles recherches sur l'ovogenese et l'organogenèse de l'ovaire des mammiferes (chat). *Arch. biol.,* T. 24, pp. 1–142, 165–276, 373–432, 627–651.

Windle, William F. 1939. Induction of mating and ovulation in the cat with pregnancy urine and serum extracts. *Endocrinology,* vol. 25, pp. 365–371.

Windle, W. F., and M. W. Fish. 1932. The development of the vestibular righting reflex in the cat. *Jour. Comp. Neur.,* vol. 54, pp. 85–96.

Windle, W. F., and A. M. Griffin. 1931. Observations on embryonic and fetal movements of the cat. *Jour. Comp. Neur.,* vol. 52, pp. 149–188.

Windle, W. F., J. E. O'Donnell, and E. E. Glasshagle. 1933. The early development of spontaneous and reflex behavior in cat embryos and fetuses. *Physiol Zool.,* vol. 6, pp. 521–541.

Wislocki, G. B., and E. W. Dempsey. 1945. Histochemical reactions of the endometrium in pregnancy. *Amer. Jour. Anat.,* vol. 77, pp. 365–403.

Wislocki, G. B., and E. W. Dempsey. 1946. Histochemical reactions in the placenta of the cat. *Amer. Jour. Anat.,* vol. 78, pp. 1–45.

Wislocki, G. B., and G. W. D. Hamlett. 1934. Remarks on synchorial litter mates in a cat. *Anat. Rec.,* vol. 61, pp. 97–107.

Young, W. C. 1941. Observations and experiments on mating behavior in female mammals. *Quart. Rev. Biol.,* vol. 16, pp. 135–156, 311–335.

9· FERRETS

T. HUME BISSONNETTE
Trinity College
Hartford, Connecticut

INTRODUCTION

The ferret of England and America, variously called *Putorius vulgaris, P. foetidus vulgaris,* or *Mustela putorius,* is believed by some authors to be a domesticated variety or albino mutant of the wild polecat, *Putorius foetidus,* of northern and central Europe, originally introduced there from northern Africa before the time of Pliny, who described its use in hunting rabbits. The tame ferret and its wild relative interbreed freely, and the hybrids are believed to give the color varieties appearing even in the same litter with white kits. The exact ancestry is still in question, however. Other authors think its ancestor came from southeastern Asia. Aside from the color phases just mentioned, the tame ferret has pink eyes and yellow-white fur, so coarse as to be of little commercial value. In England, there are two fairly common color phases; the "polecat" or "black," with creamy under fur and black guard hairs; and the chocolate brown, with brown upper parts and black under parts and a few scattered light hairs on the face. Either or both of these may be called "polecat ferret" in England, where the brown one is believed to result from crossing the white ferret with the wild polecat. Doubtless other color phases will be met with as more attention is paid to such "sports," as is the case with mink of late years.

The ferret reaches a body length of fourteen inches with tail of about five inches. In England there seems to be a larger strain. When mature, females are considerably smaller than males in both strains. Some common names for ferrets and polecats and their hybrids, other than white, are fitch, fitchew, fitchel, fichen. Tame ferrets raised in America were probably introduced from Europe, where they have been used in hunting rabbits and rats, and for taking ropes through long underground pipes.

Native or wild species of polecat are found in Europe, Asia, Africa, and the middlewestern states of America. They are:

1. *Putorius foetidus,* of north-central Europe, said to have been introduced from Africa; now rare in Britain; dark brown above, black below, with face mottled with light hairs; fur long and coarse; stronger and less active than the marten, and rarely tree climbing.

2. *P. eversmanni,* of northern Asia.

3. *P. sarmaticus,* from east Poland to Afghanistan; fur white, mottled with reddish spots above.

4. *P. nigripes,* the black-footed polecat of the midwestern plains of North America; fur creamy yellow, with brown legs and black feet and tail tip; length up to two feet in some specimens in addition to the tail. No hybrids of this species with the ferret have been reported.

Classification is based upon skull measurements in addition to size and coat color, as for weasels or other Mustelidae. All the above-listed forms have similar habits and eat small mammals up to rabbit size, snakes, lizards, frogs, fish, eggs, and birds, particularly poultry. Like weasels, ferrets and polecats will kill far more than they can eat if they get loose in a chicken coop. Wild female polecats bear six to eight kits in April or May after about a two-month gestation period. As will be seen, the tame ferret does better than that and often has two litters in one spring and summer season.

CARE AND HOUSING

Domestic ferrets are easily tamed, the white ones more readily than the dark, but show little or no real affection, even for their masters who feed them; they act as neutrals. Females often eat their young when improperly fed or excited. Neither sex shows any fear, even of a cat, but goes about its business. Ferrets are vicious biters when wild or teased and hang on like bulldogs. It is said that their grips can be loosened by pressure of the thumb just over the eyes. Until gentled, they should be handled with thick leather gloves. They may be handled by waving one glove slowly in front of them, to catch their attention and let them take hold if they will, while the other hand in the glove "sneaks up on them" from behind and grasps them quickly and firmly by the neck so that thumb and fingers meet under their chins. They may then be lifted and examined and handled to "gentle" them. When tamed by handling in this manner, they may be handled without gloves and seldom bite unless hurt or teased. They seem always to snap at a moving object, and so should not be teased by waving things in front of them. Another method of catching them is to turn the fingers of the right hand downward, spread the index finger away from the middle finger like a forked stick, and slip the fingers down over the neck from above with the nose toward the palm of the hand. The writer has not been able to perfect this technique, which he saw Dr. John Hammond, of Cambridge University, use very expertly.

When angered or excited, ferrets emit, from two scent glands under the tail, a peculiar pungent and penetrating odor like that of a weasel, very offensive and nauseating to some persons and animals. Hence the name "foetidus" for the polecat. The odor is much less powerful and does not spread as far as that of the skunk, its not too distant relative. It is perhaps caused by one of the mercaptans like that of the skunk, or a sulphide of an alcohol, in which the sulphur replaces oxygen. The odor is very persistent in clothing and in rooms.

Although wild polecats are not sensitive to cold and damp, ferrets are, and must be kept in dry, clean, well-ventilated rooms, cages, or hutches, protected from cold drafts. They are also very susceptible to "foot-rot," like that of sheep, an itch, like scabies, which causes swelling and scaling off of the toes and feet and spreads finally to the ears and most of the body. Strict cleanliness and watchful care are the only way to prevent it, and, in our experience, there is no really effective cure. Eventually foot-rot kills the animals.

Methods of housing the animals depend largely upon the type of experiments to be carried out or the numbers of animals to be kept. Some years ago we secured several ferrets, at intervals, from W. E. Cariens, Route 1, Port Clinton, Ohio, who was breeding ferrets up to 2000 or more per season for sale and for "ratting." In the little booklet which he sends to his clients, he recommends a dry clay or wooden floor for each pen, cleaned out once weekly by removing all straw, scraping the floor to get rid of all droppings, spraying floor and walls with naphtholeum, used one part in ten of water, to sweeten and disinfect the pen, and replacing plenty of clean wheat straw on the floor. At mid-week he adds more clean straw to refresh the footing. He thinks oat straw is irritating and leads to mange and scurvy. But, since many animal breeders fail to realize these are due to infection and deficiency of vitamins, Cariens may well be missing the real causes. He also runs the males and females together in large pens from August to late January, when he separates the males in individual cages and lets the females run in lots of five or six until four weeks after they are bred, in boxes 3 by 4 by 3 feet in size, without tops. These boxes must be in a light, dry room, free from cold damp drafts. At 28 days after breeding and 14 days before littering, the females are separated into individual pens, with good clean straw, and left undisturbed until after the kits are some days old. This separation keeps the females from quarreling and helps to prevent them from eating their young. In these individual nest boxes the straw or shavings used for beds should not be too deep after litters are born, lest the very young kits get buried too deep and become chilled and damp from lack of the mothers' warmth or fail to get to her to suckle. Because their eyes remain shut for about 5 weeks, they cannot find the mother by sight and she may overlook them and let them starve. This is good advice under any type of housing.

For our experimental work with ferrets as laboratory animals, we have used four different sorts of cages or pens. Among them, and slight modifications of them, the following have proved most useful and flexible in manipulation.

In Dr. F. H. A. Marshall's laboratory at Cambridge, England, where we first used ferrets in studies on photoperiodic control of sexual and breeding cycles and where we learned to take care of them, we used the same wooden pens or hutches as for rabbits. In them we kept individual ferrets or pairs, with straw, excelsior, or shavings in the rear of the pens for nesting places. The doors were of wire mesh of a convenient size and opened, on hinges, outward to permit feeding and cleaning out with a small hoe. The dimensions of these pens are not important, but they can easily accommodate one or two ferrets if one rabbit is not too large for them.

In this country, we have used three other sorts of pens. One, which we found useful for carrying over ferrets between different experiments and for breeding and rearing the few we have raised from matings, was a set of six rabbit pens grouped as a unit in three tiers of two pens each. The dimensions of each pen were 32 by 42 by 16 inches, with wooden corners and 1-inch chicken wire mesh for sides, bottom, and top. Under each pen was a space for sliding two trays of galvanized iron to catch the droppings. The front side of each pen had its door hinged at one end. In each pen we placed a wooden box, without top, 16 by 16 by 8 inches and with about 2 to 3 inches of shavings or excelsior in the bottom for nesting. Any number of ferrets up to six can easily be accommodated in these pens. Cleaning is very easy, as droppings fall through the mesh on to the trays and can be cleaned away and the trays disinfected or flamed. Small kits remain in the nest boxes until large enough not to fall through the mesh. An electric light for controlling reproductive or pelt cycles is easily placed in the pen so as to shine in both nest box and outer pen, to prevent the animals from hiding from the light whenever it is turned on.

For individual cages during controlled experiments in smaller space, we have used smaller pens grouped in three tiers of two pens each (figs, 2, 4, 5). This permits members of the two sexes to be subjected

FIG. 1. Part of ferret room, showing cages modified for weasels in foreground, exhaust fan in window, extra wire runways, and wooden dens. Metal ferret cages in four tiers, described in text, in rear of operator. Darkroom in right rear. Racks for holding den-runway systems to give motility.

at once to the same treatments. These cages may be with or without exercise cages of wire mesh attached to the rear face of each pen. The pens were 23 by 15 by 12½ inches inside, with sides, floors, and doors of ½-inch wire mesh and the back of wood. When the door is open, a galvanized-iron tray is slid into each pen, with sides and back 8 inches tall, no front, and floor or bottom 14 by 21½ inches. The door closes properly when the tray is in place. Trays are interchangeable for cleaning and disinfecting. Through the side of each tray, next the adjacent pen, and through the back are cut 3½-inch round holes corresponding to similar openings in the wooden partitions, for doorways to the other pen or to the wire runway on the back of the pen. These have metal disks attached so as to cover the holes when desired and prevent the animals from going through. The exercise cages are 21 by 13½ by 13½ inches with all sides, top, bottom, and far end made of ½-inch galvanized wire mesh. The other end, of wood, is fitted to the back of the pen above described. The cages are clean and the pens are bedded with shavings, sawdust, or excelsior to keep the animals off the metal floors when cold (fig. 2).

We also used metal cages in four tiers, with mesh sides, doors, floors, and tops (except in the top tier). The floors were removable and rested

FIG. 2. Ferret pens of wood in three tiers of two pens each, some with runways of wire mesh, used in lighting experiments. Drinking vessels in place; metal pans to facilitate cleaning; electric bulbs in place in sockets inside pens. One feeding dish is on top of runway.

above sliding removable pans for droppings and refuse. These were secured from the Geo. H. Wahmann Company, Baltimore, Md. The dimensions of these pens were 10 by 13 by 22 inches. They were bedded in the same way at the back, where a plywood platform was inserted above the wire floor (seen behind operator, fig. 1).

The most flexible combination of den and exercise cage used was an adaptation of the combination used by local mink breeders and described by us (Bissonnette and Bailey, 1940) (figs. 1, 3, 4, 5, 6) for use with weasels and other small mammals. The dimensions used for weasels were enlarged to double those given there, or more. Use of these requires more floor space than the cages previously described, because the double units of den and cage cannot well be used in vertical tiers, as was done with the others. But the arrangement is more flexible because the dens and cages are freely separable and inter-changeable for cleaning and disinfection, or for other manipulation of animals, without manual contact with them.

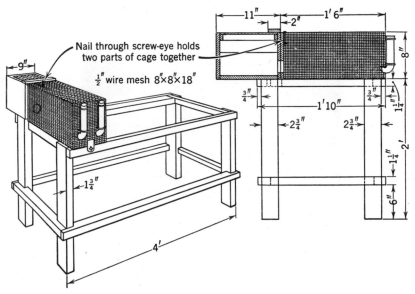

FIG. 3. Plans for den-runway units on racks, modified for weasels. Dimensions should be doubled for ferrets.

For use in rooms, under cover from rain and snow, the dens are wooden boxes, 18 by 22 by 20 inches, outside measure, with removable plywood covers which slide in and out under a cleat along each side and at the end next the runway. The covers are held in place by pushing a wire nail down through them just inside the box (fig. 5). This prevents the animals from freeing the lid. Covers may be light-tight or may admit some light around the edges, as desired (fig. 6). Under each plywood cover is a false top consisting of a rectangular wooden frame, covered by ½-inch soldered galvanized wire mesh, which fits snugly inside the box, resting on two cleats and held in place by two wire finishing nails inserted into it through two holes in the sides of the box near opposite ends, one on each side. The false tops are removable for cleaning and for handling the animals, as needed. If the animals tear at the wire of the false top, a sheet of glass slid along just under it will discourage them until the habit is broken, when the glass may be removed until needed again. For outside use exposed to rain and snow, the real top fits down over the box and is light- and water-tight.

The interior of the box is divided lengthwise, by a wooden partition half the height of the box, into a bedchamber, partly filled with shavings or sawdust for the nest, and a hallway, from which a proper-sized

Fig. 1. Runways shown in figure 9 to show drinking vessels in place and cleats to hold runways to racks. In rear, upper tiers of ferret pens shown in figure 2.

round opening for a doorway is in line with a similar hole in the near wall of the runway, also of wood.

About 3 inches from the top of the box and next the runway is centered an eyebolt with the eye projecting outside the box to pass through a slot at the same level in the wooden near-end of the runway. A suitable-sized wire nail is passed down through the wire mesh and the eyebolt (figs. 3, 5) inside the runway, with its head projecting above the wire, to hold the den and runway firmly together. (Dimensions shown in fig. 3 should be doubled as they were for use with weasels, much smaller animals.)

Of the runway, the top, bottom, sides, and one end are of ½-inch galvanized wire mesh, securely wired together with overlap, to prevent tearing apart (figs. 3, 4, 5). The end next the den is of wood, with a round opening in line with the exit from the den, and a slot to receive the eyebolt, as described above. The dimensions are 18 by

36 by 16 inches for ferrets. A short metal strip is soldered to extend vertically downward from the far wall of the runway to hook over a nail suitably placed in the rack which supports the den-runway units (figs. 3, 4, 5). This prevents the whole unit from upsetting by reason of the overhang of the den at the other side of the rack. A wire-mesh

Fig. 5. Attachment of den-runway units to racks; nail through eyebolt to hold den to runway on the rack; milk and water drinking vessels; other type ferret pens in rear.

door in the top of this runway, suitably fastened to keep the animals in, makes feeding and catching the animals by hand an easy matter.

Two drinking vessels per runway may be used to contain water or milk or both as desired. The ones we have used (figs. 3, 4, 5) consist of a 1½-inch glass tube drawn out at one end to ½ to ¾ inch diameter and bent at right angles. The large vertical part is stoppered with rubber cork, and the curved horizontal part projects through a small opening in the wire near the floor of the runway, where the animals will not defile it. In the upper side of the horizontal part is a hole large enough for drinking, and very little is spilled. Each vessel is kept in place by two fine wire springs of suitable strength stretched across it near top and bottom and firmly anchored to the wire mesh at each end. The bottom spring must be strong enough to hold the opening in place for drinking (obtained from the Geo. H. Wahmann Company, Baltimore, Md.).

These double units, of den and runway, may be placed side by side in groups up to five or more as needed, across racks or wooden frames

Fɪɢ. 6. Putting den-runway system into "darkroom" for short-day experiment. Racks on large casters; plywood roof of den in place over wire-mesh false top.

on legs with large casters for motility. The height of these racks should be about two feet; the other dimensions and the type of wood can be changed to suit the circumstances, and in keeping with the sizes of the dens and runways (figs. 1, 3, 4, 5, 6, with changes to suit).

Advantages of the system are: Grouping on racks with casters gives easy movement on cement floors in cleaning and hosing or mopping up droppings. Up to five units can be easily handled with enough flexibility for experiments on groups of animals caged individually or in pairs, but treated similarly. Multiples of these numbers do well as controls and experimental groups. The wire runways can be cleaned and disinfected by dipping into boiling water and effective solutions or by rapid flaming. Dens can be opened, sunned, dipped, aired, or painted. Animals can be easily changed from one to another combination by slipping a metal shingle between den and runway to cover the door and retain the animals in one, while the wire nail is removed from the eyebolt and the other removed and replaced. Replacing the nail through the eyebolt locks the new unit in place. Matings may be permitted in this way without handling the animals, and animals can be transferred to pens with glass sides for photographing. Daily inspection of units prevents escape. Such units can be piled in small space while not in use. They are particularly adapted for use with small quick animals, like weasels, that may be vicious, by reducing handling and injury to a minimum. Feeding is easy. The materials are easily available, and new features to suit conditions can be added for new experiments. This type of installation is not expensive.

The pens above described may be kept in convenient-sized basement rooms or others with cement floors and a sink hole, with removable perforated iron top, leading to the sewer by a wide pipe which does not tend to clog with deposits from soupy wastes. This arrangement permits easy cleaning and disinfecting of the room with hose, spray, and mop or broom. Air should be drawn out continuously during the day by an electric fan in one of the windows, controlled by a time switch to go into action early in the morning and shut off after nightfall (fig. 1). This removes odors to the outside in case there be persons in the building who resent such odors, and draws air from the rest of the building which is kept at a comfortable temperature. It also prevents dampness.

Metal containers of convenient size for storage of straw, shavings, excelsior, or sawdust for bedding, and other containers for feed, a power-driven meat chopper, and an electric water heater, for use when steam is not available from the heating plant to give scalding water to clean utensils and pens may be kept in the same or adjacent rooms. A refrigerator to keep milk, which forms a large part of the food of ferrets, and for vitamin concentrates, which are sometimes used, is also helpful.

REPRODUCTION

The ferret has a long autumn-winter period of sexual quiet or anestrus from late August or early September until March. Without artificial manipulation of lighting cycles or endocrine treatment, ferrets have not been known to breed during that period in the northern hemisphere. They show a similar period in the southern hemisphere from March to late August. By reversing cycles of day length and controlling them, the animals can be induced to come into heat or estrus and mate at any desired time of year. Two or perhaps even three litters can be produced in one normal season if conditions of lighting and nutrition are favorable and matings are secured as soon as possible in the spring and after the weaning of young.

The complete normal sexual cycles of females were studied at Cambridge, England, by Marshall (1904), Hammond and Marshall (1930), and Marshall (1933) in detail, and modifications of their sexual cycles by light control, by Bissonnette (1932, 1935c, 1936), Marshall and Bowden (1934, 1936), Hill and Parkes (1933b, 1934), and Allanson, Rowlands, and Parkes (1934). The sexual cycle of the male has been well described by Allanson (1932), and its modification by controlled lighting by Bissonnette (1932, 1935a, 1935b), Hill and Parkes (1933a), and others. The relation of this modification to the anterior pituitary gland and to pelt or hair cycles was described by Bissonnette (1935d).

Ferrets are almost ideal for study of sexual periodicity because the sexual condition in females can be learned by casual inspection of the vulva, which swells greatly and becomes turgid as heat or estrus comes on. This obviates the need to take vaginal smears to ascertain the stage of the cycle. In males, the size and position of the testes as learned by palpation give a good indication of sexual state. Willingness to mate and presence or absence of sperms in the exudate from the vulva of the female after copulation, followed by pregnancy or pseudopregnancy, are excellent criteria of hormone activity or germinal epithelium of the testis, respectively. Both sexes are relatively small and easily kept for breeding and experiment on a standard system of care and feeding, and they become tame and easily handled.

Normal Cycles. In females not mated, estrus lasts continuously from March to August, signalled by great swelling of the vulva and secretion of serous fluid in the receptive period. There is some slight, more or less cyclic variation in the size and turgidity of the vulva; but the vulva does not subside until anestrus begins late in August or during pregnancy or pseudopregnancy, which follows ovulation without fertilization and lasts as long as real pregnancy. Ovulation occurs only after prolonged coitus with males, either fertile, or sterile from incomplete activation or vasectomy. Successful coitus may last from about ten minutes to three hours, and follows a very sadistic courtship of an hour or more. Swelling of the vulva recurs with a second or even a third estrus after pseudopregnancy or pregnancy and suckling, before late August, at which time the autumn-winter anestrus comes on with shortening days. Should an animal become pregnant just before this occurs, it will complete its gestation period of 42 days (Hammond and Marshall, 1930).

Males exhibit a shorter quiescent period for the testis, accompanied by a reduced condition of accessory sex organs, lasting from October to January, during which period no sperms are produced and epididymides are empty of sperm and their linings are reduced in height. Testes reach maximum size in relation to weight in March; minimum size in October, after which germ cells increase in November and December (Allanson, 1932).

Copulation and Ovulation. In the anestrous season the males make no attempt to copulate, and the members of both sexes play together like kittens. But as the estrous period approaches, the males begin to take an interest in the females. Anestrous males in winter take no sexual interest in females, even in those brought into complete heat artificially, but play with them in the off-season manner, even when the females attempt to induce copulation. After the male has been subjected to added lighting, for only two weeks, his attitude toward estrous females changes and he is very much interested in them sexually and attempts to copulate; he may even succeed in it and become locked to the female. The males are normally ready to copulate in late February, but are not producing the necessary sperms in sufficient numbers to be fully fertile until about March 1st. The females become receptive in March and April, depending upon their previous light-history.

Courtship of a female in complete heat or estrus, by a vigorously potent male, as stated above, is a very rough and violent performance; so much so that the writer, when he first witnessed it, feared he would lose his female and asked the laboratory assistant, who was familiar with the performance, to "get the female out of there before he kills her." The reply was, "You have not seen anything yet. He will not do her any permanent damage."

The male grasps the female by the nape of the neck with his teeth, apparently in a preferred spot where it does not seem to hurt as much as at nearby places, and drags and throws her about for periods up to an hour. If he grasps the right spot, she offers no resistance to this, but when hurt emits little plaintive squeaks or whines. If she is not in full heat she fights vigorously and he finally desists. She should be removed to be tried again later. When she is limp and apparently almost comatose, with eyes nearly closed, the male, still holding her firmly by the neck, proceeds to attempt coition. Entry may take from several minutes to half an hour, and eventually the penis is inserted into the vulva and vagina and pushed home with strong convulsive thrusts while the female is held between the forefeet and partially by the hind feet. The bone of the penis (os penis) is almost as long as the femur and one-third to one-half as thick, with a hook at the distal end, and is anchored firmly to the symphysis pubis by ligaments. After the penis is firmly locked in the vagina, the male retains his grip on the neck and the animals remain locked together for as long as three hours. During this time there are periods of quiet, while the animals lie on their sides, alternating with grinding, thrusting spasms like those by which complete entry was achieved.

While the animals are locked in this manner they can be picked up by grasping the male at the shoulders and back. They remain locked unless teased or shaken, after which they separate and the coitus is over. The result, however, can be pregnancy, or pseudopregnancy, after a copulation as short as twelve minutes, if the male is completely potent and fertile. There must be an excess of sperms in the ejaculate to insure fertilization. Small amounts are ineffective. One male should not be used for matings with more than five or six females in a limited time, if certainty of large litters is desired, lest his ejaculations fall below the minimum effective quantity of semen or the sperms be imperfectly matured in the epididymides.

On separation, the male may be left with the female or he may be removed. The same female may be induced to accept copulation as often as three or four times in two days by placing her with a vigorous potent male or with successive males. One successful copulation leads to ovulation between thirty and forty hours after the animals become locked. In some cases, after thirty-four hours ovulation had not occurred, and at thirty-six hours eggs were in the oviducts. The time is variable. Ovulation results from a stimulus from the orgasm to the anterior pituitary gland, which shifts the balance between the amounts of follicle-stimulating hormone and of luteinizing hormone liberated into the blood stream. The repeated copulations within the short time before the female becomes non-receptive induce ovulation and luteinizing of the follicles into corpora lutea, as do single copulations; and they may also lead to luteinizing of additional follicles from which eggs have not been extruded. Corpora lutea of this sort have been observed containing eggs still surrounded by granulosa cells, again in turn surrounded by luteal tissue.

The eggs shed are picked up by the ostia and pass down the oviducts, surrounded by follicle cells for some time, and then sperms penetrate the eggs and fertilization is effected in the tubes. The second polar body is extruded from the egg after ovulation.

Placentation, like that of other carnivors, is of the zonary type, with the signet-ring effect due to the large dark blood-filled region at the base of the umbilical stalk. The expansion of the fetal placenta around the endometrial wall is incomplete at the meso-metrial region for a considerable time, and implantation is antimeso-metrial. In cleared specimens the placenta is seen to resemble a signet ring of which the part of the ring opposite the signet is open or in-complete. Gestation period is approximately forty-two days.

BUILDING A BREEDING STOCK

Care of Mother and Young. As stated above (under Care and Housing), females may be housed together in groups of two to six until about two weeks before littering time. They should then be housed individually to prevent quarreling and eating of young. Bed-ding should be scanty but soft, too shallow to permit loss of the little ones in the bedding, causing sweats and starvation. The mother and young should not be disturbed for a few days lest she get excited and eat her kits.

The eyes of the kits remain closed for about five weeks, but they can often be helped to find their way to the feed dish before the eyes open. When they set up a fuss when the mother leaves them to feed, they should be taught to take food in addition to suckling, particu-larly if the litter is as large as nine or ten kits. Bread soaked in fresh milk can be scattered in pieces in the nest twice a day after the mother has been fed or while she is feeding. The kits will eat the bread for the milk and eventually find their way to the feed dish with the mother. This will insure that all members of the litter get enough food to grow well, and none should starve.

As soon as the young are weaned, or soon thereafter, the female should be removed to a pen by herself or the young taken from her. She may then come back into heat and mate. She may do this also after her young die for any cause or are eaten. Her food should be rich enough in proteins, vitamins, and minerals to prevent deficiency diseases and eating of young. This is also true for the growing young ones.

The manner of mating the animals will depend upon the type of breeding stock desired. If an inbred stock of very uniform genetic background is wanted, then judicious matings of strong individuals possessing the desirable qualities and related to each other should be undertaken. If, on the other hand, it is intended to keep the stock

variable and heterozygous for hybrid vigor, unrelated males and females should be mated and new strains should be introduced from
time to time. Selection for genetic characters may be practiced as
with any other animals, such as rats and mice. Several different
strains can be developed together.

No diseased or crippled animals should be used for breeding, and
rigorous culling out should be practiced. A period of quarantine
should be followed with all introductions from other stocks. Pedigree without high physical quality will not guarantee good stocks.

SANITATION

Sanitary precautions were suggested above in the section on Care
and Housing. Regular cleaning out of straw, excelsior, shavings, or
sawdust, used for bedding or carpet, and of droppings and the clay
surface, if clay is used for floors, and replacement with new bedding
and carpet once a week or oftener, are imperative. Pens and runways should be scrubbed with boiling water, sunned, and sprayed
with a good disinfectant, not irritating to the animals, and under
precautions against poisoning. Metal parts should be flamed regularly.

All feeding and drinking utensils must be boiled or scalded and kept
in clean dry places while not in use. Clorox or another oxidizing
agent helps. Excess of food above that eaten at each feeding should
be avoided so that none stands for flies to infect and spoil between
feeding times.

Flies, fleas, and other insects must be guarded against. Spraying
of walls, floors, ceilings, and cages with DDT or some good equivalent should help keep flies down and, to a lesser degree, fleas. Screening of windows and prevention of first infestation are the most effective sanitary measures. Continual vigilance is necessary. Admission
of all cats and dogs must be prevented; they carry fleas among other
pests.

Ferrets are susceptible to "foot-rot." Like that of the sheep, it is
due to an itch mite which attacks the feet and spreads quickly to the
ears, nose, eyes, tail, and over the body. It causes the feet to swell
and scale off, the nails to become thick and long, scabs to appear on
the body and in the ears, the hair to drop out, and the animals to
become inert, to sleep all the time except for brief periods used to
eat, drink, and attend the calls of nature. They finally become moribund and die. Once started in a stock, foot-rot is passed from animal

to animal. Unless stopped at once, it may affect the whole stock, which must be destroyed. The young become infested from the nest; the mother may be caused by her itching to bite and eat her young.

To prevent foot-rot we have mixed flowers of sulphur in the beds and on the floors. As sulphur ointment, we have rubbed it well into the feet and ears. Sometimes it helped, if the infestation was slight; often not. We received one lot of animals already infested, and we had to destroy them after a quick experiment. Luckily foot-rot does not prevent photoperiodic activation of the sexual organs of ferrets. They are affected just as much by the added lighting, when almost moribund with foot-rot, as when in good health and not infested at all. But any young born would be infected from birth. General good health appears to help the animals to fight foot-rot off or remain free from it. It is suggested that some of the newer cures or treatments for scabies in man be tried, such as the one suggested by O'Loghlen (1942), in which two solutions are used: A—sodium hyposulphite, 20 per cent solution in water, sponged on and let dry; B—a 5 per cent solution of hydrochloric acid. After sponging with solution B, the feet and tail can be wrapped in bandages. Two days of treatment is effective in man and with transfer to uninfested quarters should work for the ferret. Cariens recommends application to the affected parts of an ointment made by melting lard, stirring in an equal amount of sulphur, and adding equal amounts of coal oil and turpentine. Two days later trim the nails and remove loosened scabs. Repeat two or three times a week until all signs of the infestation are gone. Quarters also must be thoroughly freed from the mites. Cleanings should be burned and pens flamed.

If the young show signs of "sweats" from being covered too deeply in bedding, reduce the bedding and air the animals until they are dry and not too warm. Otherwise they may get sticky, adhere to each other, and die.

Swellings on the head and neck occasionally develop in even the finest and apparently healthiest of ferrets. These resemble boils and result from infection with pathogenic bacteria. They may be poulticed or kept moist with Epsom salt solution on bandages until they soften or come to a head, but should never be squeezed in handling the animals, lest the infection be spread into the general circulation or lymph stream. Then lance with a sterilized scalpel or knife and gently work out the pus. It helps to make two incisions in the form of a cross and snip off the corners to prevent the incision from closing. Keep it open and well disinfected around about with Lugol's solution of iodine or some milder non-poisonous solution or ointment, to prevent reinfection from the hair follicles near by. Isolate all infected animals until fully healed. The animals generally recover, but we have had one die from one of these swellings that did not come to a head.

Ferrets are very susceptible to dog and cat distemper and to human influenza virus. They have been used a great deal since Dunkin and Laidlaw (1926) described dog distemper in the ferret, and Hinz (1931) immunized ferrets against dog distemper. Horsfall and Lennette (1940) showed that canine distemper and human influenza gave synergic effects when ferrets were subjected to them at the same time. Ferrets are being used in the Department of Bacteriology and Immunology of the University of Minnesota by the late Professor R. G. Green and his students in his studies (1945 et seq.) on the effects of passing dog distemper several times through ferrets to see if it is changed in virulence. They are also utilized to test the virulence of sera used on puppies (Slavitz, 1935). Dalling (1931) has described distemper in the fitch, and Spooner (1938) has described a distemper-like disease epidemic in ferrets.

The techniques used in these procedures fall outside the scope of this chapter, but references are appended to the bibliography at the end of the chapter for those who wish to follow up the subject. Breeders and users of ferrets as laboratory animals for other studies will need to exert every precaution to keep the stock from becoming infected lest they lose them all from epidemics, which spread like wildfire. Persons with influenza or recently exposed to it, dogs and cats, and their handlers should be rigorously excluded from the buildings in which ferrets are kept and from food stored for ferret use. If infection is suspected, immunizing inoculations or procedures should be carried out at once to forestall epidemics. A good veterinarian or some other person having experience in these diseases and procedures should be brought in to help. We have never seen such an epidemic in our experience with ferrets in England in 1931–1932 and since that time in America.

NUTRITION

Bread and milk are used by many persons who keep only a few ferrets, and that diet seems to be satisfactory. In England we used fresh unpasteurized milk as the sole food for our ferrets, as did Drs. Marshall and Hammond. Raw milk is believed to contain all the necessary vitamins, whereas some are lost or reduced by pasteurization.

Since returning to this country we have used Grade A milk, and pasteurized milk with yeast tablets or powder to increase the content of the vitamin B group and sometimes with vitamins A and D added in various forms. These last additions caused our animals to become fat. We also had a good dog food or fox food in checker form available to the animals. (Any of the advertised brands seems to be satisfactory; we used those made by the Wayne and Elmire companies.) When the dog or fox food was given to the animals for the first time in checker form in a box, they did not seem to relish it; so we tried dropping a few checkers into the milk to soften, and they took the food that way. Some salt was added, particularly with ferrets whose pituitary glands were removed, and the amount increased as the animals declined in vigor. We got the impression that this prolonged their lives, and one or two became active again after being almost moribund.

For animals carrying or suckling young it is advantageous to increase the relative amount of protein in the diet by adding these checkers or some freshly killed small animals or meat. Starlings, mice, and rabbit meat are economical.

For large stocks of ferrets for which the above regimen is expensive, Cariens (1932) recommends a mush made from boiling water with fine whole-wheat flour, like graham flour, stirred in while boiling. To the needed amounts of this add double the amount of fresh milk, and do not let the mush get lumpy when cold. Cariens also recommends fresh meat for young animals. But salty meat or pork should not be used.

Overfeeding should be avoided lest the animals become too fat and lazy for good health and breeding. Animals to be used for rabbit hunting or against rats should be hungry and fed only half rations on the day before using. They should be in good lean condition, like an athlete.

FEEDING DIRECTIONS

Some breeders or keepers of ferrets feed but once a day, in the morning, but Cariens (1932) recommends twice, morning and night, and three times for young growing kits. We have fed ours night and morning and oftener for young ones. Not more is given at a feeding than the animals will clean up completely, and the dishes should be inspected frequently to be sure there is no excess. Any left-overs should be removed to prevent spoiling and overfeeding by returning between meals. This should be adhered to with young as well as mature animals. Water should be available for them at all times in the drinking vessels described under Care and Housing.

For feeding dishes we use earthenware basins with vertical sides 2¼ inches high and 6½ inches in diameter with an overhanging inside margin to prevent spilling. These were obtained from the Geo. H. Wahmann Company of Baltimore, Md. They are scalded thoroughly every morning before being put in use. Enough are available so that one set is in use while the other is being cleaned and sunned. (See fig. 2, on top of runway.)

If meat is given it should be at mid-day and in as many pieces as there are animals in the pen, lest they get to fighting over it and wound each other. Such fights may lead to cuts and scratches which may fester and give the abscesses on the throat and head described above.

PRACTICAL SUGGESTIONS

Throughout the above account of the care and breeding of these animals, the benefit of our experiences has been given as they occurred to mind at each step, as cautions. They may be repeated here briefly.

1. Cleanliness is of paramount importance.

2. Prevention of infection is *the* way to keep free from disease. Quarantine from dogs, cats, and persons with influenza and distemper or from handlers of such vectors should be practiced.

3. The type of cage to be chosen depends upon the use to be made of the ferrets and the numbers to be kept. Adaptive changes in the types suggested above will fit them for special purposes.

4. Avoid overfeeding, but be sure vitamins and minerals are available in proper amounts.

5. Being carnivores, ferrets in nature would live mostly on meats and high-protein diets, with entrails of prey to furnish vitamins, and bones, the needed calcium.

6. Frequent careful handling from time of weaning onward will "gentle" the animals and make them easy to handle without excitement and injury to ferret or keeper. Heavy leather gloves should be used at first to protect the handler. Bites from ferrets usually fester and should be thoroughly disinfected in both man and ferret.

7. The animals make good pets, when tamed, but show little real affection even to those that feed them. Like cats, they are individualists.

8. The old superstition, that if a female ferret is not mated and allowed to become pregnant she will die as a result, is not true. Fe-

males may be kept over two years at least without mating and live a healthy life if properly fed and kept free from dirt and infection.

9. Experience will dictate changes in the above program and regimen of care, housing, sanitation, feeding, and breeding to suit your conditions. The above is only a set of tried procedures that have suited our studies with ferrets.

REFERENCES

The following list is by no means exhaustive but covers some of the studies made upon ferrets.

Housing and Reproduction

Allanson, M. 1932. Reproductive processes of certain mammals. III. The reproductive cycle of the male ferret. *Proc. Roy. Soc.*, ser. B, vol. 110, pp. 295–315.

Allanson, M., I. W. Rowlands, and A. S. Parkes. 1934. Induction of fertility and pregnancy in the anoestrous ferret. *Proc. Roy. Soc.*, ser. B, vol. 115, pp. 410–421.

Bissonnette, T. H. 1932. Modification of mammalian sexual cycles. Reactions of ferrets (*Putorius vulgaris*) of both sexes to electric light added after dark in November and December. *Proc. Roy. Soc.*, ser. B, vol. 110, pp. 322–336.

——. 1935a. Modification of mammalian sexual cycles. II. Effects upon young male ferrets (*Putorius vulgaris*) of constant eight and one-half hour days and of six hours of illumination after dark between November and June. *Biol. Bull.*, vol. 68, pp. 300–313.

——. 1935b. Modification of mammalian sexual cycles. III. Reversal of the cycle of male ferrets (*Putorius vulgaris*) by increasing periods of exposure to light between October second and March thirtieth. *Jour. Exp. Zool.*, vol. 71, pp. 341–373.

——. 1935c. Modification of mammalian sexual cycles. IV. Delay of oestrus and induction of anoestrus in female ferrets by reduction of intensity and duration of daily light periods in the normal oestrous season. *Jour. Exp. Biol.*, vol. 12, pp. 315–320.

——. 1935d. Relation of hair cycles in ferrets to changes in the anterior hypophysis and to light cycles. *Anat. Rec.*, vol. 33, pp. 159–168.

——. 1936. Modification of mammalian sexual cycles. V. Avenue of reception of sexually stimulating light. *Jour. Comp. Psychol.*, vol. 22, pp. 93–103.

Bissonnette, T. H., and E. E. Bailey. 1940. Den and runway system for weasels and other small mammals in the laboratory. *Amer. Mid. Naturalist,* vol. 24, pp. 761–763.

Cariens, W. E. 1932. *Feeding, breeding, and working ferrets* (a breeder's pamphlet). Route #1, Port Clinton, Ohio.

Hammond, J., and F. H. A. Marshall. 1930. Oestrus and pseudopregnancy in the ferret. *Proc. Roy. Soc.*, ser. B, vol. 105, pp. 607–630.

Hill, M., and A. S. Parkes. 1933a. Studies on the hypophysectomized ferret. IV. Comparison of the reproductive organs during anoestrus and after hypophysectomy. *Proc. Roy. Soc.*, ser. B, vol. 113, pp. 530–536.

Hill, M., and A. S. Parkes. 1933b. Studies on the hypophysectomized ferret. V. Effect of hypophysectomy on the response of the female ferret to additional illumination during anoestrus. *Proc. Roy. Soc.,* ser. B, vol. 113, pp. 537–540.

Hill, M., and A. S. Parkes. 1933c. Studies on the hypophysectomized ferret. VI. Comparison of the response to oestrin of anoestrous, ovariectomized, and hypophysectomized ferrets. *Proc. Roy. Soc.,* ser. B, vol. 113, pp. 541–544.

Hill, M., and A. S. Parkes. 1934. Effect of absence of light on the breeding season of the ferret. *Proc. Roy. Soc.,* vol. 115, pp. 14–17.

Marshall, F. H. A. 1904. The oestrous cycle of the common ferret. *Quart. Jour. Micro. Science,* vol. 48, pp. 323–345.

——. 1933. Cyclical changes in the vagina and vulva of the ferret. *Quart. Jour. Exp. Physiol.,* vol. 23, pp. 131–141.

——. 1940. The experimental modification of the oestrous cycle in the ferret by different intensities of light irradiation and other methods. *Jour. Exp. Biol.,* vol. 17, pp. 139–146.

Marshall, F. H. A., and F. P. Bowden. 1934. The effect of irradiation with different wave-lengths on the oestrous cycle of the ferret, with remarks on the factors controlling sexual periodicity. *Jour. Exp. Biol.,* vol. 11, pp. 409–422.

Marshall, F. H. A., and F. P. Bowden. 1936. Further effects of irradiation on the oestrous cycle of the ferret. *Jour. Exp. Biol.,* vol. 13, pp. 383–386.

Use in Immunology and Pathology

Dalling, T. 1931. Distemper in fitch. *Vet. Rec.,* vol. 11, pp. 1051–1052.

Dunkin, G. W., and P. P. Laidlaw. 1926. Dog distemper in the ferret. *Jour. Comp. Path. Therap.,* vol. 39, p. 201. (Cited from succeeding paper.)

Dunkin, G. W., and P. P. Laidlaw. 1931. Some fuller observations on dog distemper. *Jour. Amer. Vet. Med. Assoc.,* vol. 78, pp. 545–551.

Green, R. G. 1945. Temperature reactions of dogs to distemper after ferret passage. *Proc. Soc. Exp. Biol. Med.,* vol. 58, pp. 103–105.

——. 1944. Zoological and histological modification of distemper virus after ferret passage. *Amer. Jour. Hyg.,* vol. 41, pp. 7–24.

Hinz, W. 1931. Short description of immunization of ferrets against distemper by Laidlaw-Dunkin method. *Vet. Rec.,* vol. 11, p. 625.

Horsfall, F. L., and E. H. Lennette. 1940. Synergism of human influenza and canine distemper viruses in ferrets. *Jour. Exp. Med.,* vol. 72, pp. 247–259.

O'Loghlen, J. E. 1942. Treatment of scabies. *Brit. Med. Jour.,* vol. 1942, (2), p. 416.

Slavitz, C. A. 1935. Insusceptibility of young puppies to distemper virus. *Proc. Soc. Exp. Biol. Med.,* vol. 32, pp. 1225–1229.

Spooner, E. T. C. 1938. Disease resembling distemper epidemic among ferrets. *Jour. Hyg.,* vol. 38, pp. 79–89.

10. THE OPOSSUM

EDMOND J. FARRIS

The Wistar Institute of Anatomy and Biology
Philadelphia, Pennsylvania

INTRODUCTION

Of all our native animals, the opossum is the only mammal living in North America that carries its young in a pouch, and for this reason it presents unique mammalian material for research purposes, particularly for problems concerned with embryology.

The Wistar Institute conducted a series of experiments with four different species of American opossums:

1. The *Didelphis virginiana.*

2. The gray-masked or four-eyed opossum, *Metachirops fuscogriseus.*

3. The wooly or red opossum, *Metachirus nudicaudatus dentaneus (Philander).*

4. The murine opossum, *Marmosa isthmica.*

The last three types, all belonging to the order Marsupialia didelphidae verae, were captured especially for breeding experiments.

Many myths have originated about the opossum's reproductive processes, and this can easily be understood when one considers that an opossum can produce a litter of thirteen young only 12 to 13 days after conception. At birth the entire litter is so small that all the young would easily fit into a soup spoon; each opossum weighs approximately 1.3 grains.

The physiology of reproduction in the opossum is interesting. The female opossum has two uteri, and the embryo has no placenta, lying free in the uterus during its intrauterine development. During this period the embryos may be moved from one uterus to another, and still go on growing normally. At birth, the tiny, undeveloped half-inch creature crawls among the moistened ventral hairs along its mother's belly to the teats in her abdominal pouch, and there the

256

young one finds the nipples of the mammary gland. The nipples swell to form an oral plug-like attachment, and the young opossum grows

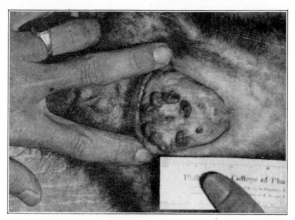

F𝚒𝚐. 1. Nine young opossums attached to mammary gland nipples in mother's pouch. Three nipples are visible. Note the undeveloped appearance of the young, in residence in the pouch for about one week.

in the pouch (fig. 1), feeding on the mother's milk for about 60 to 70 days. During this period the fetuses are particularly interesting for observation and experimentation.

When the offspring are developed enough to emerge from the pouch, but are still small enough to need the mother's attention, she carries them about on her back (fig. 2), their tails entwined about hers, their feet clutching the fur of her back.

TYPES OF OPOSSUMS

Of the four types of opossums studied at the Institute, the American opossum, *Didelphis virginiana*, is abundant and easily caught in almost any part of the United

F𝚒𝚐. 2. Litter of young opossums at about 75 days of age, enjoying the protection and transportation of their mother.

States. This particular marsupial has been studied for many years. A technique was developed for breeding it in captivity, and this has resulted in many interesting discoveries. However, since the adult opossum *virginiana* grows to about the size of a small dog, it becomes expensive to raise for laboratory work, so The Wistar Institute attempted to domesticate and breed three types of tropical opossums.

Description. Size is the chief difference between the Virginia opossum and the tropical opossums. The average weight of the Virginia opossum is 6 to 8 lb., and those raised in captivity for laboratory purposes have attained weights as high as 19 to 25 lb. at the age of 495 days. On the other hand, the usual weight of the tropical opossums is less than 1 lb.

Fig. 3. The gray-masked or four-eyed opossum, *Metachirops fuscogriseus*.

Some of the differential characteristics of the four genera, as noted before domestication, are as follows:

A. The gray-masked (fig. 3) or four-eyed opossum, *Metachirops fuscogriseus*.
   1. Tail furred basally.
   2. Vicious, hisser and spitter.
   3. Terrestrial.
   4. Seven mammae, arranged in pairs, with one in lower midline.
   5. Weight 250 to 500 grams.

The gray-masked opossum is one of the best experimental animals of the *Didelphis* family. It is about the size of a rat, inexpensive to raise, and, being terrestrial, does not climb very high. It is calm, easily domesticated, and hardy. It is amenable to being raised in dormer cages such as are used for rats (page 45 or fig. 9) and is easy to keep clean. It has been established that this variety has a life span of about 3½ years, which in our experience is about ½ year longer than the Virginia opossum.

Experiments with hormones (Nelsen and White, 1941; Farris, 1941a) suggest that the animal can be forced to ovulate and breed successfully in captivity. We have been able to breed several litters successfully at the Farm (Farris, 1941b). The average number of young in a litter was four, with a slightly higher ratio of males.

In view of the value of the gray-masked opossum as a laboratory animal, a few additional details may be of value. About eighteen females were collected in grass fields about 47 miles from Panama City, in a location that is low and damp all the year. The animals

Fig. 4. The "wooly" opossum, *Metachirus nudicaudatus dentaneus.* This opossum was bred and domesticated at The Morris Biological Farm of The Wistar Institute.

usually were found in their nests in palm trees, at the branching and in the leaves about 5 feet off the ground, where conditions were relatively dry. A stick was pushed carefully into the leaf nest where the animal was well hidden. The poking usually resulted in the blowing or hissing sound of the animal. It could then be captured by hand or by any other method.*

B. Wooly opossum (fig. 4), *Metachirus nudicaudatus dentaneus.*
   1. Wooly fur, reddish brown, with a furred tail.
   2. Aggressive and vicious.
   3. Semi-arboreal.
   4. Five mammae in pouch.
   5. Weight 300 to 400 grams.

* Appreciation is expressed to Dr. Herbert Clark, Director of the Gorgas Memorial Laboratory, to Mr. J. B. Shropshire, of the United States Army Sanitation Bureau, and to Dr. G. B. Fairchild, Entomologist, Gorgas Memorial Laboratory, for aid in capturing many of the tropical opossums.

Although the wooly opossum is described as the most vicious and aggressive of all the opossums, the young can be easily domesticated.

The wooly opossums were secured on higher ground, in trees or in hollow holes. The animal resists capture and fights severely until caught. The eyes protrude, and look as though they are about to pop out of the head, suggesting the usual thyroid exophthalmos.

The females carried pouch young when caught. These young were about 1½ to 2 cm. in length up to weaning size. These were captured during the months of August and September.

FIG. 5. The murine opossum, *Marmosa isthmica*.

The wooly opossum, also about the size of a rat, has one disadvantage. It is semi-arboreal. Therefore all its cages and quarters have to be enclosed.

C. The murine opossum (fig. 5), *Marmosa isthmica*.
1. Nude tail.
2. Quick and active.
3. Large eyed, large eared.
4. Brown cinnamon color.
5. Semi-arboreal.
6. Thirteen mammae.
7. Weight 75 to 100 grams.

The small murine opossum, about the size of a mouse, is too fast for the usual caretaker to handle and is very difficult to breed. This particular type has not been bred in captivity, and trials in other laboratories have met with failure.

D. Virginia opossum (fig. 6), *Didelphis virginiana*.
1. Slow but aggressive.
2. Primarily terrestrial.

3. Thirteen mammae usually, but have had as many as eighteen in a litter being raised in the pouch.
4. Weight 6 to 8 lb.

In the United States the opossum is found in great numbers in the wooded eastern part, from eastern New York, southern Wisconsin, eastern Nebraska, and down to the Gulf Coast.

This opossum prefers swamps and wet lowlands bordering along streams. It is found in hollow trees and in holes under roots of trees. This form has never been domesticated in the colony.

Fig. 6. The Virginia opossum, *Didelphis virginiana*. This opossum was bred and reared at The Wistar Institute farm.

PENS

An opossum colony building (fig. 7), constructed of steel and cement blocks covered with stucco, serves very well for the establishment of a satisfactory breeding colony. The building accommodates four animal rooms, one detention room for receiving of new stock, a small library, four laboratories or offices, and the usual rooms for mixing diet and preparing cages.

Located in the library is the largest and finest collection of opossum eggs and embryos ever collected. This collection was started by Dr. Carl Hartman. It includes a series of prepared microscopic slides, showing almost all the possible gradations in size to 4 mm. Another collection contains separate embryos from 4 to 13.5 mm. The opossum is delivered when it is 10 to 13 mm. long, and some of these last-mentioned specimens are pouch young.

Adjoining the laboratory building (fig. 7) are three outdoor pens, each approximately 15 by 12 by 6 feet, made of 21-inch link wire on pipe frames, with cement curbs and sand floors on a foundation of broken stone and gravel, drained by subsoil tile, and shaded by trees. This type of pen is intended to be a permanent structure, adequate

FIG. 7. The opossum colony building. Note the enclosed outdoor quarters.

for breeding or for holding non-breeders throughout the year. The sand floors are used to reduce foot injuries, of which there are many. They are easily cleaned and drained, and this aids in elimination of the intestinal parasites and fleas.

FIG. 8. Portable kennels.

Portable Kennels (fig. 8). These are 8 by 3 by 2 feet, made of wood and wire, and equipped with weather-proof shelter boxes. All outdoor pens are provided with larger weather-proof shelter boxes, allowing each animal 1 to 2 square feet of floor space. The boxes are lined with Cellotex for protection of the animal from heat or cold.

Detention Room. The new stock, or animals that were maintained outdoors and are to be brought indoors for general experimentation purposes, are carefully observed in the detention room in smaller cages (fig. 9) for approximately 10 days, or until the animal is known to be "clean" and satisfactory for research purposes.

The experimental animals are maintained indoors, in cubicles large enough to permit exercise. The floor is concrete and easily hosed, a provision essential in preventing disease.

FIG. 9. Corner of the detention room, in which opossums are conditioned until satisfactory for research purposes.

The Institute has found that opossums survive severe winters outdoors, provided they have adequate nesting material to insulate them against sudden changes in temperature.

REPRODUCTION

The breeding season of the Virginia opossum begins in late January practically throughout the country. In the colony, eggs and embryos can be secured nearly all the year round, a brief anestrous period being noted from October through December.

By employing artificial means for extending the length of daylight in the colony, the Institute has been able to breed opossum successfully during the "anestrous" period, and has had several litters born in December.

The Institute has found that the opossum breeds very well the first and second year, but that it is rare to secure litters during the third year. Contrary to the findings of certain authorities, the opossums, under colony conditions, are old in 3 years, and the author is convinced that this is approximately the life span of the Virginia opossum.

The writer has observed several opossums during copulation. The male and female lie on their sides, and remain attached for at least 20 to 40 minutes. The female is attracted by a peculiar clicking sound, a sort of metallic ring, produced by the courting male. He apparently is inspired to make this sound only when the female is coming into heat about once every 23 to 28 days. If the male insists upon mating before the female is sufficiently excited sexually, and in the proper stage of estrus, he may be killed. At such times a tiny female is fully capable of annihilating a much larger male. This behavior was particularly noted after the administration to the females of estrogenic hormones. They became extremely irritable and combative to the male for a short period before mating was permitted. When the pairs were left together overnight, it was not uncommon to find the male dead in the morning.

The birth of these animals has been described by Dr. Carl Hartman (1920), who removed unborn opossums from the uterus at term, and found that they could crawl easily on the mother's moistened abdomen. He witnessed the actual migration of opossum young, proving conclusively that they were able to pull themselves up to the pouch in spite of their immature state of development.

At birth the posterior appendages are useless paddle-shaped structures, but overdevelopment of the anterior appendages provided with deciduous claw-like structures permits the young to crawl among the ventral hairs to reach the pouch and nipples of the mammary gland.

NUTRITION

Opossums eat a great variety of foods, including meat, eggs, fruit, cooked vegetables, nuts, insects, mice, young rats, and bananas. The various dry diets advocated by commercial companies for mammals such as minks and foxes proved to be quite unsatisfactory for opossums, particularly for breeding purposes. The dry diet is satisfactory only for general maintenance of opossums during short-term experiments.

Two diets are herewith listed. The first diet was used successfully for an excellent colony in the sixth generation of inbreeding, when the diet was finally discarded for a less expensive one. The diet for a colony of 80 adult animals and feeding directions are as follows:

MONDAY. 13 lb. of liver, 5 quarts of milk, 6 eggs, bone meal (about 1 teaspoonful sprinkled into each food pan), wheat meal (1 teaspoonful sprinkled into each food pan), occasional fruits and vegetables according to season. *Note:* This is approximately 2½ oz. of meat per day per animal. The bone meal and wheat meal are sprinkled on the raw meat to soak up the meat juices so that the animals will eat it. The eggs are broken and stirred into the milk so that they are evenly distributed to all the animals.

TUESDAY. 5 quarts of milk, 6 eggs, bone meal, wheat meal, occasional fruits and vegetables.

WEDNESDAY. Same as Monday.

THURSDAY. 9 lb. of beef, 3 calf hearts, 5 quarts of milk, 6 eggs, bone meal, wheat meal, no fruits or vegetables.

FRIDAY. Same as Monday.

SATURDAY. 10 lb. of beef, 4 calf hearts, 5 quarts of milk, 6 eggs, bone meal, wheat meal, occasional fruits and vegetables.

SUNDAY. Dry diet of any variety.

Water should be available at all times.

In an effort to reduce the cost of feeding the large Virginia opossums, Dr. Herbert Ratcliffe adopted a diet which he had developed for the large mammals in the Philadelphia Zoological Garden, for the purposes of the opossum, and found it to be quite satisfactory. The breeding rate was reduced slightly.

Meat	60% (hog liver 2 parts, beef heart 1 part)
Vegetables	10% (any green vegetables)
Dry powder	30%, made up as follows:

Soybean	30%
Peanut meal	25%
Dried skim milk	25%
Oyster shells	6%
Alfalfa leaf meal	4%
Brewer's yeast	4%
Iodized salt	2%
Cod-liver oil	4%

The meat is ground twice, and the 10 per cent of vegetables added. In a large electric food mixer of about 25-quart capacity, the dry powder is added. All this mixture is combined and blended, and stored in an ice box or frozen until needed. Bone meal or its equivalent was added to the diet to avoid rickets, which proved to be a serious condition in the early days of the development of the colony.

It was also noted that with unbalanced diets 56 per cent of the young had become detached and were lost.

It should be noted that opossums are *nocturnal,* and therefore *in hot weather the food should be made available in the late afternoon, to avoid spoilage.*

DISEASES

Worms and all types of intestinal parasites are common.

The animals die as a result of spontaneous infection or infections following injuries by cage mates.

Chronic nephritis of the opossum is a common condition.

Chronic enteritis, abnormal hyperplasia of the mammary glands, and tooth decay are also common conditions.

When the humidity is low, opossums frequently develop swollen tails and sores that soon suppurate. When the humidity is adjusted to approximately 50 to 60 per cent, the swelling of the tails disappears. The open sores are treated with a sulphur ointment or a coating of medicated Vaseline.

Control and Treatment. With the routine practice of thorough cleanliness and proper diet, many of the above conditions clear up within a reasonable time. For specific instructions relative to sterilizing, using dusting powders, and insect control, refer to pages 478 to 487.

PRACTICAL SUGGESTIONS

1. Select stock carefully for the development of the colony. Mate brothers to sisters whenever possible, until inbred strains are established. Maintain an inbred stock in preference to purchase of animals from the field.

2. For short-term experiments, animals secured from the field are adequate.

3. It is advisable to check the pouches daily to ascertain the presence of newborn. For closer timing, the pouches should be examined twice daily.

4. Do not permit more than one proven male in a cage with the females at any time.

5. The cages should be sufficiently large to permit ample exercise.

6. Clean the cages daily, and incinerate the refuse.

7. Remove the straw from each room once a week, and burn.

8. Nesting boxes, pens, concrete floors, and cooking jars should be washed with soap and water once a week.

9. Food pans should be washed and sterilized daily.

10. In the summer, dusting powder or sprayed insecticides should be added to the fur of the opossums if bothered with any insects.

11. Sodium amytal or ether is good as a general anesthetic.

12. The pouch young can be immobilized on a rigid surface by adhesive tape.

13. Opossums may be identified by tattooing, with an electric needle, numbers directly on the skin of the back, after the hair has been shaved. Ear marking and amputation of toes, tags with chains, etc., have proven quite unsatisfactory because of the frequent fights of the animals, with resulting loss of ear lobes or even feet.

REFERENCES

Enders, R. K. 1935. Mammalian life histories from Barro Colorado Island, Panama. *Bull. Museum Comp. Zool., Harvard College,* vol. LXXVIII, no. 4, pp. 385–502.

Farris, E. J. 1041a. Behavior responses of the tropical opossum (*Metachirops*) to gonadotrophic hormones. Abst., *Anat. Rec.,* vol. 81, suppl. p. 105.

——. 1941b. *Tropical opossums.* Kodachrome motion picture, 400 ft., 16 mm. The Wistar Institute, Philadelphia.

Hartman, Carl G. 1920. Studies in the development of the opossum, *Didelphis virginiana* L. V. The phenomena of parturition. *Anat. Rec.,* vol. 19, pp. 251–261.

——. 1023a. Breeding habits, development, and birth of the opossum. *Smithsonian Inst. Rept. for 1921 (Publ.* 2689), pp. 347–363.

——. 1923b. The oestrous cycle of the opossum. *Amer. Jour. Anat.,* vol. 32, no. 3, pp. 353–421.

——. 1928. The breeding season of the opossum (*Didelphis virginiana*) and the rate of intra-uterine and post-natal development. *Jour. Morph.,* vol. 46, pp. 143–215.

McCrady, Edward, Jr. 1938. The embryology of the opossum. *Amer. Anat. Mem.* 16. The Wistar Institute, Philadelphia.

Moore, Carl R., and David Bodian. 1940. Opossum pouch young as experimental material. *Anat. Rec.,* vol. 76, pp. 319–327.

Nelsen, Olin E., and Elizabeth Lloyd White. 1941. A method for inducing ovulation in the anoestrous opossum (*Didelphis virginiana*). *Anat. Rec.,* vol. 81, pp. 529–535.

11· THE DOMESTIC FOWL

F. B. HUTT
Department of Poultry Husbandry
Cornell University
Ithaca, New York

INTRODUCTION

As experimental animals for various kinds of research, birds have many features to commend them to biologists. For centuries the favorite source of material for embryologists, they have in more recent times proven ideal for various studies in endocrinology, genetics, immunology, and certain fields of physiology.

Of all the birds that have been used, none has proven more adaptable to various kinds of research than the domestic fowl. It has many advantages, especially for those who must keep their experimental animals in small confined quarters or in cages. In comparison with ducks, fowls are much cleaner under such conditions. In comparison with the pigeon, which is also frequently used for various kinds of investigation, the fowl has the big advantage that when properly managed it will lay almost continuously, whereas the pigeon lays only two eggs before incubating them. Another point for consideration is that the fowl has only rarely been incriminated as transmitting any disease to man, whereas parrots, budgerigars, and pigeons are occasionally responsible for the transmission of psittacosis. For all these reasons, and because each species of birds requires different management, the present account is limited to the fowl. Those interested in pigeons are referred to the detailed account of that bird by Levi (1941).

The domestic fowl, *Gallus gallus*, is really a pheasant. Its native habitat is India and parts of southwestern Asia. However, it has proven its adaptability to confined quarters. This applies both to chicks of all ages and to adults. With modern equipment, it is quite feasible to maintain a colony of 100 adult birds in a laboratory of average size without any difficulty. This is done by keeping them in the laying cages, or "hen batteries," which have proven so suitable for such work. Similarly, several hundred young chicks may be kept in a very small floor space by utilizing an electric brooder that has four to six decks. So long as the ventilation and sanitation are adequate, unpleasant odors should be even less of a problem with chicks than with other laboratory animals.

One advantage of the fowl is that it is available in all sizes from the small Bantam weighing less than 800 grams to the big Jersey Black Giants and Langshans weighing over five times that amount. Some of these breeds have been shown to have certain physiological peculiarities, and the research man aware of these may choose a breed with some assurance that it is suitable for his kind of work. This is discussed further in a later section of this chapter. The genetic variability is now better known for the domestic fowl than for any other bird. Similarly, its anatomy, physiology, and endocrinology have been so thoroughly investigated that a great volume of factual information in these fields is available.

Another advantage of the fowl is that stock is readily available anywhere in the country. One may get at comparatively low cost birds of improved strains that are capable of rapid growth and high egg production.

One of the reasons for using the domestic fowl is that with a little care one may time operations to get eggs as soon as they are laid and embryos at any stage desired. If a hen with a laying cycle of three or four eggs on successive days should lay her first egg of that cycle on a Monday, one may usually count on eggs for Tuesday, Wednesday, and Thursday, each coming an hour or two later than on the previous day. Some hens lay every day, and at about the same hour. By timing the period of incubation properly, one can get embryos of desired stages with a very slight margin of error.

An advantage of the chick for some types of experimental work is its extremely rapid growth. Ordinarily, a chick just hatched from a two-ounce egg will then weigh about 38 grams but will more than double that weight in the first 2 weeks of its life. At 4 weeks it will weigh from 150 to 200 grams, if properly fed. Because of this rapid growth, the chick is well suited for quantitative tests of diets deficient in certain essentials, for example, in vitamin D. If the rate of growth were less rapid, deficiencies would not be revealed quite so readily.

The domestic fowl has a special advantage for the endocrinologist in that it shows remarkable secondary sex characters (fig. 1). The large comb and wattles of the male have proven ideal for testing for the presence or absence of male hormones, and it has been shown that capons will respond to very small doses of androgens by growth of the comb. Another secondary sex character is the plumage, which shows sex dimorphism in the neck, the wing-bow, and the saddle. The males have long, narrow lanceolate feathers, whereas those of the females in the same areas are rounded. Poulardes (ovariectomized females) show a plumage similar to that of the capon. When poulardes or capons are treated with female hormones and the plumage plucked, the new feathers will be of the female type.

Fig. 1. Typical specimens of the White Leghorn, one of the most popular breeds in the United States. As in all fowls, sex dimorphism in size is genetic, but the differences in comb, wattles, and structure of the plumage are determined by secretions of the gonads.

Finally, the reader is reminded that whereas the disposal of rats, mice, and guinea pigs at the end of an experiment may sometimes be a problem, the disposal of unneeded chickens and eggs is more likely to be a pleasure.

HOUSING AND CARE

Few workers in biological laboratories will be able to maintain their chicks and adult birds under conditions comparable to those found on the farms. It will be necessary at most laboratories to keep the stock in confined quarters, not to allow it to range out on an open field, or in a fenced yard, as is the common practice. For the brooder house, the colony house, and the laying house of the poultry farmer, the laboratory biologist will have to substitute the electric brooder, the unheated broiler battery, and the laying cages for adult birds.

Perhaps it will be easiest to follow through the kinds of housing and care needed for chickens at various ages by starting with day-old chicks just as they come from the hatchery. Before their arrival it will be necessary to have the brooder all warmed up with the thermostat controlled and a supply of food and water in the containers provided for them. Many types of chick brooders are available on the market. The kind most suited for the laboratory will be

FIG. 2. An electric brooder with twelve compartments, each capable of accommodating up to fifty chicks. The opened door shows the heated section. Feed troughs are placed on both sides and a water pan at each end. (Courtesy of Petersime Incubator Company, Gettysburg, Ohio.)

those that have several decks (fig. 2). Important things to ensure in selecting a brooder are that there be some space available for the chicks away from the heated compartment, that there be ample hoppers for feeding, and that the whole may be easily cleaned. The ease or difficulty with which chicks may be caught, removed, and returned should also be considered. If the dropping pans under the hardware cloth that forms the floor of the brooder are cleaned out daily, there should be little smell about the chick room.

In general, chicks will do better in small flocks of 25 to 50 than in larger ones. The commercial poultry farmer who has special brooders for raising chicks on the floor in houses approximately 12 feet square can manage there as many as 200 to 250 in one flock. In the 6 to 8 square feet of the average compartment in an electric brooder, best results will be obtained with groups of 50 or less. There should be plenty of space for most of the chicks to feed at the same time should they wish to do so.

The thermostat should be adjusted so that when the chicks go into the brooder the temperature is approximately 90 to 95° F. at the level of the chick's back, or slightly more if the bulb of the thermometer is higher up. After the first week the temperature can be dropped

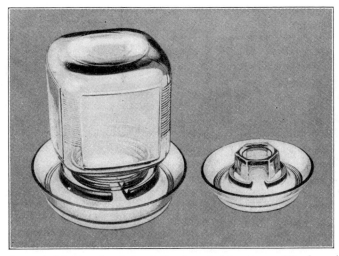

FIG. 3. A good drinking fountain of the inverted-jar type. The base is shown separately at the right.

down to 85° and gradually reduced somewhat, although the chicks will still need heat up to about 6 weeks of age. With a little practice, the observant biologist can readily tell when the chicks are too hot or too cold and can adjust his thermostat accordingly even without considering the thermometer reading. When the chicks crowd away from the source of heat, they are too hot; if they huddle together, they are too cold. At all costs, chilling must be avoided in the critical first week.

When chicks are first put in the brooder, fresh water should be supplied in two or three drinking fountains spaced apart near the edge of the heated compartment, so that the chicks will learn to drink with a minimum of delay. A favorite water container for young chicks is the inverted glass jar in a shallow pan or saucer. When refilled regularly these containers provide a constant supply of fresh water, but one not deep enough to let any chicks get drowned. An ordinary pint jar inverted in a saucer will serve very well, but good drinking fountains of this type are available in a variety of sizes (fig. 3). Directions for feeding are considered later on.

When the chicks are about 3 to 5 weeks old, it will be necessary to thin them out and to put in two brooder compartments the number that was originally started in one. At about 6 weeks or a little later (the exact age varying inversely with the temperature of the room),

Fig. 4. The hen battery provides for each bird a separate compartment supplied with feed, water, and a sloping floor on which the egg rolls gently to a receptacle (*lower front*), where it cannot be pecked or eaten. The rollers at the right are used in cleaning the pans under the birds. (Courtesy of L. M. Hurd.)

they may be put in another type of cage, sometimes called a broiler battery, where they do not have heat supplied. When these cages are used, the temperature of the room should not drop below 60° F. Birds can be carried along in these broiler batteries until they are 10 or 12 weeks old, and by that time it will probably be necessary to move them into the kind of cages that are commonly used for adult fowls.

In these laying cages, sometimes called hen batteries (fig. 4), there is a separate compartment for each bird. However, since they can be built up to a height of three or four decks, the number of birds that can be maintained on a small floor space is remarkable. One of the big advantages of these laying cages is that each bird is by itself and thus relieved of any social pressure. There is no loss from cannibalism, which is sometimes serious when birds are kept in crowded quarters in laying houses. Another advantage of the laying battery is that it automatically provides a record of the egg production of each hen. In each compartment there is some arrangement for the egg, when laid, to roll to the front of the cage and fall down to a position where the hen cannot peck it. By collecting these eggs at regular intervals and recording the cages from which they come, one easily gets a record of the egg production of all the hens.

Fig. 5. Wire exhibition coops, useful for birds that must be kept under observation or isolated. Note metal cups for feed and water.

With battery brooders and with battery laying cages one may use some absorbent litter like sawdust on the pans for collecting the feces, but this is not essential. In some of the laying cages there is an arrangement whereby a big roll of Kraft paper is pulled through daily and the feces with it. Another common practice is simply to put newspapers on the pans to facilitate cleaning.

One advantage of adult fowls compared with other laboratory animals is that they can tolerate room temperatures over a rather wide range. In general, a hen is likely to be most comfortable at about 60 to 70° F., but will not experience any hardship should the temperature drop to 40°, or even below. At temperatures above 80° there is likely to be some discomfort, and it has been shown that temperatures of 70° or above cause the laying of smaller eggs than one gets at lower temperatures. The hens will not be seriously injured if the temperature of the laboratory should be very low some morning because something went wrong with the heating equipment. Many poultrymen keep birds in uninsulated hen houses in which the water freezes on cold nights.

An important point to remember is that the hen breathes rather rapidly and gives off approximately 0.5 liter of carbon dioxide per kilogram per minute. Furthermore, the hen's breath is heavily laden with moisture, which in cold rooms or buildings is likely to condense

FIG. 6. A chick box of the type used for shipping day-old chicks is useful equipment in the laboratory. In this illustration, the cover is removed to show the 4 sections, each large enough for twenty-five chicks.

on the floor. Adequate ventilation to ensure good air movement but without draft is essential. As in any animal room, it is advantageous to have the floor smooth so that it can be easily cleaned out. If it be provided with a floor drain so that it can be washed out, so much the better.

Apart from the larger units for holding chicks or adult hens, a few pieces of extra equipment will be useful. Single or double "exhibition" coops (fig. 5) of the kind that one sees in poultry shows are ideal for holding one or two birds that need special observation or more space after an operation or some special treatment. For handling young chicks it is convenient to have commercial chick boxes of various sizes (fig. 6). These are made in capacities to hold 25, 50, or 100 day-old chicks. Most of them have an arrangement whereby the number of ventilating holes can be increased in warm weather. These boxes are useful for carrying chicks to classes and for holding them if the electric current should be shut off. When the chicks are huddled together in the compartments of these boxes, they can maintain a temperature high enough to prevent chilling.

One may see samples of other equipment at almost any good hatchery and descriptions of it in any poultry magazine. While most

FIG. 7. The "dubbed" male (*right*) had his comb and wattles removed at six weeks of age; the undubbed male (*left*) may find his large head furnishings a nuisance in a wire cage. When not needed for studies in endocrinology, the comb and wattles are better removed, particularly from Leghorns.

of the laying cages and battery brooders provide ample equipment for feeding and watering the birds, it may be useful to have on hand some extra vessels for supplying water and also small feeding cans or cups with which to look after single birds in special circumstances.

Fowls confined to cages indoors tend to develop unusually large combs and wattles. The combs lop over, particularly in females of the Leghorn and Minorca breeds, but also in males of all breeds. Sometimes these enlarged appendages interfere with feeding and drinking. This trouble can be easily avoided by cutting off the combs and wattles (fig. 7) with scissors, preferably before the birds are six weeks old. Some poultrymen do it on the first day after hatching.

Finally, it is appropriate to say in this section a few words about the proper way to hold a chicken. It is not desirable to hold a bird by the legs and with the head down. It should be held with the body resting on the forearm and head to the rear or to the side, with one finger between the legs and three fingers or more holding the legs securely by locking around the "hock" joint or above it. With a little practice, one learns to hold the tip of one wing in the same hand that grasps the legs, and thus to prevent flapping. The same result is attained by tucking the head under the arm, so that it points to the rear.

For further information about poultry husbandry in all its aspects, one should consult such useful books as those by Lippincott and Card (1946), Winter and Funk (1947), Jull (1938), and Rice and Botsford (1949).

REPRODUCTION

In modern domestic fowls, the natural cycle of reproduction in the domestic fowl has been almost entirely eliminated. A peak in egg production can be induced in fall or winter by providing artificial light. Chicks can be supplied by many hatcheries all the year round. Accordingly, the fowl is suitable for laboratory work in which animals may be wanted for tests at all seasons of the year. The original system of reproduction devised by nature has now been perfected to the point at which later stages of that process can be nicely handled by artificial incubation and artificial brooding, without nests, and without catering to the caprices of broody hens. Many strains of poultry have now been developed to the point at which few hens go broody during the year, but the annual molt has not been eliminated. Some hens begin this replacement of feathers in July, others not till November.

Egg Production. Leghorn pullets will ordinarily begin to lay somewhere between 150 and 200 days of age. The exact time will depend to a great extent upon the amount of light available when the birds are nearly mature. Under the more or less natural conditions of the rearing range, Leghorns hatched in early March will begin to lay in August, but birds of the same stock not hatched until May will not begin to lay until November. Sexual maturity is retarded by the shortening days of the fall months. Under laboratory conditions, the onset and rate of egg production can be accelerated by providing artificial light after the birds are 150 days old. With respect to the heavy breeds, such as Rhode Island Reds and Barred Plymouth Rocks, the usual thing is for females of these breeds to come into production at 6 to 7 months. Strains especially bred for early maturity will start laying somewhat earlier.

The number of eggs that the hen will lay on successive days varies with the individual, but is usually characteristic of the bird. Some hens will lay for about 3 days and then miss 1 day. Others may lay for 4 or 5 days and then miss a day, and some hens have been known to lay for several months without stopping. Some lay on only 1 day in 2 or 3. By considering the records of each bird, the laboratory man will learn exactly when to expect eggs and when not to expect them. The size of eggs increases with laying for about the first 3 to 5 months of the laying period and thereafter remains fairly constant. A pullet may start with an egg of 38 gm. in August, and by December be laying eggs of 58 gm. The standard egg size preferred on the market is 2 oz., or 58.7 gm. Pullets starting to lay at 160 days or less begin with small eggs and take several months to attain the full size, whereas those not beginning until over 200 days of age are more likely to start with large eggs and to reach final egg-size quickly.

Fertility. To obtain fertile eggs in the laboratory one must either provide a small breeding pen or coop in which the male and female can be placed, or resort to artificial insemination. Details for carrying out the latter process are given by Burrows and Quinn (1939). Ordinarily, fertile eggs may be expected from 30 to 40 hours after the first mating. Under more natural conditions, in which one male is placed in a pen with fifteen to twenty females, poultrymen do not expect normal fertility from the pen as a whole until about 7 days after the male is introduced. Similarly, once fertility has been well established it will be maintained for about a week after the male has been removed. In some cases, however, fertile eggs have been obtained as long as 3 weeks after removal of the male. Few laboratories are likely to have space available for large mating pens of this sort, and the best that can be done, apart from artificial insemination, is to put the pair of birds in some pen, box, or coop about 4 to 6 feet square. With such "stud matings" better fertility is likely to be obtained if the birds are mated in the afternoon rather than in the morning, when some of them carry eggs in the uterus.

Incubation. Small incubators of the old-fashioned type in which all the eggs are held in a single tray are still to be found in some laboratories, but they are gradually being replaced by more modern incubators of the cabinet type (figs. 8 and 9) in which several trays are arranged one above the other. Such incubators are now available in a variety of sizes to accommodate from 500 eggs up to many thousand. Since the conditions of operation will vary somewhat according to the type of incubator and the regulating devices, the directions of the manufacturer should be closely followed.

The optimum temperature for incubation in cabinet incubators in which the air is constantly in motion is from 100° F. down to 99.5°. The relative humidity should be about 60 per cent, and carbon dioxide

FIG. 8. A modern incubator of the cabinet type, suitable for the small laboratory. It will hold 600 eggs, which may be set all at once or at intervals. (Courtesy of American Electric Incubator Company, New Brunswick, N. J.)

0.5 per cent or even somewhat less. Eggs will hatch at humidities other than 60 per cent, but when the humidity is above 70 per cent or below 30 per cent the hatchability is markedly reduced. It is important that the passage of air through the incubator should not be too rapid. Sometimes the poultryman can gauge this by measuring the evaporation as indicated by the air cell in the large end of the egg. Under optimum conditions, at 8 days of incubation that air cell should be approximately ¾ inch deep. If it is less or more, there has been either too little evaporation or too much, respectively. Formerly it was thought that eggs had to be cooled once a day to simulate the natural conditions when the hen got off the nest to feed, but it is now known that such cooling is unnecessary. It is important that the eggs be turned three or four times daily, preferably at widely spaced intervals. This is particularly necessary during the first 2 weeks of incu-

FIG. 9. A larger incubator, capacity about 2000 eggs, showing how the eggs are turned by tilting the trays. (Courtesy of American Electric Incubator Company, New Brunswick, N. J.)

bation but should *not* be done after the seventeenth day of incubation. Most of the cabinet-type incubators have mechanical devices for tilting the trays (fig. 9) and thus ensuring that the eggs be turned. Best results seem to be obtained when the eggs lie on the tray with the large end raised at about 30° from the horizontal. They should never be incubated with the small end up.

The eggs should be tested at 8 to 10 days of incubation to detect the infertile ones and any in which the embryos have died. Such testing can easily be done in a darkened room by holding the egg up to a "candling" device such as is used by commercial egg candlers. If one of these is not available, a home-made one can easily be made by putting an electric light in a closed tin can with a hole about 1 inch in diameter in one side. With little practice one will learn to identify the clear eggs that are infertile and also the ones in which embryos have died. The latter will usually show a "blood ring" or streak of blood. A second test at 17 days of incubation permits removal of more eggs with dead embryos before the eggs go into the hatching compartment. If the identity of the parents is to be preserved, each

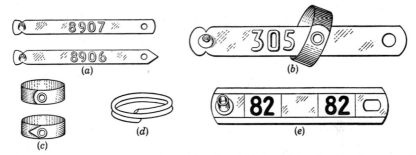

FIG. 10. Types of bands used to identify chickens: (a) wing bands to go in anterior border of wing; (c) the same when sealed (the pointed ones need no slit to be cut in the wing for insertion); (b) numbered aluminum leg band for adults; (d) spiral Celluloid leg band, available in several different colors; (e) leg band using colored numbers under a Celluloid cover. (Courtesy of National Band and Tag Co., Newport, Ky.)

hen's eggs should be hatched in a separate "pedigree basket," and the chicks must be wing banded as they are removed. Various types of identifying bands are shown in figure 10.

Broodiness. Broodiness has been almost entirely eliminated from Leghorns and other Mediterranean breeds, but in some of the heavier breeds, particularly Rhode Island Reds, New Hampshires, Barred Rocks, and Wyandottes, one may expect a certain proportion of the flock to go broody. This is less likely to happen when the birds are in laying cages than when they are in floor pens. The poultryman's usual procedure for breaking up broody hens is to put them in slatted crates with plenty of ventilation. If it is desired to let the hen set as a demonstration for classroom purposes, she should be transferred in the dark to a suitable nest and kept in darkened quarters until she has become adjusted. Food and water should be supplied, and it may be necessary to remove her from the nest daily to make sure that she does get off to eat.

Care of Eggs for Hatching. An important point to remember, when eggs must be held to get a number big enough to incubate at one time, is that the hatchability declines in eggs over a week old. Under most conditions, ordinary laboratory temperatures are too high for holding eggs that are to be incubated. The ideal temperature is from 50 to 55° F., preferably in a room with a relative humidity of around 60 to 70 per cent. When the eggs are kept thus, some poultrymen manage to get good hatches even with eggs held as long as 2 weeks before incubation. However, it is desirable to turn eggs that are held longer than 7 days. This can be done by keeping them in an ordinary egg crate or carton and tilting it at least once, and preferably twice, each day during the holding period. It should be remembered, in selecting the eggs to be incubated, that those of 50 to 58 gm. in weight will hatch better than those over 60 gm. or less than 45 gm.

Finally, let no biologist expect the impossible. Most poultrymen are more than satisfied if the proportion of infertile eggs is no higher than 10 or 15 per cent, and if they hatch about 75 to 80 per cent of the eggs incubated. Some embryonic mortality is inevitable. If it can be kept to as little as 10 per cent of the fertile eggs, that is unusually good. For excellent reviews of factors affecting hatchability, the surveys by Landauer (1948) and Taylor (1949) should be consulted.

DEVELOPMENT OF BREEDING STOCK

While many biologists try to maintain their own colonies of hamsters, rats, or other animals, this is hardly necessary with the domestic fowl because fowls are now available everywhere in large numbers.

To begin with, there are well over a hundred different breeds or varieties of the domestic fowl to be found in this country. They range in size from little Bantams weighing no more than 600 to 800 gm. up to the big heavy breeds weighing well over 4000 gm. for males. While many of these are fanciers' varieties, and not available in quantity, there are at least eight or ten good breeds which can be had in almost any quantity desired.

It seems questionable whether the biologist is warranted in developing his own strain of Leghorns or Plymouth Rocks or Rhode Island Reds, when there are already available so many highly improved strains of these breeds. Furthermore, many hatcheries can supply chicks at almost any season of the year.

Physiological Differences between Breeds. An important point not to be overlooked by anyone doing research with the domestic fowl is the fact that, apart from the more obvious differences in size, color, plumage, and comb which go to make up the breeds, there are some remarkable physiological differences between breeds. These must be taken into consideration in the interpretation of results of research. The best known of such physiological characters are those which distinguish White Leghorns, and probably Leghorns of other colors as well, from the so-called "heavy breeds" like Rhode Island Reds, Plymouth Rocks, and Wyandottes. Some of the differences between them known to date are as follows:

Fig. 11.   Rapid feathering (*left*) and slow feathering (*right*) shown at 10 days of age in development of wings and tail. The difference indicates two of several genetic variations in physiology to be considered in the selection of stock for the laboratory and in the interpretation of experiments.

The Leghorns have genetic rapid feathering, in contrast to the slow feathering of heavy breeds. The difference is best seen at 8 to 10 days of age, when the Leghorn chicks have little tails about half an inch long and wings almost as long as the body (fig. 11). At the same time chicks of the heavy breeds have no tails and comparatively short wings. The recessive sex-linked gene causing rapid feathering also induces slightly more rapid growth up to 10 or 12 weeks of age than is found in the slow-feathering chicks.

The White Leghorns, though not really non-broody, as they are sometimes said to be, show much less broodiness than do the heavy breeds.

The Leghorns lay white-shelled eggs; the heavy breeds named lay eggs with shells that are brown and not quite so thick.

White Leghorn chicks are much more resistant to pullorum disease, an infectious disease caused by the organism *Salmonella pullorum*, than are chicks of the heavy breeds.

Adult Leghorns are better able to tolerate extreme heat than are the heavy breeds, but are apparently more susceptible to intestinal roundworms.

A remarkable difference is that the Leghorns can get along on less thiamine than is needed by the heavy breeds. Similarly, the thiamine content is about 60 per cent higher in Leghorn eggs than in eggs of heavy breeds. White Leghorns also get along with less vitamin D than do the heavy breeds and have a lower requirement of manganese.

It is entirely probable that there are other physiological differences between the Leghorns and the heavy breeds yet to be discovered. Even within the heavy breeds as a class there are differences to be found. Infertility and poor reproduction are sometimes more of a problem in White Wyandottes than in some of the other breeds. The New Hampshires seem to be characterized by more rapid growth than is usually found in other heavy breeds.

These differences among breeds have here been emphasized merely to ensure that the biologist should remember, when interpreting his data, that invisible genetic differences in physiology often accompany the more evident differences among breeds.

Strains, Hybrids, and Crossbreds. Of the more common breeds, such as the Leghorns and Rhode Island Reds, many different strains are available, some of them only slightly related. Such strains may differ in nutritional requirements, in resistance to disease, and in other respects, all of which must be considered in the interpretation of data obtained in experiments.

Because inbred strains are now in demand for crossing, a number of poultrymen are now trying to develop highly inbred strains. These may prove advantageous for the research worker who prefers to use stock with a minimum of variability, such as is most likely to be found in inbred lines. The inbred strains are now being crossed to produce chicks with exceptional hybrid vigor. In so far as this is reflected in greater viability, and perhaps in more rapid growth, the interstrain hybrids may have advantages for some kinds of research.

Crossbred chicks are also available from the intercrossing of distinctly different breeds. A popular one is the cross of Barred Plymouth Rock females (fig. 12) with Rhode Island Red males, which produces chicks that show no barring in the females, but do show barring in the males. This difference is recognizable at hatching, so that from such crosses the hatcheryman can supply chicks guaranteed to be all pullets or all cockerels. Apart from this means of identifying sex, many hatcheries are now having the chicks' sexes determined by examination of the cloaca and can supply males or females in which the identification of sex will be accurate up to 95 per cent or better. Sometimes the hatcheryman has a surplus of Leghorn cockerel chicks, for which the demand is slight. From such a source one may be able to get an abundant supply of chicks at very little cost, if White Leghorn males are suitable for the research to be undertaken.

Breeding a Distinct Strain. While the writer hesitates to encourage this proceeding, there is no reason why it should not be attempted by some courageous spirit who wishes to try it. The fact is that to make improvement by selection for any quantitative or physiological character, one would need to test a great many breeding animals each season to find the best families. It is doubtful if the average laboratory could provide space for such an enterprise. On poultry farms, one needs from ten to twenty breeding pens in order to find each year the few superior sires by which progress can be made. In laboratory work, similar results might be accomplished by using artificial insemination, but it would require a great deal of time and effort to make

Fig. 12. A Barred Plymouth Rock female. The breed is available anywhere, and in some regions is much in demand for sex-linked crosses.

the necessary matings. For all these reasons, and particularly because of the limitations of space, it would seem that the biologist may be better advised to find some strain of birds suitable for his purposes and then to arrange for a supply of hatching eggs, of chicks, or of stock as he may need it.

For a detailed review of genetic variations in the fowl and of principles of breeding, the book by Hutt (1949) may be consulted.

SANITATION

On the whole, the biologist who keeps a few score of birds in his laboratory has much less occasion to worry about disease than has the poultry farmer with several hundred. This is so for several reasons. Day-old chicks taken from the hatchery directly to a laboratory where there is no adult stock have a fair chance of keeping free of disease. Another reason why disease should be less in the laboratory is that there most of the birds are kept in wire-bottomed cages. This automatically dispenses with litter and provides less chance for the harboring of germs and the eggs of parasites. The wire cages in most cases are easily cleaned so that, even if some disease should break out, they can be readily disinfected. However, lest any biologist feel too secure about this matter, it is well to mention some of the risks.

In the first place, young chicks should not be kept in the same room with old birds if that can be avoided. In many cases the old stock has survived various diseases but still carries the infectious organisms, even though resistant to them. Such association of old stock with young chicks is likely to result in infestation with parasites and to spread infectious diseases like lymphomatosis and colds.

Before bringing chicks into the laboratory for the first time, the whole place should be thoroughly disinfected. This process should be repeated after the chicks are cleaned out. For this purpose one may use lye, at the rate of 1 ounce to the gallon, or some of the cresol disinfectants, particularly the saponified cresol. With these it is well to follow the manufacturer's recommendations as to the strength. Most of them call for a solution of 3 to 5 per cent of cresol in water. With such a disinfectant, walls, floors, and particularly the cages should be thoroughly cleaned.

Some diseases are spread in the incubator, and it is well to disinfect that piece of equipment occasionally. This is particularly so if the chicks develop omphalitis, or navel ill. The organisms causing this sometimes accumulate in the incubator. They can be destroyed by fumigating it with formaldehyde at the rate of 20 cc. of formalin (40 per cent formaldehyde) per 100 cubic feet to be disinfected. The formalin should be soaked up in cheesecloth, which is then suspended in the incubator while it is in operation. Another procedure is to pour formalin on potassium permanganate, using 35 cc. of formalin and $17\frac{1}{2}$ gm. of potassium permanganate per 100 cubic feet. When this is done, the mixture should be made in a dish somewhat larger than the materials to be mixed. The gas will do the job thoroughly and quickly.

Equipment for feeding and watering should be of such a kind that it excludes the droppings. This is not always possible with the first feeders and fountains for young chicks, which go right on the floor of the brooder. The mash hoppers should have a lip so that there is little feed spilled on the floor. What is spilled should be cleaned up if mice and cockroaches are not to be encouraged. Damp or moldy feed should never be left in the hoppers, and the drinking vessels should be cleaned daily.

It is a good plan to have the room in which the birds are kept completely rat proof. This precaution is valuable not only because the rats will steal feed out of the mash hoppers, but also because they will eat young chicks, and may carry diseases communicable to chickens. One of the best ways to reduce the accumulation of mice in the bird room is to keep all feed in metal containers. The ordinary precautions that would be taken with any laboratory animals should be observed, namely, the elimination of diseased birds, particularly those that can be recognized as carrying infectious diseases. It is also well to isolate temporarily any birds brought in (other than day-old chicks) to see if they are going to exhibit some disease that was not evident upon their arrival.

Respiratory diseases are more likely to cause trouble in the laboratory than are some others. To avoid them, one should make sure that the birds are not exposed to drafts and that no birds are introduced to the laboratory without first ascertaining that the flocks from which they come are not infected with respiratory disease. Against some of these, vaccination is some protection, but that is discussed in a later section.

NUTRITION

Fortunately for the biologist, the nutrition of the fowl presents no special problems, as the species is remarkably adaptable to a varied diet. The nutritional requirements of fowls of various ages have been so thoroughly investigated that it is now possible to formulate diets which should maintain the animals in perfect health. Such diets are now put up by many feed manufacturers and are available in almost any part of the country.

Essential Nutrients

Carbohydrates. The most important constituent of any diet for fowls is a source of energy. Fortunately, carbohydrates meeting this need are abundantly available. The common grains are used, and some of their by-products, which are cheaper than the grains themselves, are also utilized in mashes. Wheat and corn are the favorites for this purpose. Ordinarily, grains and their by-products comprise from 75 to 90 per cent of the total ration. What fat the bird needs can be manufactured from the carbohydrates, so there is no need to supply fat as a separate constituent of the diet. Fiber may serve some useful purpose by providing bulk and by loosening the mass of feed to provide better access of digestive juices. However, most normal diets will contain from 5 to 8 per cent of fiber and more than that is not desirable.

Water. Water is one of the most important things necessary for fowls. It comprises from 55 to 75 per cent of the weight of the body and about 80 per cent of the weight of the albumen in the fresh egg. Apart from this, it serves a very useful function as an aid in controlling body temperature. Like other birds, the fowl lacks sweat glands and is well insulated by its plumage. Its best cooling device is the expiration of water from the lungs in the form of water vapor. When one remembers that the normal temperature of the adult fowl is from 105 to 107° F., the importance of the cooling mechanism can be better appreciated. It is essential to maintain a constant supply of fresh water before the birds at all times. Any biologist wishing to give his birds a special treat in the winter may be amused to find that even when water is abundantly available they will relish a bit of snow if it be put in the container.

Some idea of the importance of water in the diet is conveyed by the fact that 100 hens will normally consume about 3½ gallons of water per day. In hot weather, and especially if the birds are laying well, the intake may run as high as 5 to 6 gallons. However, one cannot measure out the daily ration because a certain amount of water is always spilled. The only safe plan is to make sure of an abundant supply at all times.

Proteins. Protein is an important constituent of the diet, as it makes up a large part of the body tissues and of the egg. Any rapidly growing organisms, such as chicks, need a high content of protein in the diet. One must not assume, however, that the total protein content indicates the adequacy of the diet. The requirement is not really of protein as such, but rather of the amino acids which in varying amounts make up different kinds of proteins. At least ten of these amino acids have been shown to be indispensable for chicks, and others may be needed under some conditions. The exact quantitative requirement of each amino acid can hardly be satisfactorily given, and the usual practice is simply to feed proteins from different sources and thus provide from all of them the balance that may be needed. Some good proteins come from vegetable sources, but a diet must also contain some animal proteins to provide a proper balance of amino acids. Most common grains will contain from 9 to 12 per cent of protein, and additional vegetable proteins can be provided from soybean oil meal, which is now available in quantity. When these are balanced by fish meal, beef scraps, or animal protein from some other source, the result is usually adequate for all purposes.

While most analyses of mashes for chicks and hens will show that the protein content runs from 18 to 22 per cent, it must not be assumed that birds at different ages all have the same requirements. The actual amount consumed varies according to the amount of grain fed as a supplement to the mash. Thus, young chicks will usually start out on an all-mash ration in which the protein content may be from 20 to 22 per cent. After about 4 or 5 weeks, a certain amount of cracked wheat and corn or other "scratch feed" is given, and, since this has a lower content of protein, it causes the total intake of protein to be reduced below the 20 per cent shown for the mash. Laying hens are commonly given a laying mash containing about 20 per cent of protein and a mixture of grains for which the protein content may run about 11 per cent. When the intake of these two kinds of feed is about equal, they provide together about 15 to 16 per cent of protein, which has been found to be adequate for normal egg production and reproduction.

Minerals. While the fowl needs many different mineral elements for normal physiology, the chief ones that have to be supplied in the diet to ensure an adequate amount are calcium and phosphorus. The optimum Ca:P ratio is approximately 2:1. Although a certain amount of these elements comes from other feeds in the diet, the usual practice is to supply bone meal at a level of 1 per cent in rations for chicks and 2 per cent in mashes for adult birds. The requirement of calcium is met by supplying ground limestone in the mash or, better yet, by providing ground limestone or oyster shells for consumption *ad libitum*. For birds in cages, where hopper space is limited, it may be fed right on top of the mash. Calcium is particularly important for laying hens because the shell of the egg is almost entirely calcium carbonate and it forms about 10 per cent of the weight of the egg. This means that an ordinary egg will carry from 5 to 6 gm. of calcium carbonate.

Common salt is provided in most mashes at a level of 0.5 per cent. Other minerals need not be supplied with the exception of manganese, which seems to be required in abnormally high amounts by some birds of the heavier breeds. It is particularly necessary for birds in cages, as they seem more subject to slipped tendon (perosis) than birds on the ground. This condition (fig. 13) is more frequent with diets low in manganese. Most mashes now contain the quarter pound of manganese sulphate per ton that ensures adequate protection. The Leghorns do not need this supplement, but fortunately it is cheap. In comparison with mammals, the fowl seems to have an unusually low requirement of iodine. Although goitre is extremely rare in chickens, some nutritionists recommend the feeding of iodized salt.

Vitamins. The vitamins which must be supplied in adequate amounts in the diet are vitamin A, vitamin D, and riboflavin.

Fig. 13. An extremely bad case of slipped tendon (perosis), a condition more common in caged chicks than in those raised on the floor. In milder cases, only a swelling of the tibiotarsal joint may be evident.
(Courtesy of G. F. Heuser.)

The amount of vitamin A needed in the diet for chicks is 1350 International units per pound of feed. It will be supplied if the mash contains 0.5 per cent of cod-liver oil, or other equivalent fish oil, or 25 to 30 per cent of ground yellow corn, or the usual amount of good alfalfa leaf meal. For laying hens the requirement of vitamin A is somewhat higher, but is usually adequately met if the grain mixture is about half yellow corn. The usual symptoms of a deficiency of vitamin A include white pustules on the larynx and esophagus and deposits of urates in the kidneys and in other organs. However, it takes some 4 or 5 months on a diet low in vitamin A to deplete the reserve of adult birds that have been properly fed, and to show these symptoms of deficiency.

Vitamin D is also needed, but is ordinarily provided by 0.5 to 1 per cent of any good cod-liver oil in the mash. The exact amount of vitamin D required has been estimated as 180 AOAC units per pound of feed for chicks and about $2\frac{1}{2}$ times that amount per pound of feed for adult birds. For birds exposed to direct sunlight that contains ultraviolet light, no vitamin D supplement will be necessary. For birds confined in the laboratory, however, it will be safest to ensure an adequate supply of vitamin D by feeding throughout the year either fish-liver oil or the synthetic vitamin D now available. Most mashes now contain adequate amounts of vitamin D.

Riboflavin is needed by chicks for growth and by adult hens for reproduction. Apparently it need not be given for egg production alone, but, if one wishes to ensure good hatchability of those eggs, then supplementary riboflavin should be provided. This is done by including alfalfa meal, milk products, or other riboflavin supplements now available to feed manufacturers.

Apart from these three vitamins, the requirements of the others seem to be met by ordinary normal diets. For details of the amounts needed of thiamine and other vitamins, one may consult the publication *Recommended nutrient allowances for poultry* of the National Research Council. Anyone noting therein horrible pictures of what happens to chicks that do not get enough vitamin E, vitamin K, thiamine, nicotinic acid, pantothenic acid, or biotin need not be unduly perturbed by them. Unfortunately, the legends do not make it clear that these abnormalities were induced on special experimental diets, and are not likely to be encountered when normal diets are fed.

For detailed information on nutritional requirements and feeding of poultry, the books by Heuser (1946) and Ewing (1941) may be consulted.

DIRECTIONS FOR FEEDING

To show how these nutrient requirements are met when the feeds are compounded, the mashes prepared by one manufacturer for chickens of different ages are shown in table 1.

It should be noted that the chick starter is supplied usually without any supplementary grain or anything else. By the time the chicks are 6 to 8 weeks old, a considerable amount of grain is being fed. Birds getting the laying mash or the breeder mash shown in table 1 are supposed to be getting approximately half of their total feed intake from some mixture of whole grains. A common one is the mixture with equal parts of wheat and corn, but other grains, such as oats, barley, and buckwheat, may be used. If older birds are given an all-mash diet, formulae different from those in the table would be necessary.

It will hardly pay the biologist to compound his own ration. So many ready-mixed feeds of high quality are already available, that it will save him much time to use what has been prepared in large quantities for the trade. Of course, there may be exceptional circumstances, such as occurred during the war when some British biologists kept their stocks going on table scraps collected from several homes.

Some mashes are also available as little compressed pellets simulating kernels of grain. The advantage of these is that the birds eat them more readily than the ground feed, and hence food consumption

TABLE 1

Typical Diets for Chickens of Different Ages as Made by One Feed Manufacturer

Ingredient	Chick Starter (per cent)	Growing Mash * (per cent)	Laying Mash * (per cent)	Breeder Mash * (per cent)
Yellow corn meal	43.23	17.5	30.13	23.28
Wheat bran	. . . . .	16	14	12.2
Wheat standard middlings	. . . . .	14	15	10
Wheat flour middlings	10	. . . . .	. . . . .	. . . . .
Ground oats	5	10	5	10
Alfalfa meal	3	5	. . . . .	7
Soybean oil meal	20	23	23	15
Meat scrap	3	4	3	4
Fish meals	5	2	2	4
Corn distillers' solubles	4	. . . . .	. . . . .	4
Riboflavin supplement	0.4	0.4	0.4	0.5
Dried whey	4.0	3	3	4.0
Ground limestone	1.0	2	. . . . .	. . . . .
Iodized salt	0.25	1	1	1.0
Manganese sulphate	0.02	. . . . .	0.02	0.02
Dicalcium phosphate	1.0	2	3	3.0
"D"-activated animal sterol	0.05	0.05	. . . . .	1.0
Vitamin A feeding oil	0.05	0.05	0.35	1.0

* To be fed as part of a diet that also includes whole grain.

is stimulated somewhat. One must remember, however, that, if laying hens are allowed to eat only pellets, they are getting an unnecessarily high amount of protein because there is no low-protein grain to balance the materials in the pellets. Another method of stimulating production when the birds are off feed is to give some of the mash slightly moistened, so that it is in a crumbly state. Birds will eat moist mash readily when they cannot be tempted by dry feed.

Finally, one should remember that there are some protective foods, such as milk, whether skimmed milk, whole milk, buttermilk, or skimmed-milk powder, also liver meal, and green feeds. When these are provided, they tend to overcome any deficiencies that may possibly exist in the diet otherwise supplied. The birds will relish an occasional bit of lettuce, fresh lawn clippings, cabbage, or other green feeds.

The amount of feed the birds will need at different ages varies considerably according to the efficiency of the feed containers and the amount wasted, but the following figures may be some guide. Up to 6 weeks of age White Leghorn chicks will eat just slightly over 2 pounds of feed per bird. To raise a Leghorn female requires 5.3 pounds to 10 weeks of age, and about 16 pounds to 20 weeks. Males of this same breed will eat slightly more, and chickens of the heavier breeds will eat still more. With respect to adult birds, one may expect that a Leghorn female of average weight (about 4 pounds), and laying 15 to 20 eggs a month, will eat about 7 pounds of feed per month. A 6-pound Barred Plymouth Rock hen will eat approximately 8½ pounds per month. However, if the bird is not laying well, the feed consumption will be less, and if it should be an exceptional layer, the feed consumption must be higher.

An important point is to make sure not only that the feed is in the containers, but also that these are readily accessible to the birds, and that there is ample feeding space. For day-old chicks, one should provide at least 1 inch of feeding space per chick. This means that a hopper 2 feet long with feed available from both sides will suffice for fifty chicks. When water is first given, it should be in containers right on the floor of the cage so that the chicks can dip down to get it. At about 3 weeks of age the fountain may safely be put on 2-inch blocks, or even higher, to reduce contamination. Any observant biologist should be able to tell by watching the birds for a few minutes whether or not they can get at the feed and water readily.

With the kind of equipment that one will have to use in a laboratory, the containers for the feed are usually provided. Thus, the chick brooder has ample trough space, usually one for the water as well as for the mash. It will be necessary to provide both mash and water in containers right on the floor of the brooder for the first day or two, until the chicks learn to use the larger feeders. The first supply of mash may be given on flat boards or in shallow hoppers. The batteries for older chickens usually have ample accommodation for both feed and water.

DISEASES

In the brief space here available it is possible only to mention a few of the more common diseases likely to be encountered in the small laboratory. For a full discussion of these and other poultry diseases the reader is referred to the books by Biester and Schwarte (1948), Blount (1947), and Barger and Card (1943).

Diseases of Young Chicks

Pullorum Disease. The bacterium, *Salmonella pullorum*, is responsible for one of the more common diseases of young chicks. Affected birds may show a white diarrhea, appear stunted, and frequently die soon after onset of the disease. They do not eat well and stick close to the heat, but, since other diseased chicks show similar symptoms, one cannot always be sure merely from inspection that the disease is pullorum. Accurate identification can be made by bacterial cultures from affected chicks.

This disease is transmitted from surviving carrier hens through the egg to their offspring. One chick thus infected put out in the brooder may subsequently infect many others. The situation is aggravated if the chicks are chilled in any way, particularly if that happens during the first 3 or 4 days after hatching. After 10 days of age the chicks are comparatively resistant. White Leghorn chicks are much less likely to have trouble with pullorum disease than are those of the heavy breeds. The best protection against loss from this source is to ensure that any chicks bought come from flocks of hens that have been blood-tested by the agglutination test, so that carriers of the organism have been detected and removed. With this precaution there should be little trouble from pullorum disease. Since most hatcheries now make sure that their supply flocks are blood-tested, there should be little difficulty in securing a supply of chicks from tested flocks and thus avoiding trouble.

Coccidiosis. The domestic fowl is subject to infection by at least eight different kinds of coccidia, all species of the genus *Eimeria*. It is true that loss from these is avoided to some extent when the birds are raised on wire, as in the chick brooders and hen batteries, but coccidia cannot be entirely avoided. The more serious species are *Eimeria tenella* and *Eimeria necatrix*. The former invades the ceca, and its presence may be diagnosed by severe hemorrhage in those organs. *Eimeria necatrix* affects the small intestine more than the ceca.

With poultry raised on commercial farms, it is safest to allow the chicks out on the ground at an early age, so that they may thus obtain some slight exposure and thereby develop resistance. Within recent times a very effective drug for the control of coccidiosis has been discovered. This is sulfaquinoxaline. Where trouble with coccidiosis is expected, sulfaquinoxaline fed to 8 weeks of age at a very low level permits the birds to resist infection and to develop immunity. This applies to birds on the floor or on the ground, where exposure is inevitable. With birds in cages, it will probably suffice, if coccidiosis appears, to feed the drug in the mash at a level of 0.05 per cent for 3 or 4 days. Mashes containing sulfaquinoxaline are now generally available.

Ordinary sanitary precautions, such as ensuring that the feed does not get contaminated, and the avoidance of overcrowding in the brooders, together with regular cleaning of the wire floors, will help to lessen any risk of a serious outbreak of coccidiosis.

Infectious Bronchitis. This disease, which is caused by a filterable virus, may attack both chicks and adult birds, but chicks once exposed are likely to be immune thereafter. Until very recently there has been no protection against it, but it is now known that by vaccination with a live virus an immunity may be established which apparently lasts for a long time, if not for life. The disease is to be suspected if the birds have difficulty in breathing and gasp for air. However, since other diseases produce similar symptoms, it might be safest to have the condition diagnosed by someone familiar with the different respiratory diseases of poultry.

Diseases of Adult Birds

Lymphomatosis. In birds on poultry farms the most serious disease at the present time is lymphomatosis, or avian leucosis. It has many symptoms, including paralysis of legs or wings, tumors in the ovary, liver, or other visceral organs, and even a type of blindness called "iritis." However, this disease should not bother the laboratory flock as much as it does the commercial farm. If the chicks are obtained from a hatchery as soon after hatching as they can be dried off, and then taken directly to the laboratory, there is a fair chance that they may never be exposed, particularly if there are no adult birds in the laboratory. It has been shown that if infection or exposure can be avoided until the birds are about 5 months of age, most of them are then highly resistant. Presence of the disease is to be suspected if young birds become paralyzed between 6 weeks and 8 months of age. The visceral forms of the disease (the tumors) are more likely to be found in older birds. In those that are paralyzed, swellings of the brachial and sciatic nerves may be found, but sometimes smaller nerves in other parts of the body are affected without lesions in these larger ones.

Chicken Pox. This disease is not so likely to be found in the laboratory as on the large poultry farms. In one form of this disease, encrustations like scabs form on the comb and wattles; in the other, the so-called wet type, there are yellowish patches in the throat around the larynx. Some of these may become so bad as to cause death by choking. The latter form is sometimes called avian diphtheria, or diphtheritic roup. Where the disease has occurred for several seasons and is to be expected again, protection can be ensured by vaccination at about 3 months of age. If an outbreak is detected before many birds are affected, some protection can still be ensured by vaccinating with pigeon pox vaccine.

Laryngotracheitis. This respiratory disease is characterized by an extreme difficulty in breathing. The affected birds hold the head extended apparently at full length and with the mouth open, as they gasp for air at each inspiration. These attempts to breathe are sometimes accompanied by loud cries as the birds try to clear out the blood and mucus in the trachea. It is possible to vaccinate against the disease, but, as with other vaccines, care must be exercised not to introduce the live virus where the disease has not been prevalent. If the first bird affected in the colony should be noticed promptly, it may be possible to lessen the ravages of the disease by vaccinating the others.

Newcastle Disease. This respiratory disease has recently been identified in this country from coast to coast. Even an expert may have difficulty in distinguishing it from other respiratory diseases without the appropriate tests of the blood. Newcastle disease frequently causes malfunctioning of the nervous system, so that some birds walk with a peculiar gait and others show twisting of the head and neck. When an outbreak strikes a flock of hens for the first time, it causes an almost complete cessation of laying, and the birds may take some 3 weeks or more to get back into production, the exact interval varying with the individual. In this country Newcastle disease has not been quite so serious as in some others, so far as adult birds are concerned, but it has been highly disastrous with young chicks. Infected hens that recover pass through the egg to their offspring a passive immunity that seems to last for about 3 weeks. Vaccines effective against Newcastle disease are available but are not likely to be needed in the small laboratory.

Colds. Like the common cold in man, the common cold in the chicken is a perennial nuisance in any flock. In recent years it has been designated as infectious coryza and attributed to the causative organism, *Hemophilus gallinarum*. While this may be the primary cause of the condition, the infected sinuses are usually invaded by other organisms, particularly those of the *Pasteurella* group, and these set up an infection that may become chronic. In some cases the linings around the eye become distended and the whole eye may be completely closed. Treatment of the individual is not very effective. The safest precaution against the disease is to eliminate any birds that have had the disease, or that show symptoms of it, as soon as they get it. Among poultrymen the usual practice is to dispose of the adult flock and to leave the buildings vacant for a period of several weeks before the young stock is brought in. This breaks the chain of infection and gives the new population a chance to start clean.

Parasites. Birds kept in isolated laboratories are not likely to have much trouble with parasites, but it is well to be continuously on the lookout for them. The nematode, *Ascaridia galli,* is serious in some cases, and, where it has been found, other birds in the colony should be individually treated with capsules especially made for the purpose. These contain nicotine, the most effective drug against this worm. The caecum worm, *Heterakis gallinae,* is likely to be found in most birds, but it apparently does little or no harm. Birds in the laboratory are not likely to have infestations of tapeworms because they are less likely to eat the insects and slugs which in the outdoor world serve as intermediate hosts for the cestodes to which the chicken is subject. The birds should be examined at regular intervals to see if they have acquired an infestation with body lice or feather lice. These may be found throughout the plumage, but usually there is a concentration of them just below the vent. If they are present, they can easily be controlled by dusting the birds under the wings, under the vent, on the back, and in the fluff feathers of the abdomen with a pinch of sodium fluoride.

The common red mite, or roost mite, may possibly cause some trouble, but can easily be controlled by soaking the crevices in roosts, walls, or battery with DDT in some vehicle such as kerosene. Mites are not likely to be as serious with birds kept in metal cages as they are in wooden poultry houses. Another mite that sometimes causes trouble is the northern fowl mite, a small grayish organism which may be found all through the feathers, with accumulations at the base of the tail or in the fluff feathers of the abdomen. A fairly effective way of controlling this is to dust the birds thoroughly with free sulphur, such as is used in spray mixtures, or to paint on the fluffy feathers of the abdomen a little of the tobacco preparation, Black Leaf 40, that is used for aphids.

Other Conditions. Apart from infectious diseases, one must expect to lose some birds from tumors of spontaneous origin and from various disorders of reproduction. Birds kept in batteries sometimes seem more subject than others to kidney diseases of various kinds. Young chicks are likely to develop the bad habit of feather pulling or toe picking, especially if they are overcrowded. Once the birds are big enough to be put in individual compartments in hen batteries there is little trouble on that score.

OTHER SOURCES OF INFORMATION

The reader is reminded that at most agricultural colleges one may find specialists in the feeding, breeding, and management of the fowl, and also in the diagnosis and control of its diseases. Most veterinary colleges now include specialists in diseases of poultry. Many of the feed manufacturers have well-trained poultry specialists who can be helpful. The hatcheryman who supplies the chicks has undoubtedly heard of every kind of trouble that can befall them, and is usually both able and willing to give helpful advice when it is sought.

REFERENCES

Barger, E. H., and L. E. Card. 1943. *Diseases and parasites of poultry.* 3rd ed. Lea and Febiger, Philadelphia.

Biester, H. E., and L. H. Schwarte, eds. 1948. *Diseases of poultry.* 2nd ed. Iowa State College Press, Ames, Iowa.

Blount, W. P. 1947. *Diseases of poultry.* Williams and Wilkins, Baltimore.

Burrows, W. H., and J. P. Quinn. 1939. Artificial insemination of chickens and turkeys. *U. S. Dept. Agr. Circ.* 525.

Ewing, W. R. 1941. *Handbook of poultry nutrition.* W. R. Ewing, Upper Montclair, New Jersey.

Heuser, G. F. 1946. *Feeding poultry.* John Wiley & Sons, New York.

Hutt, F. B. 1949. *Genetics of the fowl.* McGraw-Hill, New York.

Jull, M. A. 1938. *Poultry husbandry.* 2nd ed. McGraw-Hill, New York.

Landauer, W. 1948. The hatchability of chicken eggs as influenced by environment and heredity. *Storrs (Conn.) Agr. Exp. Sta. Bull.* 262.

Levi, W. M. 1941. *The pigeon.* The R. L. Bryant Company, Columbia, S. C.

Lippincott, W. A., and L. E. Card. 1946. *Poultry production.* 7th ed. Lea and Febiger, Philadelphia.

Rice, J. E., and H. E. Botsford. 1949. *Practical poultry management.* 5th ed. John Wiley & Sons, New York.

Taylor, L. W., ed. 1949. *Fertility and hatchability of chicken and turkey eggs.* John Wiley & Sons, New York.

Winter, A. R., and E. M. Funk. 1947. *Poultry science and practice.* J. B. Lippincott, Philadelphia.

12· REPTILES

CLIFFORD H. POPE *
Chicago Natural History Museum
Chicago, Illinois

INTRODUCTION

There is no "laboratory" reptile comparable, for example, to the laboratory rats and mice among mammals. This chapter, therefore, deals with reptiles in general, and merely gives the principles of their care. Reptiles constitute a class of vertebrates that includes snakes, lizards, turtles, crocodilians, and *Sphenodon*, a "living fossil" of New Zealand on the point of extinction. Snakes and lizards together form the order Squamata, the turtles belong to the order Chelonia (Testudinata), and the crocodilians to the order Loricata (Crocodilia).

The initial difficulty in the care of reptiles lies in an adjustment to the slow reactions of most of these animals. Nearly every human being has had some experience with mammals of some kind and therefore knows in a general way what to expect of them. A hungry mammal reacts to food quickly, whereas the hungry reptile may hesitate for many moments before seizing its prey. A rat or other small mammal will detect an object from a distance, seize, and gobble it in a matter of seconds. In contrast, a caged reptile might not even sense the presence of food at a distance; it often has trouble identifying food lying near or even on it. There probably is no substitute for experience in dealing with reptiles, since no amount of description will cover all the types of apparent indifference to the environment. One promising experiment was abandoned because the experimenter did not know that snakes feed infrequently; the labor of collecting earthworms every day proved to be too great.

The little effort required to care for reptiles more than makes up for any difficulty in understanding their behavior. A collection of reptiles probably can be looked after in much less than half the time

* I am indebted to Dr. Bernard Greenberg of Roosevelt College, Chicago, for reading the manuscript and giving much constructive criticism.

MIDDLEBURY COLLEGE LIBRARY

needed to care for a comparable lot of mammals or birds. The very fact that daily attention is not necessary and that an occasional three- or four-day lapse can do no harm makes reptiles much less of a responsibility than the warm-blooded vertebrates.

Reptiles hibernate during the cold months of a temperate climate, but those kept warm in confinement remain active all the year. Some individuals feed less regularly or even refuse to eat during the winter. This should cause no concern, since any reptile in good condition readily survives a fast of many weeks or even months. A little extra feeding throughout the autumn is a good precautionary measure.

In spite of their requirement of relatively high temperatures, captive reptiles often succumb to summer conditions, a fact proved by numerous reports of zoological gardens. This hot-weather mortality is partly an aftermath of the fresh zoo acquisitions of early summer, but temperature extremes and feeding difficulties of hot weather are also accountable (Schroeder, 1939). Tropical reptiles often tide over the hot, dry periods by estivating.

The problem of the reptile's thermal requirements is being carefully investigated by Cowles and Bogert, who published a preliminary study in 1944. It is becoming apparent that each species has an optimum temperature requirement, which it satisfies by moving from a warm to cool place or vice versa. Changes in body temperature are relatively rapid, a fact that accounts for the quick death of captives exposed to direct sunlight from which there is no escape. This is the first lesson that a reptile keeper must learn.

House rats will attack and kill small captive reptiles; this rodent menace must be carefully guarded against.

An ultraviolet sun-lamp should be used regularly on species that sun themselves when wild unless they have access to direct sunlight under laboratory conditions.

Two papers (Walker, 1942; Hoopes, 1936) give much valuable information on the care of reptiles. The former is written from the point of view of a zoo man, the latter from that of a reptile fancier.

SNAKES

There are more than 2000 species of snakes, the majority of which are terrestrial; an appreciable number of tropical forms are arboreal. Semi-aquatic to fully aquatic species are numerous and widely distributed in both tropical and temperate regions. The sea-snakes

(Hydrophiidae) are strictly marine animals confined largely to the East Indian region. In contrast to lizards, snakes are not abundant in very arid regions.

Snakes are essentially specialized lizards. That snakes branched off at an early date is shown by fossils found in the Cretaceous. The extreme modifications that the snakes have undergone set them off sharply from other reptiles. Their specialized feeding habits and loco-motion, both correlated with a loss of limbs, are among their most interesting features.

In spite of the universal fear of snakes, reptile houses are among the most popular parts of zoological gardens. Consequently, much information on the handling and care of snakes has accumulated (Stemmler-Morath, 1938), and this group is treated more fully than any other.

Handling and Transportation

The handling of snakes presents a special problem because of the widespread fear of them, an elongate shape, and the venomous nature of certain kinds. All these difficulties can be overcome: the fear by using a little self-control, and the other two by special methods de-scribed below.

In general, a large snake should be supported at several points even though the neck is seized first; a snake held by the neck alone is prone to thrash about. Rough handling undoubtedly tends to keep any excitable individual wild and unmanageable and may injure it. For reasons that are not evident, very small snakes die if closely confined in the hand for more than a few minutes. This applies only to individuals less than a foot long.

The bites of harmless species less than five feet long are painless and seldom become infected. Large, aggressive individuals may be treated like venomous kinds. Since a snake's teeth slant inward, the reflexive jerk that the bitten person invariably gives works to the advantage of the snake. If it does not let go, the head should be seized and pushed forward at the same time that lateral pressure by the fingers spreads the jaws apart. This will force the teeth out of the victim's flesh and allow the bitten part to be removed from the gaping jaws.

The dangerous snakes of this country have, with the exception of the coral snakes, a deep pit or cavity anterior to the eye and at a somewhat lower level. These are the pit-vipers, which include the copperhead (*Agkistrodon contortrix*), the water moccasin or cottonmouth (*Agkistrodon piscivorus*), and the numerous species of rattle-snakes (*Crotalus* and *Sistrurus*). The coral snakes are recognized by the black snout and bands of red, yellow, and black that *completely encircle the body*. A few entirely harmless kinds, the so-called mimics, are strikingly similar in general appearance. The water moccasin and coral snakes are confined to the southern part of the United States; the former are absent from the southwestern states, the latter excessively rare there.

Venomous species should be first seized close behind the head and supported at one other point before being raised. This procedure is facilitated by a snake stick (fig. 1) with which the head is pinned

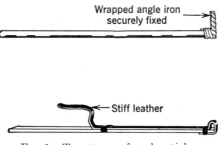

Fig. 1. Two types of snake sticks.

to the ground. Sometimes the snake stick may be used to raise the snake by the middle of the body and place it where desired, thus avoiding actual contact with the reptile.

The method illustrated in figure 2 has the advantage of safety for experimenter and snake alike, since the former does not even have to touch the snake and the latter is little if any annoyed and will not be injured. The basic piece of equipment is a tube of appropriate length made by rolling a few sheets of cellulose acetate and holding them in position by rubber bands. A glass tube may be used but is, of course, non-adjustable in bore and cracks readily. One end of the tube is closed by a loosely fitting stopper controlled by a U-shaped piece of wire; the other end remains open. A cloth sack is then placed in a carton and the sack's mouth held open by turning its border outward over the rim of the carton. A loop of string must be placed around the sack before arranging it. A corner of the sack with a hole in it is pulled through a small hole in a lower corner of the carton. The protruding extremity of the sack is then secured by adhesive tape around the open end of the tube. The snake is gently lifted with the snake stick and lowered into the sack, which is at once closed by the string. The snake will soon crawl into the tube, where its position is controlled by the stopper. The tube is now separated from the sack. The snake can be handled in any way; being strongly thigmotactic, contact with the tube will keep it quiet for long periods. Once inside the tube, the most dangerous individual becomes innocuous and may be minutely examined. A hole in the tube will allow scrutiny of any part of the snake.

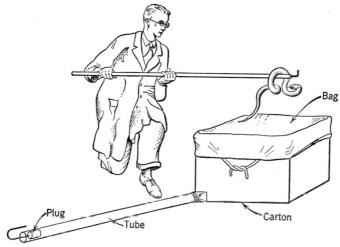

FIG. 2. Equipment for handling dangerous snakes and method of introducing specimens.

Jaros (1940) has described a simple operation by which a true viper or a pit-viper is rendered harmless. The venom duct is laid bare where it passes beneath the skin at a point directly below the eye. Electrocoagulation of the duct at this point prevents the future passage of venom to the fang. It has also been suggested that the duct could be tied off rather than severed by electrocoagulation. Occlusion of the duct apparently has no bad effect on the general health of the snake.

It is well to keep a snake from ever striking at or biting a hard object. A snake stick should never be held so that it can be bitten; a snake in a box should not be teased or induced to strike the glass. Even slight injury to the mouth may develop into infection that is especially hard to control.

Keeping track of individuals of a series housed together is no simple matter when size differences do not help. Painting symbols on an animal that so frequently sheds its entire body covering is often

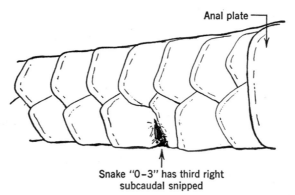

Anal plate

Snake "0-3" has third right
subcaudal snipped

FIG. 3. Ventral view of the basal region of a snake's tail marked by the Blanchard-Finster method.

worse than useless. The most feasible way to insure recognition of specimens has been described in detail by Blanchard and Finster (1933). They snip with sharp scissors one or more of the subcaudal

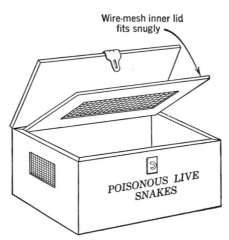

Wire-mesh inner lid
fits snugly

POISONOUS LIVE SNAKES

FIG. 4. Shipping box for snakes.

plates or scales. The fact that most snakes have a double row of these plates makes it easy to devise a system that will allow a large number of specimens to be individually marked (fig. 3). If the cut is made deep enough to expose the underlying muscles, the resulting scar will last for at least two or three years; eventually it may nearly or entirely disappear (Conant, 1948).

The transportation of snakes presents a problem because of post office prohibition and express company regulations. The latter are fulfilled by a wooden or metal box with a handle, a lock, and a double top, the inner one made largely of wire screening or hardware cloth (fig. 4). Marking the box either "poisonous live snakes" or "harmless live snakes," as the case may be, will caution against unnecessary delay and rough handling. When snakes are thus shipped it is well to put them first into bags. These prevent injury and also keep the larger snakes from devouring the smaller, a mishap that may occur even among non-ophiophagous species when they are closely confined together. It is not necessary to provide food and water, since snakes are capable of such long fasts and do not readily succumb to thirst.

Bags are also the most practical containers for snakes being carried by hand; drawstrings should not be used because a snake can escape from a bag unless a string or rubber band is wound tightly around the mouth. They invariably escape from cardboard boxes or any other container that is not securely and completely covered.

Housing

Snakes thrive in boxes with a smooth inner surface and relatively small cubic content. They do not seem to require exercise. Wire cages are not suitable because restless snakes will injure their noses in their efforts to escape. A sliding glass panel in front provides the most convenient access. Water deep enough to cover the occupant's head should be supplied in shallow containers that cannot readily be overturned. It is imperative to keep the boxes dry even though semi-aquatic species are involved. Therefore, if wood is used the inside should be thoroughly painted or water-proofed in some other way. Direct sunlight should never be allowed to flood the box because a snake will absorb heat and quickly succumb. The temperature should be maintained between 22 and 27° C. (about 72 to 80° F.), especially after feeding. Cleanliness is essential. Most individuals will defecate two or three days after feeding, but the giant snakes may not do so for many days or even for several weeks. The urine is solid and often passes out with the feces. Because snakes feed infrequently and swallow their food whole, their boxes can be kept clean with little effort. The better the ventilation the less the trouble in keeping the boxes dry. Snakes like to hide under any suitable flat object, but it is not necessary to supply most of them with such.

Although there is no objection to arranging a box "naturally" with earth, branches, and even water for swimming, snakes in general live as well without such elaborations. Providing a few branches for arboreal kinds is a simple matter, but supplying earth merely adds to the difficulty of cleaning. Water snakes thrive with only enough water to drink. Some of the small burrowing snakes as, for example, species of *Leptotyphlops,* must be supplied with moist sand (Burt, 1935). The sand-swimming desert species like those of the genus *Chilomeniscus* require dry sand.

Food and Feeding

All snakes are carnivorous and swallow their food without chewing it. The pieces thus consumed are relatively large. In fact, a snake often finds difficulty in swallowing a small piece of food, especially if it is an active animal, which may literally leap or crawl out of the predator's throat.

An object with a diameter several times that of the snake's neck is readily swallowed, although giving such large objects is not recommended. In general, the diameter of a suitable meal should not greatly exceed that of the snake's head; where the head and neck are about equal in girth, the food object may have a diameter three times that of the head. The length of the food is of less consequence; a snake can easily swallow an object 10 to 25 per cent as long as its body. In fact, it is not unusual for a snake to devour another snake equal to it in length and girth, the tail being zig-zagged in the predator's throat.

Whole animals are the better food although strips of muscle may be used in the absence of or to supplement whole animals. A compromise can be effected by introducing a strip of meat into the mouth just as the snake is about to finish its meal. The swallowing reflex will usually continue until the meat has disappeared. This method may induce a snake to eat raw meat independently.

Contrary to general belief, dead animals will be accepted sooner or later by most snakes; an occasional individual may steadfastly refuse them. Pit-vipers have a heat-sensitive organ in the head (the "pit") and therefore take warm, freshly killed mammals much more readily than they take cold ones. Dunking a dead rat or mouse in hot water will often arouse a feeding reflex in a reluctant individual.

Snake keepers use many other stratagems. The food object may be manipulated to simulate the action of a living animal, or a small area of skin may be removed to give better olfactory perception. A more elaborate device is to place the object in a box with one small opening and allow the snake to crawl in. Encountering the food thus in the dark will occasionally arouse the desired reflexes. Lowe (1943) has described such a feeding box. A tube will serve the same purpose. Leaving the food in the cage overnight may help in feeding some nocturnal species.

Snakes should be fed three or four times a month but can live on much less frequent feedings. They are often indifferent eaters; in selecting specimens this point should be borne in mind. Food can be forced down their throats, but such a procedure should be resorted to only as an extreme measure after weeks or even months of starvation, and, in doing so, great care must be exercised to avoid injuring the mouth. The temperature of the boxes should never be allowed to become low soon after a meal; lowering of the temperature appreciably slows up digestion. Needless to say, rough handling will sometimes cause a snake to eject a recent meal.

Snakes often eat other snakes. Before mixing specimens it is wise to determine the food habits of the kinds involved. Care must be exercised at feeding time or else two snakes may start to swallow the same object and meet at its middle; as often as not, the larger snake will consume the smaller one.

A snake with cloudy eyes usually refuses to eat and should not be coaxed to do so because the period before shedding is a critical one, during which the snake should be let alone. At such a time it is apt to strike wildly and thus injure itself or bite its keeper.

The disadvantages of having to use live animals for food are many: rats, mice, and other small mammals are prone to attack snakes and may even seriously injure them; a living animal may readily escape both the keeper and reptiles.

Rodents and other small mammals and birds are taken by many of the larger snakes, insects by some of the smaller. The natural food of species found in the United States and Canada is given by Schmidt and Davis (1941). The raising of insects for reptile food is discussed below under the lizards.

Garter snakes may be induced to eat raw hamburger by first mixing the meat with earthworms and then slowly reducing the amount of earthworms until the snakes are eating only the meat.

Health and Disease

A snake in a good state of health should have no injury or inflammation in or about the mouth; the skin should be free of blisters and other blemishes (excepting old scars) and parasites. Ecdysis should take place regularly and completely. The readiness with which food is accepted may be used as an additional criterion even though some snakes in good health are inconsistent in this respect; they will feed readily at one time and less so at another. If ample material is available, ready feeders should be selected in preference to diffident ones.

A snake's eyes become cloudy about eight days before ecdysis but clear up two or three days before the molt is actually cast off. During the interval between the appearance of the cloudiness and ecdysis, a great deal of moisture is lost through the skin (Bogert and Cowles, 1947), and this loss is a health hazard that may be overcome by warm baths of a temperature about four Centigrade degrees higher than that in which the snake is constantly kept. If the snake insists on crawling out of the water, it is probably too warm. Very large individuals may be covered with wet bath towels instead of bathed. When molting is imminent, something heavy should be put in the cage: a stone or small brick wrapped in inner-tubing rubber will do. A light object will slide about when the snake pushes against it in its effort to molt.

When the skin is shed incompletely, the remaining parts of it may be peeled off by allowing the snake to crawl through the hand against a moderate amount of resistance, or by using tweezers for the removal of the eye caps and other fragments. A record of each shedding, giving the date of clouding and clearing of the cornea and of actual ecdysis, will be useful for future reference. The interval at which ecdysis occurs varies not only from species to species but from individual to individual; it may be expected to take place once in five to ten weeks. It should also be complete, although the molt may come off in pieces.

Snakes caught in a wild state are frequently infected with both ecto- and endoparasites. These parasites often have little effect on the health of wild specimens which may, however, develop symptoms after some time has been spent in captivity. Obviously, it is best to secure individuals relatively free of infection.

The following data show how commonly helminths occur in wild snakes. Harwood (1932) examined 72 Texas snakes representing 17 species and found helminths in all but one. Nematodes were taken from 12, trematodes from 8, cestodes from only 3 species. Mueller (1937) found 90 per cent of Florida water snakes (*Natrix*) infested with cestodes. Fowler (1941) reports 83 flukes in one specimen of the common water snake (*Natrix sipedon*). These parasites inhabited the lung, the trachea, and the intestines. Another individual had 41 trematodes in the mouth alone. All parts of the reptiles may be invaded, including the subcutaneous connective tissue in which nematodes have been seen to thrive and produce external sores over much of the body (Richmond, 1945).

The degenerate arachnids known as linguatulids live as adults in reptiles, chiefly snakes; a fish or a mammal is usually the intermediate host although the whole life history may be carried out in the reptile. More than fifty species have been described. These endoparasites are recognized by their large size, elongate, ringed bodies, and the two pairs of chitinous hooklets that lie on either side of the mouth. There is little information regarding the effect that linguatulids have on their hosts; so large a parasite must certainly be an unpleasant guest in the reptile lung, where the linguatulid usually lives. Some species are more than three inches long in the adult stage.

Among ectoparasites mites are the only very troublesome animals that attack snakes. These pests are sometimes found on specimens taken wild, and it is therefore wise to examine fresh material carefully. A bad infection of mites will cause severe losses to a snake collection. It is a simple matter to detect these parasites crowded under the edges of the scales. Since shedding will often rid the snake of mites, a quarantine system should be effective. Any snake isolated until it has shed will be free of mites unless it is so heavily infected that the slough splits at the point of infection and leaves the mites behind.

Zoological garden curators have tried various methods of fighting mites, but none of them is entirely satisfactory. Giving the snakes baths in warm water is one simple and effective method. The victims must be forced to remain almost completely submerged for at least twenty-four hours. A cage harboring mites must be washed with very hot, soapy water, and disinfected with lime-sulphur dip or bichloride of mercury. Conant and Perkins (1931) give further details. Mattlin (1947) had some success with flea powder. After sprinkling the powder inside a bag he puts the snake in it for about fifteen minutes. The cages are thoroughly dusted with the same powder. A serious infection was checked by this easy time-saving method, which other zoological gardens also have found very effective. Perkins now uses *fresh* powdered pyrethrum and keeps the patients in the bag for from one to several hours or, in the case of very large individuals, even days.

Geimann and Ratcliffe (1936) have studied the life cycle and morphology of a protozoan, *Entamoeba invadens,* that produces amoebiasis in snakes and other reptiles. This species proved to be so much like *Entamoeba histolytica* that the investigators were tempted to call it a variety of that well-known species. They compare *invadens* to the several other amoebae that parasitize reptiles. Since reptiles suffering from amoebiasis often die, this disease must be considered a real menace.

Various bacteria invade the digestive tract of snakes and produce enteritis, which is frequently fatal. In fact, the zoological gardens often sustain greater losses from maladies of the digestive system than from any other single cause. Lovell (1930) has reported on bacteriological studies of fatal cases of enteritis that occurred in the London Zoo but he was able to identify only a few of the bacteria isolated. It is well to note his statement that animals must be examined shortly after death, or else putrefaction makes isolation of the lethal bacteria all but impossible. This is partly caused by the high temperatures in which reptiles are kept.

Spontaneous tuberculosis in snakes has been discussed by Aronson (1929), who found that the bacillus involved differs in antigenic and cultural character and staining reaction from the acid-fast bacilli of warm-blooded animals. Cultures were made from the lungs and livers of the snakes. It is possible that the germ found in reptiles is merely a modified form of the one that attacks man and other warm-blooded animals (Glidden, 1936).

Pneumonia and other diseases of the respiratory system always account for a large percentage of deaths among snakes living in zoological gardens. Such causes can be prevented in small collections by keeping the snakes in well-constructed, uniformly heated boxes.

Mouth-rot, so frequently fatal, is seen only in snakes that are captives. This disease is much feared by snake fanciers because it is especially hard to treat. The victim succumbs in a few days to the acute form, whereas the chronic one persists for months and may also end fatally. The symptoms are refusal of food, reluctance to use the tongue, and inability to close the mouth completely. White spots may be seen on the edematous mucous membrane of the mouth, and the head often becomes twice its normal size. Although primarily an infection of the oral lining, mouth-rot may attack the jaw bones and terminate in a generalized sepsis.

Burtscher (1932) investigated a series of cases and was able to isolate a bacillus that he believes plays an important rôle in the development of the malady. He was unable to work out any specific cure but recommended total warm baths, or painting or washing the affected parts with a good disinfectant such as a 3 per cent tincture of iodine. Exposure to ultraviolet rays and use of vitamins are also methods that have been tried with more or less success.

Snakes are subject to various other diseases, but so little is known about them that even their listing is scarcely advisable. The chances that they will appear in small collections are slight; if they do, the recording of data on their nature and effect is highly worth while.

Reproduction

Anyone interested in experimental work that involves the breeding of snakes should, if possible, select a viviparous species because the care of reptile eggs is troublesome. This is in strong contrast to the care of the youngest snakes, since reptiles are born, or hatched, fully prepared to fend for themselves. Although many of our native snakes lay eggs, three large groups, the garter snakes (*Thamnophis*), the water snakes (*Natrix*), and the pit-vipers (*Agkistrodon*, *Crotalus*, and *Sistrurus*) bear their young alive. The question of viviparity versus oviparity in snakes and lizards is discussed below under the reproduction of lizards.

The sexing of snakes presents certain difficulties because external sexual characters seldom occur. If a series of adults is at hand, the tail of the females (viewed from below) will be found to taper more or less abruptly near the vent; in contrast, the tail of the males will maintain a uniform width for a short distance. Pressure on the tail a little caudad to the vent will force the hemipenes of the male into view, especially when the pressure is directed toward that orifice. If hemipenes do not appear, it may be hard to decide whether the pressure is insufficient or the specimen is a female. The matter is not simple even when dissection can be made: there is a striking, though superficial, similarity between the anal glands of females and the hemipenes.

Some hardy species will breed in captivity. Courtship patterns vary so much that an understanding of them is helpful. Davis (1936), Noble (1937), and Blanchard and Blanchard (1942) discuss them in sufficient detail. The last reference concerns garter snakes only. After mating, there are no special precautions that must be taken except to isolate females with advanced embryos in a place suitable for laying or giving birth.

The period of gestation in snakes is hard to determine because the sperm may live for weeks in the oviducts, and the rate of development varies with the temperature of the mother. Under natural conditions, mating takes place in the spring, and the young appear in the summer or early autumn. Autumn matings produce young at the usual time. In the egg-laying forms, the considerable development that may take place before deposition shortens the incubation period.

A female about to lay must be watched constantly so that the eggs can be promptly removed before desiccation kills the embryos. The eggs are elongate in shape, almost or entirely unpigmented, and have a parchment-like shell. Development will usually take place if the eggs are placed on thin stones that in turn rest on a layer of moist sterile sand in a covered glass container. This will keep the humidity relatively constant. A high temperature (24 to 27° C.; 75 to 80° F.) will facilitate rapid development. Any absorptive substance such as sphagnum moss that can be rendered sterile may be used instead of sand. The container must be relatively large; the eggs should not occupy more than 5 or 10 per cent of its internal volume. Since the eggs are quickly spoiled by mold, there is some advantage in keeping them in the light rather than in the dark. Shriveling of a fertile egg is a sign that greater humidity is needed; the eggs normally increase in size and remain smooth and firm to the touch. The period of incubation varies from species to species and with the temperature. Schmidt and Davis (1941) give the normal periods for our native forms.

The live-bearing female should produce her young without trouble or special attention. The newborn soon escape from the birth membrane and crawl about. The first ecdysis occurs in a few days. Investigators who want series of living embryos from the same female may use the technique described by Clark (1937) and modified by Franklin (1945).

All young snakes can escape through the smallest cracks and holes so they must be kept in tight containers. Some of them probably have food requirements unlike those of adults, but little information on this subject is available. They will live for weeks if not months without food, but they should, of course, be fed as frequently as the adults, and supplied with water.

LIZARDS

Among more than 2000 species of lizards there are numerous terrestrial and arboreal types. Aquatic species are few in number, truly marine ones unknown. Deserts and other arid regions abound in lizards. Lizards did not flourish until the Cretaceous, although fossil forms are known from the Jurassic.

Since lizards create no strong emotional reaction in man, and therefore do not make such popular zoo exhibits as snakes, it is not surprising that information on lizard care is relatively hard to find. Moreover, lizards are comparatively easy to understand and care for because of their less specialized form and behavior. The legless lizards with fragile tails, such as the glass snakes (*Ophisaurus*), are exceptions.

Lizards of the genus *Anolis*, commonly known as "chameleons" and "anoles," swarm in the New World tropics and are often sold as pets by circuses. They have been used in the zoological laboratory because many of the numerous species are readily secured in quantity through dealers. The anoles perhaps come as near being suitable laboratory lizards for American workers as any other saurians. They are not to be confused with the true chameleons (Chamaeleonidae) of the Old World, chiefly Africa. The true chameleons are really aberrant agamids, whereas their American namesakes are iguanids. The confusion arose because both are arboreal groups with a striking ability to change color.

Handling and Transportation

The handling of lizards presents little difficulty because few species are venomous and most of them cling to any available object. The two species of beaded lizards, or Gila monsters (*Heloderma*), are the only venomous kinds. Many others will bite, but they can do little harm. Heavy gloves should be worn when lizards more than twenty inches long are handled because their teeth can inflict ugly wounds. Even the beaded lizards become docile in captivity and can be handled with little danger.

So many lizards have fragile tails that a good rule is always to seize the body and allow the tail to go unrestrained. The greatest precaution may be ineffective in keeping an active skink from losing the tail. The loss of this appendage has no bad effects on the health of the owner; a new but inferior tail will slowly replace the original one.

A very small lizard should never be enclosed by the hand for more than a few moments, since doing so may kill it.

Lizards are conveniently carried about in cloth bags or paper boxes although they are prone to escape from the latter. They can be shipped in simple boxes or sent via the post office in mailing tubes. Sphagnum moss or sticks will keep them from being damaged in transit. Small individuals can now be sent to almost any part of the world by air. Like snakes, they live for weeks without food or water and therefore need not be provided with such while in transit.

Marking individual lizards for future recognition may be simply accomplished by cutting off the toes in combinations. Attaching skin clips to the side of the neck and painting numbers on the body are methods that may also be used. Painted numbers, of course, are lost at each molt.

Housing

Lizards may be kept in many types of cages, the size of which is of no great importance. Exercise is not required.

The ground-living forms have some tendency to rub their noses against a rough surface, but, in general, wire screening may be used in spite of this. Arboreal species should be provided with branches on which to climb, terrestrial forms with flat objects such as boards or strips of bark under which to hide. Skinks and certain burrowing lizards, the amphisbaenids, for instance, apparently lose moisture rapidly through the skin (Bogert and Cowles, 1947) and therefore should have in the cage damp sand covered with pieces of bark or wood.

Lizards that will not drink out of a vessel can be given water by sprinkling leaves of plants kept in the cages. There is remarkably little information on this aspect of lizard behavior. Direct sunlight may kill a lizard that cannot escape from it, so some shade is needed at all times. The temperature should be maintained between 22 and 27° C. (72 to 80° F.). The cages should be kept clean and dry. Lizards swarm in warm, arid regions, and in captivity many of them require a sandy substratum. Apparently, a lizard does well on sand, whereas many snakes thrive on a bare floor. Few lizards are semi-aquatic, none truly aquatic.

Food and Feeding

Like other reptiles, lizards are sometimes slow in their reactions to food, and therefore feeding them may call for great patience; indifference to food may be of limited duration. Movement of the food object is often necessary for carnivorous forms; in such cases, live animals supply the proper stimulus. As a rule, lizards should be fed at least four times a week. However, skipping a meal now and then will do no harm, especially if hearty feeders are involved. The live-food eaters will chase their prey about the cage. Vessels containing flying prey may be uncovered in a tightly closed cage.

The vast majority of lizards subsist on insects and other small invertebrates; a few are omnivorous, and a few others entirely herbivorous. In the United States (and Canada) all but the crested lizard (*Dipsosaurus*) and the chuckwalla (*Sauromalus*) are carnivorous. Both occur in the extreme southwestern part of this country, where they subsist on the more tender parts of desert plants. In captivity, the chuckwalla will eat lettuce, watermelon, cantaloupe, bananas, flowers, and similar plant matter. The few omnivorous lizards belong to the rough-scaled group (*Sceloporus*); since they also eat invertebrates as well as plants, they may be fed as if they were wholly carnivorous. It is common practice to feed certain large forms such as the Old World monitors (*Varanus*), the Gila monsters (*Heloderma*), and the giant stump-tailed skink of Australia (*Trachysaurus*) raw eggs mixed with chopped meat, which is presented in a low dish. The giant skink will also take small lizards and snakes, soft fruits, and flowers; the monitors, rats, fishes, and birds.

Specific feeding habits of any of our native lizards may be determined by reading the complete accounts of life histories given by Smith (1946). Ratcliffe (1941, 1942) describes a diet that has been used with success in feeding large and small carnivorous lizards in the Philadelphia Zoological Garden. His mixture contains so many ingredients that it would be practical only when large numbers of lizards make the raising or purchase of live food difficult.

Food for insectivorous lizards is easily procured during warm weather in meadows and open, grassy fields by sweeping a strong butterfly net back and forth through high grass. A quick turn at the end of the last stroke will close the net.

It is, of course, more satisfactory to have a permanent supply of food. Several kinds of insects lend themselves to laboratory culture.

Among them, the meal worm is probably the easiest to keep in stock. This is a black beetle larva (*Tenebrio molitor*) that lives on bran and thrives in trays or boxes. An initial supply is easily secured, since pet shops handle meal worms. They have hard bodies and therefore should not be fed too steadily to the same lizards. Blowflies (*Cynomyia*) are second only to meal worms as food for lizards. The usual method of rearing these flies is both time consuming and filthy, but Frings (1941) has described a simple method that largely obviates these difficulties.

Although wax worms and wax moths (*Galleria mellonella*) are much more difficult to raise and to handle than meal worms and blowflies, they are excellent food for delicate lizards. The worms are a pest in apiaries, whence they may be procured. Their food is old brood comb; if this is not available, a mixture of one part each of fine corn meal and dried powdered yeast and two parts each of whole wheat flour, skimmed-milk powder, and standard wheat middlings may be substituted. The proper consistency can be attained by adding honey and glycerine in equal amounts (Hydak, 1936). Fruit flies (*Drosophila*), used so often for experiments in zoological laboratories, may be fed to very small hatchling lizards. These flies are commonly bred in milk bottles. During the summer months exposed meat will soon give an abundant supply of house-fly maggots. Methods of raising the foregoing and other kinds of live invertebrates suitable for lizard food are dealt with in detail by Snedigar (1939) and Walker (1942).

Health and Disease

A lizard in good health will feed regularly and have an alert manner and a skin free of blemishes. It should not be emaciated. Food may not always be accepted at once; due allowance must be made for the slow reptile reactions already discussed.

Ecdysis is not so clearly related to health in lizards as it is in snakes, and does not therefore serve as well in allowing one to judge the general condition. The process occurs at either regular or irregular intervals, which may vary greatly from species to species. It may take place as often as twice a month or as seldom as once a year. The juveniles molt more often than the adults. The molt does not come off in one piece; it may be shed at one time or different sections may be cast at different times, each section having its own individual cycle. Parts of the molt are sometimes devoured. The lizard often rubs itself against some object to get the edges free, and therefore something suitable should be provided. Ecdysis is aided by a mechanism that enables the lizard abruptly to increase the size of its head and thus loosen the skin (Bruner, 1907). Wetting a lizard just before it sheds or while it is shedding is often beneficial. Some terrestrial species take advantage of a shallow vessel of water.

A wild-caught lizard may be host to a large number of external as well as internal parasites without showing any ill effects. Since these parasites often assert themselves after captivity has lowered the host's general vitality, it is desirable to secure stock relatively free from infection.

Helminths are extremely common in wild lizards, as the following facts will illustrate. Harwood (1932) examined 8 species of Texas lizards and found nematodes in 6, cestodes in 4, and trematodes in 2 species. A single species was uninfected, but it was represented by only one specimen. A skink, *Leiolopisma laterale*, harbored no fewer than 2 species of cestodes, 2 of trematodes, and 4 of nematodes. Three of these species were found in from 20 to 37 per cent of the 11 specimens examined.

Lizards are much more commonly parasitized by mites than are snakes. No doubt the former's possession of limbs gives the parasites better places for concealment, and the habit of walking rather than crawling keeps them from being scraped off. In an intensive study of the relationship between mites and lizards, Lawrence (1935) found that the absence of imbricated scales, which afford convenient and more or less protected points of attachment, is the chief factor correlated with absence of parasites. Serpentiform bodies, limb reduction, and aquatic habits are rated as minor factors. The monitors (Varanidae), the true chameleons (Chamaeleonidae), and the ringed lizards (Amphisbaenidae) do not serve as hosts for mites. At the opposite extreme are certain geckos with deep pockets in the armpits that seem to serve as brooding sites for larval mites: Loveridge (1926) removed fifty-nine larvae of the genus *Trombicula* from a single pocket of *Gymnodactylus lawderanus*.

Since mites become a pest in collections of captive lizards, it is well to avoid introducing them. It is, in fact, harder to rid lizards of mites than it is to get them off snakes. This is partly due to a difference in the manner of shedding, partly because the lizard body affords better concealment, and partly because lizards do not lend themselves so readily to almost complete submergence in a bath. The methods of fighting mites described above for snakes apply as well to lizards except for the cases in which a lizard's shape makes the warm bath impractical. The bag-and-powder method is probably more suitable for lizards. A box may be substituted for the bag if the lizard is not easily placed in the former.

Lizards, like snakes, suffer from amoebiasis (Geiman and Ratcliffe, 1936) and other disorders of the digestive system. They also die of respiratory diseases, especially pneumonia. The discussion of the digestive and respiratory maladies of snakes will apply in general to lizards, a fact that is not surprising in view of the close relationship of these groups.

Lizards in a wild state are frequently infected with malaria parasites (*Plasmodium*). Thompson and Huff (1944) list fourteen species of lizards in the blood of which these parasites have been found. The distribution of the parasites is very sporadic: in certain areas more than half of the lizards are infected; in others, not very distant and with a similar fauna, none. The natural incidence in 1136 North and Central American lizards proved to be 9.4 per cent. Species of the Iguanidae are most frequently infected. There is evidence that these lizard parasites attack a limited number of hosts. Some lizards succumb to the infection. Other blood parasites of lizards are the coccidians of the genus *Karyolysus*.

Reproduction

Anyone doing experimental work that involves the breeding of lizards would do well to select viviparous species and thus avoid the difficulties of incubating eggs. Although lizards in general lay eggs, the numerous live-bearing kinds are widely distributed over the earth. For example, the snake lizards and girdled lizards of Africa (Cordylidae) are entirely, the cosmopolitan skinks (Scincidae) mostly, viviparous. In North America, viviparity occurs among species of the iguanid genera (*Phrynosoma* and *Sceloporus*) and the alligator lizards (*Gerrhonotus*); all the night lizards (*Xantusia*) and the shovel-snouted legless lizards (*Anniella*) bear the young alive. However, the skinks seem to be the only large family that is predominantly viviparous.

Although most textbooks explain the difference between reptile oviparity and viviparity (often called ovoviviparity) as merely one of duration of egg retention, it has been shown that in some species of the Squamata there is a placental connection between embryo and parent. This condition will no doubt prove to be much more widespread than is now realized. Weekes (1935) has written a review of reptile placentation.

The sexing of lizards is a simple matter in the numerous species that have marked differences in color or external form. Many iguanids and some skinks are examples. Much help can be derived from the descriptions of Smith (1946). In species without such convenient characters, the hemipenes can often be forced out by pressure just behind the vent. This method is not satisfactory in skinks, geckos, and other lizards with fragile tails.

Lizards may be bred in captivity. Learning what breeding behavior to expect is relatively easy because a single pattern of courtship and copulation is usually found throughout a natural group. Noble and Bradley (1933) have discussed saurian mating behavior and included an extensive bibliography. Greenberg and Noble (1944) give a detailed account of the social behavior of the American chameleon (*Anolis*). The little that is known about our native lizards is summarized species by species by Smith (1946).

The varied laying and post-laying habits of lizards make the selection of viviparous species strongly advisable. Lizard eggs are almost or entirely unpigmented, are round to elongate in shape, and usually have a thin, parchment-like shell. In geckos the shell is brittle and noticeably calcareous. Gecko eggs consequently cannot increase in size during incubation as do those of nearly all other lizards.

Lizard eggs are deposited in different kinds of situations and hatch under a variety of conditions. As a rule, however, they are deposited in the ground or under some stone or other object on the ground.

Not infrequently the female coils around her eggs and shows interest in them in other ways. This is especially true of skinks. Noble and Mason (1933) have discussed lizard brooding habits in some detail, and summarize the literature.

Lizard eggs in general may be incubated somewhat like snake eggs if due allowance is made for the specialized habits of many lizards. There is very little specific information on this subject.

Hatchling or newborn lizards require no special attention; they are able to fend for themselves upon arrival and will thrive for several months. After this they are prone to die of what may be vitamin deficiencies.

TURTLES

There are only about 250 species of turtles. Among these, one finds a great range of size from the gigantic marine types weighing hundreds of pounds to small species less than a pound in weight. Many English-speaking people use the word "turtle" only in referring to marine forms; in the United States and Canada it is an all-inclusive term applied also to the "terrapins" and "tortoises."

The turtles, oldest of living reptiles, were already encased in a shell by the Triassic, and have showed little tendency to change their way of life in 175,000,000 years. The rigidity of the shell has forced them in respiration to rely on a method analogous to diaphragm breathing (Hansen, 1941). Although the shell is a most effective armor for the adult, it does not develop early enough to protect the hatchlings, which are therefore especially vulnerable to attack from many predators.

No other reptiles are found in homes of the United States as commonly as are turtles. These are nearly always hatchlings that have been sold through pet shops or as souvenirs to tourists. They seldom eat and grow but, when they are nearly dead, frequently find their way to aquaria and reptile houses of zoological gardens just as young alligators do. In spite of this unfortunate traffic, there is little information available on the care of turtles in captivity.

Handling and Transportation

There is little difficulty in the handling of most turtles, since they can be picked up by the shell without damage to them or danger to their captor. Turtles apt to bite should be carried by the tail if that appendage is long as in the snapping turtles (Chelydridae). The soft-shelled turtles (Trionychidae) usually strike with surprising speed and vigorously scratch with their sharp claws. It is safer to use heavy gloves when dealing with large individuals. Not infrequently, the difficulty is to get the turtle to cooperate: it may simply withdraw into the shell and thus make work with it impossible. In such a case, the turtle should be exposed to the heat of a small electric stove or that of a flood light. When thus slowly warmed the most stubborn individual will eventually emerge and walk away. This simple method has been used with great success in photography; until heat is applied some turtles will remain withdrawn for hours.

Turtles are easily shipped in boxes or even in mailing tubes. Food and water need not be provided, but some padding is helpful in keeping them from banging about too freely. Moisture should be supplied for the soft-shelled turtles; damp moss or grass is suitable.

No reptile is more easily marked than a turtle. Strips of waterproof adhesive tape will stay on the shell for long periods; they should be no larger than necessary. Holes may be bored or notches made along the rear border of the carapace. Cagle (1939) has described a system whereby large numbers can be permanently marked. If few individuals are involved, the notation of an asymmetrical arrangement of the horny shields or a scar on any part of the animal will often suffice, turtles being likely to have irregularities in the shell.

Housing

Turtles occupy types of environment that range from arid, waterless wastes to the high seas; in captivity they must be provided for accordingly. Adolescents and adults of the marine species can be kept in tanks of either fresh or salt water, the very young in sea water, which must be frequently changed. Artificial sea water is made by adding 3.5 pounds of Turks Island salt to 100 pounds of water. Water temperatures should be above 18° C. (64° F.), preferably from 20 to 24° C. (68 to 75° F.). Landing places do not have to be provided although wild individuals do bask on beaches. Deraniyagala (1939) has written the most complete general account of marine turtle biology.

The majority of turtles are aquatic to semi-aquatic inhabitants of fresh waters and their environments. In North America the musk and mud turtles (Kinosternidae), the snappers (Chelydridae), and the soft-shells (Trionychidae) are the more aquatic groups. All these should be provided with vivaria made up largely of water even though, with the possible exception of the soft-shelled turtles, they will live on land. Strict rules of cleanliness should be observed.

There are several genera of less aquatic North American turtles belonging to the family Testudinidae. Most of them should have access to water; some almost invariably refuse to swallow food unless the head is submerged. There is, however, no apparent mechanical reason why they cannot swallow when out of water.

A large genus (*Testudo*) of terrestrial "tortoises" is widely distributed but does not reach the confines of the United States and Canada. In the former country, the gophers (*Gopherus*) are the only strictly terrestrial turtles, although the box turtles are largely so. All these terrestrial kinds will live in virtually any type of terrarium and need only water for drinking, except some of the box turtles that soak in shallow water when the atmosphere is hot and dry. The general rule that exposure to strong sun is a reptile's greatest danger applies to turtles. They also require high temperatures, 22 to 27° C. (72 to 80° F.). Drinking water is best given in a dish just deep enough to allow submersion of the mouth; water can even be dispensed with if a large amount of succulent food such as lettuce and melon is provided.

Uncovered earth in a vivarium will adhere to the feet of the turtles and keep the water muddy. This difficulty can be avoided by omitting the earth or covering it with flat stones or pieces of wood. Small pebbles should also be covered because they often cause injury. A female ready to lay must, of course, be supplied with earth in which to dig.

Food and Feeding

Turtles do not have varied or specialized methods of eating. They use their horny, sharp-edged jaws to bite out mouthfuls, which are swallowed without being masticated. Claws of the forefeet assist in separating a mouthful from the main mass. Since turtles are thus apt to make a "mess," it is often expedient to take them out of the vivarium before feeding.

Good feeders are prone to eat too much and become very fat. There is no need to feed more than three times a week. For obstinate eaters, movement of live food helps, although turtles perceive a food object more readily than do snakes. Individuals of any semi-aquatic kind that refuse food on land should be given it in the water; a few species habitually refuse to swallow unless the head is submerged.

Specialization in choice of food is also rare in turtles. Many species are omnivorous, and a number will devour carrion. Fruits, melons, and other tender, succulent parts of plants are most suitable because of the mechanical difficulty the turtle has in biting through tough, fibrous material. Small invertebrates, such as earthworms, slugs, and snails, bivalves with thin shells, crustaceans, and insects (larval and otherwise), constitute most of the animal food; adult or larval cold-blooded vertebrates and, less frequently, small mammals and birds are also eaten. Obviously, turtle food is easily procured. The feeding habits of the species found in the United States and Canada are given by Pope (1939).

In general, it may be said that the snappers, the musk and mud turtles, and the soft-shells are entirely, or at least predominantly, carnivorous, whereas the gopher turtles are herbivorous. Most of the others seem to be omnivorous. Among those well known to be omnivorous are the painted turtles (*Chrysemys*) and the box turtles (*Terrapene*).

A paper by Pearse, Lepkovsky, and Hintze (1925) gives some interesting facts about turtle diets. Painted turtles are dealt with in greatest detail.

Health and Disease

A healthy turtle does not drag itself about but lifts the body off the ground when walking. The shell is firm and without sores or soft areas. There should be no external parasites on either shell or skin, and the eyes should be clear and in good condition.

A deficient diet causes the eyes to become inflamed and the lids to stick together. The eyes of very young turtles fed on such food as the "ant eggs" commonly sold by pet shops are sure to develop inflammation that is quickly cured by an abundance of fresh, natural food. The shell of hatchlings receiving a deficient diet and too little sunlight will remain soft. Loveridge (1947) suggests that the shell malady may be prevented by keeping a lump of plaster of Paris in the water of aquatic species or mixing crushed chicken bones with the food of the terrestrial ones.

The results of the examination by Harwood (1932) of 64 Texas turtles representing 8 species show how commonly these reptiles are infested with helminths. Nematodes were found in one or more specimens of all the species, trematodes in 5. In 1941, Hughes, Baker, and Dawson published a list of 8 species of turtles from which cestodes have been taken. That helminths sometimes produce acute pathological conditions is shown by two papers, one by Smith, Coates, and Nigrelli (1941), the other by Smith and Coates (1939). The former describes a disease of the gall bladder associated with a fluke, the latter fibroepithelial tumors on a turtle of the same marine species, *Chelonia mydas*.

Aquatic turtles are frequently attacked by leeches, which adhere to the shell as well as to the skin and may apparently affect the health. Terrestrial species are sometimes infested with mites; Ewing (1926) found the common North American chigger (*Trombicula*) on the box turtle, *Terrapene carolina*. The larva of a bot fly, *Sarcophaga cistudinis*, develops under the skin and in the shell of terrestrial forms, such as species of *Terrapene*, and produces large swellings.

Enteritis is the most common cause of death in captive turtles.

Reproduction

All turtles are oviparous and deposit their eggs on a sunny bank or beach or, for that matter, any other available place exposed to the sun. The female buries her clutch in the earth or plant débris and leaves it to be hatched by solar heat or that of decomposition. The depth of the nest is determined by the length of the hind limb of the layer, a point to be considered in providing laying sites for gravid females. The forelimb is never used in digging the nest. The clutches vary greatly in size from species to species, some containing only a few eggs, others many score. The eggs are round to elongate in shape and have an unpigmented shell. Details of the laying habits of the species found in the United States and Canada are given by Pope (1939).

External sexual differences are present in many kinds of turtles. The most common and convenient difference to use in sexing is the relatively long tail of the male, the length taken as the distance from the vent to the tip of the tail. This character is especially marked in the musk and mud turtles (Kinosternidae) and in the sea turtles, but differences can be detected in nearly all kinds. The male musk and mud turtles have two patches of horny scales on the inner side of each hind limb. An interesting sexual character is found in the painted turtles (*Chrysemys*): the nails of the male's forelimb are much longer than those of the hind limb, and two to three times as long as any of those of the female. In some species of box turtles, the plastron of the male has a concavity that contrasts sharply with the flat to slightly convex plastron of the female.

Age determination is so difficult for most reptiles that the subject has not been dealt with in other groups. Such is not the case with the turtles that retain growth rings in the horny shields of the carapace. Each ring indicates a year of growth, and, if the series is complete, the original, relatively rough-surfaced shield can be seen occupying an off-center position. In individuals well past maturity the first shields are worn off, and age can no longer be told with accuracy. The box turtles (*Terrapene*), the gophers (*Gopherus*), and species of the large genus *Testudo* have shells with easily counted growth rings.

In the United States and Canada, turtles usually lay during June and early July. Incubation requires several weeks, and its duration is reduced by higher temperatures; in more northern regions, the hatching, or at least actual emergence from the nest, of some species, may be delayed until spring.

The most remarkable thing about turtle reproduction is the ability of the diamond-back "terrapins" (*Malaclemmys*) to lay fertile eggs for several years after a single period of copulating. The second season there is little diminution in percentage of fertility as compared with that of the season of mating; thereafter the percentage rapidly decreases but some evidence of fertility persists for four seasons. This fact was well established by the artificial propagation experiments carried on for many years at Beaufort, N. C. (Hildebrand, 1930, 1932). The biological implications have not been investigated, nor have the reproductive habits of other turtles been sufficiently studied to determine how widespread among turtles is this unusual attribute of the diamond-backs.

Cagle (1944) has described a technique for obtaining fertile turtle eggs without having to provide the female with a nesting place. The main difficulty is to determine just when the eggs are ready to be removed; their dissection from the oviducts after the female's plastron has been cut away is simple enough. The female is supported or held in a vertical position, and its abdomen palped from each side with the forefingers. Thus the eggs may be felt and their condition determined. If the species lays an egg with a flexible shell, the ripe egg will tend to resume its original shape after being indented by pressure. Hard-shelled eggs when mature cannot be indented.

CROCODILIANS

The recent crocodilians, a mere remnant of a glorious past going back to Cretaceous times, include but twenty-five distinct species all save two of which are tropical or subtropical; the American and Chinese alligators inhabit the temperate region. The Chinese alligator is on the verge of extinction. The crocodilians are descendants of Triassic, thecodont reptiles that also gave rise to the dinosaurs. The term "crocodilian" is used herein to include all members of the order Crocodilia or Loricata: alligators, caimans, gavials, and crocodiles.

Although the crocodilians are much too large ever to become favorite laboratory reptiles, their interesting structure, as well as their relationship to the dinosaurs, some of which reached gigantic proportions, makes them of special interest to students of evolution. All crocodilians are essentially aquatic and lay eggs, which they bury near the water in sand or vegetable débris. No breeding directions are included because of the impracticality of dealing with such a large creature in the laboratory. Crocodilians are rarely, if ever, bred in zoos; although not bred in zoos, they are always exhibited in them, sometimes in great numbers.

Handling and Transportation

Young alligators and crocodiles can be boxed and shipped by express without food and water much as other reptiles of similar size are shipped. Greater care must be taken in guarding against low temperatures, since crocodilians may lose their appetite after exposure to cold and refuse to eat again.

An alligator only a few feet long can be handled safely if the jaws are first seized and held shut. This method is even used by the experienced on large alligators, but, in the individuals with great weight, the tail is a formidable weapon that can easily knock the feet from under a man. Most crocodiles are more active and aggressive than alligators and must be handled with this in mind. A crocodilian in the water should never be allowed to seize a hand or leg because the reptile may revolve on its long axis and thus give the limb a terrific twist. Biting, striking with the tail, and twisting are the crocodilian's only means of inflicting damage.

Housing

Above all else, young crocodilians must be kept warm. The most suitable temperature is about 27° C. (80° F.), although an occasional drop to 24° C. (75° F.) will do no great harm, especially in winter.

These reptiles live well in a vivarium that is equally divided between dry land and water; the water must be deep enough to float the inmates. Each section should have a diameter equal to the length of the largest specimen. It is advisable to have the sun shine on enough of the dry area to allow voluntary sunbaths; sunlight from which there is no refuge can be highly dangerous. The vivarium should be kept clean. Especially troublesome in this regard are bits of unconsumed food.

Food and Feeding

Crocodilians are carnivorous; in a wild state they eat a great variety of animals from small invertebrates to large vertebrates. In captivity, a diet of liver, raw lean beef, or fish and fish cleanings will suffice. Cod-liver oil may be added if more vitamins seem to be needed. The cleanings do not have to be included more than two or three times a month.

Food should be given three or four times a week; the omission of a feeding once in a while will do no harm.

For very small individuals, the meat or fish should be cut into strips two to four inches long having the diameter of a pencil. A strip should then be dangled near the crocodilian's jaws and released as soon as it takes hold. About six of these strips will satisfy such a specimen. In contrast to most other reptiles, crocodilians will learn to come when called at meal times. Many very young individuals will not feed voluntarily. Food may be forced into them by gently prying the jaws open wide enough to allow a strip of meat to be pushed well into the throat by a blunt object the size of a pencil.

Health and Disease

Countless thousands of alligators have been shipped from the southeastern into the northern states as souvenirs; many of these eventually have found their way into aquaria and zoo reptile houses, where they almost invariably arrive in a bad state of health. The New York Aquarium is one of the unwilling victims of this traffic. Curator-aquarist Coates has written detailed directions for the care of baby alligators. He gives the following criteria for judging the infant alligator's state of health: one in good condition has a firm skin, a slight swelling at the base of the tail, and no teeth that are loose; the manner is alert, and any slight disturbance will cause it to hold the mouth ajar; the eyes are clear and follow movements of nearby objects; it should show some desire to move about when released. Sick specimens are inactive, but quick to snap; very ill ones will not exhibit even this much temper.

Healthy young alligators will grow about a foot a year under the best conditions.

Reese has devoted many years of work to the crocodilians. His early book (1915) gives a detailed account of their anatomy and development, with emphasis on the American alligator. In 1947, he published a bibliography of the group. The natural history of *Alligator mississipiensis* is described by McIlhenny (1935). Deraniyagala (1939) has written an account of the natural history of crocodilians, chiefly those of the Old World. References to the literature on this small group are thus made available, whereas those on the remaining reptile groups, especially the lizards and snakes, are far too scattered and numerous to be brought together here.

REFERENCES

Aronson, Joseph D   1929. Spontaneous tuberculosis in snakes. *Jour. Infect. Dis.*, vol. 44, pp. 215–223.

Blanchard, Frank N., and Frieda Cobb Blanchard. 1942. Mating of the garter snake *Thamnophis sirtalis sirtalis* (Linnaeus). *Pap. Mich. Acad. Science, Arts, Letters,* vol. 27, pp. 215–234.

Blanchard, Frank N., and Ethel B. Finster. 1933. A method of marking living snakes for future recognition, with a discussion of some problems and results. *Ecology,* vol. 14, pp. 334–347.

Bogert, Charles M., and Raymond B. Cowles. 1947. Results of the Archbold Expeditions. No. 58. Moisture loss in relation to habitat selection in some Floridian reptiles. *Amer. Mus. Nov.* 1358, pp. 1–34.

Bruner, Henry L. 1907. On the cephalic veins and sinuses of reptiles, with description of a mechanism for raising the venous blood-pressure in the head. *Amer. Jour. Anat.,* vol. 7, pp. 1–117.

Burt, Charles E. 1935. Contributions to Texan herpetology. II. Some observations and an experiment on the worm snake, *Leptotyphlops.* .*Ecology,* vol. 16, pp. 530–531.

Burtscher, J. 1932. Mouth-rot in snakes. *Bull. Antiven. Inst. Amer.,* vol. 5, pp. 59–65.

Cagle, Fred R. 1939. A system of marking turtles for future identification. *Copeia,* 1939, pp. 170–173.

——. 1944. A technique for obtaining turtle eggs for study. *Copeia,* 1944, p. 60.

Clark, Hugh. 1937. Embryonic series in snakes. *Science,* n.s., vol. 85, pp. 569–570.

Coates, Christopher W. 1941. The care of turtles and small alligators. *Bull. N. Y. Zool. Soc.,* vol. 44, pp. 107–113. (Reprinted by New York Zool. Soc. 11 pp.)

Conant, Roger. 1948. Regeneration of clipped subcaudal scales in a pilot black snake. *Nat. Hist. Misc.* 13, pp. 1–2.

Conant, Roger, and R. Marlin Perkins. 1931. This mite question. *Bull. Antiven. Inst. Amer.,* vol. 5, pp. 36–39.

Cowles, Raymond B., and Charles M. Bogert. 1944. A preliminary study of the thermal requirements of desert reptiles. *Bull. Amer. Mus. Nat. Hist.,* vol. 83, pp. 261–296.

Davis, D. Dwight. 1936. Courtship and mating behavior in snakes. *Field Mus. Nat. Hist. Zool. Ser.,* vol. 20, pp. 257–290.

Deraniyagala, P. E. P. 1939. *The tetrapod reptiles of Ceylon.* Vol. I of *Testudinates and crocodilians.* Colombo Mus., Ceylon.

Ewing, H. E. 1926. The common box-turtle, a natural host for chiggers. *Proc. Biol. Soc. Wash.,* vol. 39, pp. 19–20.

Fowler, J. A. 1941. Water snakes as a source of flukes for the laboratory. *Turtox News,* vol. 19, p. 105.

Franklin, Malcolm A. 1945. The embryonic appearance of centres of ossification in bone of snakes. *Copeia,* 1945, pp. 68–72.

Frings, Hubert. 1941. Rearing blowflies in the laboratory. *Jour. Econ. Entomol.,* vol. 34, p. 317.

Geiman, Quentin M., and Herbert L. Ratcliffe. 1936. Morphology and life-cycle of an amoeba producing amoebiasis in reptiles. *Parasitol.,* vol. 28, pp. 208–228.

Glidden, H. Spencer. 1936. Diseases of reptiles. *Fla. Rep. Inst. Publ.,* pp. 1–7.

Greenberg, B., and G. K. Noble. 1944. Social behavior of the American chameleon (*Anolis carolinensis Voight*). *Physiol. Zool.,* vol. 17, pp. 392–439.

Hansen, Ira B. 1941. The breathing mechanism of turtles. *Science,* vol. 94, p. 64.

Harwood, Paul D. 1932. The helminths parasitic in the amphibia and reptilia of Houston, Texas, and vicinity. *Proc. U. S. Nat. Mus.,* vol. 81, Art. 17, pp. 1–71.

Hildebrand, Samuel F. 1930. Review of experiments on artificial culture of diamond-back terrapins. *Bull. U. S. Bur. Fish.,* vol. 45, pp. 23–70.

——. 1932. Growth of diamond-back terrapins. Size attained, sex ratio, and longevity. *Zoologica,* vol. 9, pp. 551–563.

Hoopes, Isabel. 1936. Reptiles in the home zoo. *New England Mus. Nat. Hist. Spec. Pub.* 1, pp. 1–64.

Hughes, R. Chester, John R. Baker, and C. Benton Dawson. 1941. The tapeworms of reptiles, 2. Host catalogue. *Wasmann Collector,* vol. 4, pp. 97–104.

Hydak, Mykola H. 1936. Is wax a necessary constituent of the diet of wax moth larvae? *Annals Entomol. Soc. Amer.,* vol. 29, pp. 581–588.

Jaros, Duval B. 1940. Occlusion of the venom duct of Crotalidae by electro-coagulation: an innovation in operative technique. *Zoologica,* vol. 25, pp. 49–51.

Lawrence, R. F. 1935. The prostigmatic mites of South African lizards. *Parasitol.,* vol. 27, pp. 1–45.

Lovell, Reginald. 1930. The bacteriological findings in certain fatal cases of enteritis occurring in the Gardens during 1928. *Proc. Zool. Soc. London,* 1929, pp. 623–632.

Loveridge, Arthur. 1926. A mite pocket in the gecko. *Proc. Zool. Soc. London,* 1925, p. 1431.

——. 1947. Bone-making material for turtles. *Copeia,* 1947, p. 136.

Lowe, Charles H., Jr. 1943. An improved method of snake feeding. *Copeia,* 1943, p. 58.

McIlhenny, E. A. 1935. *The alligator's life history.* The Christopher Pub. House, Boston.

Mattlin, Robert H. 1947. Mite control in reptile collections. *Herpetologica,* vol. 3, p. 172.

Mueller, Justus F. 1937. Spargana in Natrix. *Science,* n.s., vol. 85, pp. 519–520.

Noble, G. K. 1937. The sense organs involved in the courtship of *Storeria, Thamnophis,* and other snakes. *Bull. Amer. Mus. Nat. Hist.,* vol. 73, pp. 673–725.

Noble, G. K., and H. T. Bradley. 1933. The mating behavior of lizards; its bearing on the theory of sexual selection. *Annals N. Y. Acad. Science,* vol. 35, pp. 25–100.

Noble, G. K., and E. R. Mason. 1933. Experiments on the brooding habits of the lizards *Eumeces* and *Ophisaurus. Amer. Mus. Nov.* 619, pp. 1–29.

Pearse, A. S., S. Lepkovsky, and Laura Hintze. 1925. The growth and chemical composition of three species of turtles fed on rations of pure foods. *Jour. Morph. Physiol.,* vol. 41, pp. 191–216.

Pope, Clifford H. 1939. *Turtles of the United States and Canada.* Alfred A. Knopf, New York.

Ratcliffe, Herbert L. 1941. Nutrition. *Rep. Penrose Res. Lab.,* pp. 29–37.

——. 1942. Diets for carnivorous lizards. *Rep. Penrose Res. Lab.,* pp. 26–27.

Reese, Albert M. 1915. *The alligator and its allies.* G. P. Putnam's Sons, New York.

——. 1947. Bibliography of the Crocodilia. *Herpetologica,* vol. 4, pp. 43–54.

Richmond, Neil D. 1945. The habits of the rainbow snake in Virginia. *Copeia,* 1945, pp. 28–30.

Schmidt, Karl P., and D. Dwight Davis. 1941. *Field book of snakes of the United States and Canada.* G. P. Putnam's Sons, New York.

Schroeder, C. R. 1939. Report of the hospital and laboratory of the New York Zoological Park, 1938. Mortality statistics of the society's collection. *Zoologica,* vol. 24, pp. 265–283.

Smith, G. M., and C. W. Coates. 1939. The occurrence of trematode ova, *Hapalotrema constrictum* (Leared), in fibro-epithelial tumors of the marine turtle *Chelonia mydas* (Linnaeus). *Zoologica,* vol. 24, pp. 379–382.

Smith, G. M., C. W. Coates, and R. F. Nigrelli. 1941. A papillomatous disease of the gallbladder associated with infection by flukes, occurring in the marine turtle, *Chelonia mydas* (Linnaeus). *Zoologica,* vol. 26, pp. 13–16.

Smith, Hobart M. 1946. *Handbook of lizards of the United States and Canada.* Comstock Pub. Co., Ithaca.

Snedigar, Robert. 1939. *Our small native animals, their habits and care.* Random House, New York.

Stemmler-Morath, Carl. 1938. Das Halten und Züchten von Schlangen. *Handbuch der biologischen Arbeitsmethoden,* Abt. 9, pp. 729–768.

Thompson, Paul E., and Clay G. Huff. 1944. Saurian malarial parasites of the United States and Mexico. *Jour. Infect. Dis.,* vol. 74, pp. 68–79.

Walker, Ernest P. 1942. Care of captive animals. *Smithsonian Rept.,* 1941, pp. 305–366.

Weekes, H. Claire. 1935. A review of placentation among reptiles with particular regard to the function and evolution of the placenta. *Proc. Zool. Soc. London,* 1935, pp. 625–646.

13. AMPHIBIA

R. CRANFORD HUTCHINSON

Daniel Baugh Institute of Anatomy
Jefferson Medical College
Philadelphia, Pennsylvania

Amphibia are commonly used in laboratories for three purposes. Their primary use is as an inexpensive and easily procured example of a simple vertebrate which serves to acquaint the beginning biology student with the main features of vertebrate anatomy. Frogs are the most common form used for this purpose though some teachers feel that mud puppies or axolotls are much better examples of a "generalized" vertebrate than are the more specialized anura.

Amphibia may also be used to introduce the biologist or medical student to the study of physiology. Ever since the accidental discovery of a nerve-muscle preparation by Galvani and the first observation of blood flowing through capillaries by Malpighi, frogs have been used to demonstrate these phenomena in laboratories the world around.

The third scientific use to which amphibia are well suited is that of providing abundant embryonic material which may be directly observed or experimented upon from the time of fertilization until the embryo becomes a free-swimming larva. This often requires special care for prospective breeding adults and then care of the developing offspring.

It is possible that the currently popular "male frog test" for pregnancy (Galli Mainini, 1948; Wiltberger and Miller, 1948) may become firmly established, in which case a fourth use would have to be added to those listed above. At the present writing, it is too early to determine whether this test can be considered thoroughly reliable, and further work is necessary in standardizing such variable factors as size, species, seasons, etc.

Most schools procure the material needed for morphological or physiological studies just before it is to be used, so no particular problems arise in connection with the care of such animals. Securing examples of all the early developmental stages, however, often presents difficulties. If one wants living embryonic material merely for classroom work, the easiest procedure is to arrange that part of the laboratory schedule to correspond with the local spring breeding season and then collect egg masses of frogs or of salamanders as needed. The season usually starts with the rains that follow the spring thaw, and, in the east, this is heralded by the bell-like peeping of *Hyla crucifer*, the spring peeper.

The first common amphibia (in the east) to spawn are the wood frog (*Rana sylvatica*) and the spotted salamander (*Amblystoma punctatum*). In regions where the tiger salamander (*A. tigrinum*) is present, it spawns a little earlier than the two forms just mentioned. A short while after these forms have spawned, one may expect to find egg masses of the leopard frog (*R. pipiens*) and then eggs of the pickerel frog (*R. palustris*). After these eggs have had a chance to hatch and start on their careers as tadpoles, one may find egg rafts of the green frog (*R. clamitans*) and finally in the late spring huge masses of eggs from the bullfrog (*R. catesbiana*).

If local ponds and swamps are watched carefully for one season, keeping records of dates when eggs are found and the general weather conditions, it will be possible to guess pretty closely in subsequent seasons when one may expect to find eggs. Very cold weather delays the progress of the season, while unseasonably warm weather may speed it up. A soaking warm rain at night will almost always stimulate spawning if it occurs within a week before or after the average spawning date. Hence, the best time to collect mating adults is on rainy nights, and the best time to find eggs in the one-cell stage is on mornings after a night of rain. Most of these early breeders collected during their spawning season will spawn when placed in suitable laboratory aquaria, but after they have done this they should be returned to their native habitat, for they rarely continue to live well in captivity.

If one wishes to have live amphibian eggs at certain definite dates, regardless of the season of the year, one must resort to artificially induced spawnings or maintain a colony of animals which may be relied on to mate when properly stimulated. The first method is much easier and less expensive.

Wolf (1929) was the first to report the effects of pituitary stimulation on reproduction in frogs, and this work was followed by similar studies by several investigators. Rugh (1934, 1935) reports techniques whereby anuran eggs may be secured at any season of the year by means of proper pituitary stimulation. All that is needed is a supply of gravid frogs and a cool place (10 to 12° C.) in which to keep them until such times as the eggs are desired. For details of techniques, see papers by Rugh.

Frog's eggs are admirably suited for classroom demonstrations of living vertebrate development, and they are also useful for many types of experimental procedures. However, many investigators in the field of experimental embryology prefer to use salamander eggs because both the eggs and their individual cells are larger in size. Salamanders cannot be secured commercially as easily as frogs, so one cannot apply the same technique to them for procuring eggs. In many localities, however, some species of *Triturus* may be found occurring in large enough numbers to make them a suitable source of embryonic material. If females can be collected before they have spawned, they will usually deposit eggs shortly after being brought into the laboratory. When these animals are kept at room temperature for a period of time they no longer spawn spontaneously, and, if spawning is induced by pituitary stimulation, a large percentage of the eggs thus secured may develop abnormally. Normal spawning may be postponed by refrigerating freshly caught animals and holding them at low temperatures until the eggs are desired. As far as the author knows, these difficulties are encountered with all our native urodeles. The Japanese newt (*Triturus pyrrhogaster*) and the Spanish salamander (*Pleurodeles waltl*), however, both seem to adapt themselves to laboratory life and may remain healthy and produce viable eggs for several seasons.

When one has no outdoor source from which to collect material, an alternative solution is to maintain a colony of breeding adults, but this has the disadvantages of taking up space, requiring constant care, and necessitating the expense of maintenance. Also there is the problem of finding species amenable to laboratory conditions. Let us consider this last point first.

The salamander whose eggs have been most commonly used for experimental purposes, *Amblystoma punctatum*, is rather difficult to raise in the laboratory. It requires a good deal of personal attention and patience in feeding and hence becomes too expensive to raise except in certain special cases (albinos, etc.). Also, it will not spawn in the laboratory unless specially treated.

Amblystoma tigrinum is much more amenable to laboratory conditions and learns to accept food when it is offered. At the Morris Biological Farm of The Wistar Institute, the author demonstrated, to his own satisfaction, that it would be possible to maintain a colony of *A. tigrinum* with very little personal work, providing one had the rather unusual facilities there available. The scheme, which was carried out for three consecutive seasons, was to place the salamanders, after they had spawned in the spring, in a large enclosure which simulated their natural swamp-wood habitat. This enclosure, surrounded by deeply sunken cement walls and covered over with wire netting, served to prevent the animals from escaping and, at the same time, kept out their natural enemies such as snakes and birds of prey. In the fall these animals were collected and put in a cold room (36 to 42° F.) to hibernate. When eggs were desired a few animals were removed and placed in a spawning tank, where eggs were usually deposited during the third night after removal of the animals from the cold room. In this way it was possible to get eggs when desired, and eggs were obtained as late as June, though the normal breeding season for *A. tigrinum* in this part of the east is about the end of February.

The urodele of choice to live under ordinary laboratory conditions seems to be the Mexican axolotl, which has been used in laboratories ever since the time of the early Spanish conquerors, who are said to have introduced it into Europe. This neotonous form is like an overgrown *A. tigrinum* larva. It adapts itself easily to varied conditions and may be induced to spawn quite readily. A single female usually lays several hundred eggs, so a few breeding adults can produce an abundance of embryonic material. A "sport" white form of this animal appeared and was widely cultivated in the European laboratories. These two varieties have been widely used by experimentalists for following the fate of embryonic transplants, white grafts on pigmented embryos and vice versa.

The number of axolotls which one needs to keep as a colony is best determined by individual needs. In general, one should have two or three pairs of adults for every planned spawning. Fewer animals per successful spawning will be needed as the size of the colony is increased because some of the animals, particularly males, may be used several times in one season. A colony of any size may be obtained by securing just one pair of breeding adults and then raising the desired number of their offspring, though it may take two years for the axolotl to reach sexual maturity. The colony is best maintained by keeping a few larvae to raise to adults each year. If a colony stems from a few original breeders, it is a good procedure to exchange with other laboratories such larvae as are to be raised to maturity. In this manner one avoids too close inbreeding, which is a dangerous practice.

The Mexican axolotl will live in practically any type of water-holding container, exhibits a wide range of temperature toleration, and will even survive long periods of neglect. It will live in a space a little larger than itself providing the water does not become fouled, though the animals do not grow when kept in cramped quarters. In the author's experience about six adults to an aquarium (3 by 1 by 1 feet) was a satisfactory arrangement for feeding and for keeping the animals in a good state of health. When one finds an animal whose limbs or gills have been bitten, it is well to suspect that the animals are either too crowded or are not receiving adequate food.

Some people like to have their aquaria fixed with a sand-gravel bottom, aquatic plants, snails, etc., and this is all right provided one has the time to hand-feed the number of animals he wishes to keep. However, with a large number of animals, the time consumed in feeding them and in cleaning the tanks can be greatly reduced by keeping the tank bottoms bare. Axototl are carnivorous and eat bits of raw meat, earthworms, etc. Liver seems to be the easiest and best all-round food to use in keeping either a colony or a few animals in a good state of health. The animals will snap at small pieces of meat waved in front of their noses from the end of a pair of forceps, or they can be trained to pick up pieces of meat from the bottom of their aquaria. The latter method is much quicker than the former and is the reason for keeping the bottoms of the aquaria clean. If there is sand or gravel in the aquaria, the animals, in their clumsy lunging for the meat, often miss and get stones or gravel in their mouths instead and then stop feeding.

Cutting up liver for feeding may be a tedious task. If only a few animals are to be fed, one simple method is to hold a piece of meat and snip off pieces as needed with a pair of scissors. If there are many animals to feed, the easiest procedure is to have as much liver as will be needed for feeding, frozen solid in a block ahead of time. The meat can then be thinly sliced and quickly cut into pieces of any desired size while it remains frozen. The cut-up pieces should be put in a bowl, flooded with water, and stirred, and the water then poured off. This process should be repeated two or three times or until the water becomes only slightly cloudy when the meat is stirred. (This rinse water should be saved and used for food for *Daphnia* cultures.) If the meat is not washed before feeding, the water in which the animals are kept is likely to become cloudy and then foul and will have to be changed. By washing the meat ahead of time, it is possible to keep animals in the same water for weeks or months. The meat must not be washed too much, however, for if it is it will not stimulate the animals to feed and they will pay no attention to it.

The frequency of feeding may be regulated by the amount of help availiable for such procedures and by the temperature. At cool temperatures feeding every other day or even less frequently seems to work well. However, with animals that are still growing or adults kept at room temperatures, daily feedings seem to produce the best results. Animals can be maintained with less frequent feedings, but when the interval between feedings is long, the animals naturally become quite hungry. This often results in the larger animals biting the limbs or the gills of the smaller ones. Another result is that when they are fed, certain animals will eat more than they should and a short time after feeding will regurgitate all they have taken, fouling the water as well as losing their nourishment.

The animals start feeding as soon as the meat has been put in their tanks, so it might seem possible to do the necessary daily cleaning almost immediately after feeding. However, it is better to postpone this for an hour or so because some food may be regurgitated and feeding sometimes stimulates defecation. Axolotl feces are passed in the form of pellets, and if these can be removed from the aquarium before they become broken up by the movement of the animals in the tanks, it helps to keep the water clean. When animals are kept in bowls or small aquaria, a Banta-type pipette with a large rubber bulb is one of the best means of removing uneaten food and fecal matter. Where aquaria or tanks are used, a rubber hose, employed as a siphon, will prove to be a satisfactory method. Since some water is lost every time the tanks are cleaned, it is necessary to add some fresh water from time to time. Whenever the water of a container becomes foul smelling or cloudy, it should be renewed and the container rinsed and then refilled with fresh water.

Perhaps a parenthetic note should be inserted here to point out that salamanders, like fish, are susceptible to the effects of "conditioned" and "unconditioned" water in the sense used by Allee and his students. These effects are most noticeable with young growing larvae and should be taken into account by anyone attempting to raise these animals. From a practical point of view, it may be said that fresh tap or spring water ("unconditioned") is slightly toxic, and such water which has stood 24 to 48 hours is less toxic. Water in which other animals have lived for a short time ("conditioned") is non-toxic (Hutchinson, 1939).

A further word might be added for those who may be raising these animals for the first time. If your laboratory is in a large building or if the water supply comes a long distance through pipes, do not change the water on any of the animals early in the morning, for water which has stood for a long time in pipes is likely to be toxic. Wait until late in the day when the pipes have been well flushed out. If the tap water appears to be toxic at times anyway, the safest procedure is to "physiologically test" the water by filling a large container with all the water you may need for a day or two. Place a small test animal in this water; if it still appears healthy after 24 hours, you know that particular water may be safely used and is non-toxic. When one attempts to raise a colony by feeding the larvae on *Daphnia*, the toxic effect of fresh "unconditioned" water becomes at once apparent, for when *Daphnia* are put in a bowl or aquarium filled with fresh water, they promptly die before the animals in the bowl have a chance to eat them. It has been discovered empirically that fresh water may be conditioned so as to be non-toxic to the *Daphnia* by adding a few drops of "liver water" to each bowl—"liver water" being the cloudy emulsion obtained when a small amount of ground-up liver is stirred into some water.

In maintaining a large colony it is advisable to keep each tank or aquarium isolated from every other or to keep groups of tanks containing similar animals separate from other groups. By doing this, if an infection should appear in one aquarium, it would not immediately be spread to the rest of the colony and might be checked. To maintain isolation, a set of hoses, nets, pipettes, or any other instruments which are dipped into the aquaria should be kept with each aquarium or group of aquaria and used only in that one place. The person who feeds the animals must be trained not to dip his hands into the water when putting in the cut-up meat. The washed meat tends to clump; in order to break this up and still not get one's hands into the aquarium water, it is necessary to throw the meat with a swift flick into the aquarium, so that when the meat hits the surface of the water, the force of the impact breaks the clump into its small component pieces.

When a death occurs in the colony, the animal should be autopsied at once, and, if any internal lesion or pathology is found, the other animals in the same container should be sacrificed, everything associated with the group sterilized or destroyed, and adjoining groups carefully watched. However, if no pathology is found, the aquarium should be cleaned and the remaining members of the group closely watched. If other deaths follow shortly in the same group, it is best to play safe and sacrifice the remaining animals, even if the cause of death remains obscure.

It is extremely difficult to maintain isolation between groups which are placed close together. Visitors come in and dip their hands in different tanks or get the nets mixed up. And it is difficult to remember to wash one's hands every time when going from one group to another while working with the animals.

In raising one's own colony, the danger of too close inbreeding is most evident when an infection occurs. Related animals seem highly susceptible to the same disorders, while a different stock may remain unaffected even though exposed. For this reason in a large colony it is well to have animals from diversified origins. The best source of a healthy stock is probably animals collected from the wild in Mexico.

There are numerous descriptions of techniques for inducing spawning in the Mexican axolotl. Perhaps the best and most recent of these is the paper by Humphrey (1946), which also gives a useful description of methods for the care of adults and larvae. The animals sometime spawn spontaneously without any obvious external stimulus, or they may respond to any one of several applied stimuli. But since the usual purpose in maintaining an axolotl colony is to obtain eggs whenever they are desired, each person must determine which techniques most frequently produce the desired results within the limitations of the materials and equipment that are available.

Today one naturally thinks of pituitary stimulation as the easiest method of producing spawning, but unfortunately this does not produce the desired results with axolotls. These animals are quite sensitive to pituitary stimulus and will discharge their sexual products after injection with amounts similar to or less than those used with frogs. However, although stimulated males and females are in the same tank and discharge their products at the same time or with the males preceding the females, the eggs produced are infertile. This method is useful in cases where artificial insemination is desired, for the females may be injected, and then, after eggs first appear at the cloaca, they may be "milked" (i.e., squeezed from the female's abdomen) from time to time directly into a sperm suspension; fertilization can then take place without the sacrifice of the female. But for true spawning some other technique must be used.

In the various methods proposed to induce spawning, the one common feature seems to be change. This may involve just one factor, such as temperature, or several, such as temperature, size of container, and amount of light.

When one wishes to get eggs on any particular date, he has to start preparations well in advance. The first step is to segregate by sexes the animals to be used, keeping the males together in one tank and the females in another.

The two sexes are easily distinguishable, for the males, besides tending to be long and slender, have an elongated swollen lip on either side of the cloacal opening. The gravid female has a plumpish abdomen, and there is no swelling about the cloacal vent, which is smooth. Sometimes in a female just ready to spawn one can detect a slight swelling of the cloacal lips. In selecting animals to spawn one should always select those females which appear to be most gravid.

About two weeks before spawning is to be attempted, the animals are brought into a warm room, if they have been previously kept cool, and are fed well daily. They may be kept in rather cramped or crowded quarters for the period, and their water should be kept clean but not changed. Three days before the eggs are desired the animals should be placed together in the largest tank available, and the water should be kept cold, if possible 10° C. ± 2 or 3°. It is best to introduce the animals to the spawning tank in the late afternoon or evening and then to keep them as free as possible from disturbing influences during their spawning period. Do not feed them. Usually on the first morning spermatophores will be seen on the floor of the tank, and on the third night the females usually start depositing their eggs.

The technique just described was found by the author to work well at the Morris Biological Farm. In a different laboratory with different facilities, the technique would have to be modified. If the breeding animals are kept at warmer temperatures, spawning may occur shortly after the animals are put together. One element of change which should always be incorporated is to keep the animals in well-conditioned water before the attempt to induce spawning and then to have the spawning tank filled with fresh unconditioned water. This one change by itself will often induce spawning.

The females will deposit their eggs on any aquatic plants or light twigs which may be furnished and thus make the collection of eggs much easier than when the eggs are deposited on the walls or floors of the tanks. If the eggs are to be used for experimental procedures, it is wise to take care in handling them, particularly if they are just beginning to segment or to gastrulate. The eggs seem to be particularly sensitive at these stages of development, and rough handling often results in atypical development. By rough handling is meant such procedures as lifting up a stem to which eggs are attached in order to transfer the eggs from one bowl to another or trying to remove the eggs from their jelly. Such procedures as these usually cause no apparent ill effects to the eggs, but if done at these sensitive stages may cause the subsequent development to be slightly atypical, or in some cases markedly so.

When one wishes to start an axolotl colony by raising his own animals, the basic problem is not one of procuring the embryos, which is relatively easy, but rather of securing an adequate food supply with which to raise the larvae. The fully developed embryo will snap only at living food, and since his length is only 10 to 12 mm., the food must be small. It is not within the province of this chapter to go into details on the various possible types of food and methods for raising them. However, a few words may be said about the more common techniques.

The food that is selected to be raised will probably depend on the space and facilities available. Where these are ample, *Daphnia* would probably be the food of choice, but where space and equipment are limited, *Enchytraeus*, white annelid worms, make an easily raised source of food. These may be raised in finely pulverized humus or dirt and kept in any type of container—covered boxes or dishpans which may be stacked in piles to save space work very well. When some worms have been obtained to start a culture, any one of various foods may be added to keep it going. Cooked oatmeal seems to be as satisfactory and as easy to supply as any. Small lumps of the food should be introduced in different spots throughout the culture and covered with dirt. The culture should be kept cool, dark, and moist (but not wet), and in a few days clumps of worms will be found at the points where the food was. This is the time to collect the worms which are to be fed to the salamanders, as they may be removed in clumps. The culture should then be stirred to redistribute the worms which are left, and fresh food introduced. Several rich cultures should be obtained before attempting to raise many axolotls. Axolotl larvae grow rapidly at room temperatures, and as they increase in size, their food requirements increase greatly. Hence a time soon comes when the worm cultures will be depleted unless there is a sufficient number of cultures available so that they can be used in rotation with sufficient time interval between uses to allow the cultures to replenish themselves. It is also an advantage to have many cultures when the animals first begin to feed, for at this time it is necessary to start the axolotls with only the tiniest of worms, and sometimes even these have to be cut into smaller pieces with scissors to make them small enough for the animals to handle. If one has several cultures it is usually possible to find one which will have clumps of these tiny worms, and this saves much time which will otherwise have to be spent hunting through cultures for single small worms.

The literature on raising *Daphnia* is probably more extensive than that on raising amphibia, so one can easily find numerous detailed descriptions for preparing *Daphnia* cultures. In the author's opinion it would be unwise to contemplate raising an axolotl colony, with *Daphnia* as the main source of food, unless one had tank space for the *Daphnia* which would be, at least, the equivalent of a small room. Numerous small cultures of *Daphnia* in tanks or tubs are extremely useful, particularly in starting a colony, when the tiny larvae need an abundance of small food. However, when the larvae begin to get large and demand quantities of food daily, small *Daphnia* cultures soon become depleted and are unable to replenish themselves rapidly enough to keep up with the appetites of the axolotls. However, large *Daphnia* cultures under optimum conditions can produce unbelievable quantities. The author has been able to raise more than 1000 larvae at one time from beginning feeding to sizes of approximately 100 mm. exclusively on *Daphnia* obtained from two large tanks patterned after those found in some of our state fish hatcheries. This, however, would be possible only under favorable circumstances, though it is a simple matter to raise 100 to 200 larvae with large *Daphnia* cultures to draw on.

A large *Daphnia* culture is a complex microcosm, and the proper balance which continuously favors production must be maintained; otherwise the number of *Daphnia* will start to fall off or some other form will become dominant and the culture ruined. The author has tried and successfully used various of the techniques given in the literature: sheep manure, lettuce, hay emulsions, etc. However, the method by which the best results were obtained consists of making an emulsion of ground-up liver and then distributing this sparsely throughout the culture. Only small amounts of liver should be added at any one time, and one has to learn by experience just when to add more liver and how much to add. Too much liver will foul the water and spoil the culture, while too little does not create the proper balance favoring *Daphnia* production. Plenty of sunlight may be a necessary factor in making this liver technique work.

In beginning a culture, start with as little water as possible, i.e., 1-inch depth. When *Daphnia* are well established in this, add more water once or twice a week until the tank is full. It is very difficult to start a culture in a large body of water. When beginning, obtain a sample, preferably *Daphnia magna*, from another laboratory or a supply house. Do not use material collected from the wild, for some contamination, inadvertently introduced, may later be extremely difficult or impossible to eradicate. (See chapter 14, section on Live Foods, for other methods.)

Tubifex, Artemia eggs which may be hatched, and sometimes *Daphnia* or other live foods may be obtained commercially when relatively small amounts are needed or when ample funds are available.

The standard laboratory practice of raising young salamander larvae individually in finger bowls has much to commend it. This concentrates the food in the vicinity of the animal where he can easily find it and assures each animal of a fair share. It furthermore protects the slower-growing individuals from being bitten by those which are larger. These points indicate the disadvantages of trying to raise large groups of small larvae in aquaria or large tanks. When *Enchytraeus* are used as food, the worms often clump in balls, making it difficult for the animals to feed on them, and even *Daphnia* tend to collect in one corner.

Before deciding to start a colony of axolotls, careful consideration should be given to the advantages offered by a similar colony of the African horned toad, *Xenopus laevis*. This form, like the axolotl, has been a laboratory pet of long standing, but lately it has received attention from the clinicians and from the commercial laboratories because its extreme sensitivity to pituitary hormones makes it a possible pregnancy-test animal. This use was pointed out some years ago by Crew (1939), but the animals have not been commonly used for this purpose, at least in this country. Weisman (1942), among others, suggested that *X. laevis* offered many advantages over other forms as a pregnancy-test animal. Later the Camerons (1947) showed that these toads may be quite easily raised in large numbers and that they offer some outstanding advantages for the biologist as well as for the clinician. They find that fertile eggs may be obtained from adult toads at any season of the year by injection of a pituitary-like hormone.

The main difficulty in starting an axolotl colony is in procuring an adequate supply of live food for the growing larvae, but this would not obtain with *X. laevis* since it is easy to feed during this period of development, growing well on a diet of plankton and ground-up liver. The animals metamorphose when they are about 12 weeks old but remain aquatic throughout their lives. They may be fed underwater like axolotls, though they feed in a different manner, pushing the food into their mouths with their forefeet. The adults are air breathers and are best kept in aquaria with water about 3 inches deep, since this enables the animals to rest with their feet on the bottom and their noses above the surface of the water. It is possible that the male of this form may prove to be the most satisfactory amphibian for pregnancy tests because its breeding habits seem less rigidly seasonal than those of most of our common anura. This appears to be an important factor in the reliability of this test which is currently in vogue.

Unfortunately, the author has had no experience with the young stages of this form and so is unable to state whether or not the eggs and embryos are well suited for use in embryological experiments.

In a brief survey of this sort many things have to be omitted. Amphibian embryos, larvae, and adults should, in the future, be used in a much wider range of experimental procedures than they are at present. Further field work may reveal that there are unknown forms admirably suited for certain experiments, as was discovered by Twitty (1935). Continued study of the large numbers of amphibia available is bound to lead to important findings. With the use of new forms and with the development of new procedures, old techniques and methods will have to be modified and changed, but it is hoped that some of the suggestions indicated here may help in this process.

REFERENCES

Allee, W. C. 1934. Recent studies in mass physiology. *Biol. Rev.*, vol. 9, pp. 1–48.

Cameron, T. W. M., and Mrs. Cameron. 1947. The developmental stages and anatomy of *Xenopus laevis*, African horned toad. Abst., *Anat. Rec.*, vol. 97, p. 412.

Cameron, S. B. 1947. Successful breeding of *Xenopus laevis*, the South African clawed toad-frog. *Amer. Jour. Med. Tech.*, May.

Crew, F. A. E. 1939. Biological pregnancy diagnosis tests: a comparison of the rabbit, the mouse, and the "clawed toad." *Brit. Med. Jour.*, vol. 1, p. 766.

Galli Mainini, C. 1948. Pregnancy test with male *Batrachia*. *Endocrinology*, vol. 43, p. 349.

Humphrey, R. R. 1946. The Mexican axolotl as a laboratory animal. *Ward's Nat. Sci. Bull.*

Hutchinson, C. 1939. Some experimental conditions modifying the growth of amphibian larvae. *Jour. Exp. Zool.,* vol. 82, pp. 357–369.

Rugh, R. 1934. Induced ovulation and artificial fertilization in the frog. *Biol. Bull.,* vol. 66, pp. 22–29.

——. 1935. Pituitary-induced sexual reactions in the anura. *Biol. Bull.,* vol. 68, pp. 74–81.

Twitty, V. C. 1935. Two new species of *Triturus* from California. *Copeia,* July, no. 2.

Weisman, A. I., H. F. Snyder, and C. W. Coates. 1942. The frog test (*Xenopus laevis*) as a rapid diagnostic test for pregnancy. *Amer. Jour. Obstet. Gynec.,* vol. 43, p. 135.

Wiltberger, P. B., and D. F. Miller. 1948. The male frog, *Rana pipiens,* as a new test animal for early pregnancy. *Science,* vol. 107, p. 198.

Wolf, O. M. 1929. Effect of daily transplants of anterior lobe of pituitary on reproduction of frog (*Rana pipiens* Shreber). *Proc. Soc. Exp. Biol. Med.,* vol. 26, p. 692.

14· FISHES AS LABORATORY ANIMALS

MYRON GORDON

New York Zoological Society
New York, New York *

INTRODUCTION

Zoologists have demonstrated repeatedly that fishes possess behavior patterns, physiologies, and diseases, counterparts of which are found among warm-blooded animals, including man. Several species of fishes are now standard laboratory animals in many biological and medical research institutions, but until the advent of the tropical fish hobby, the goldfish was the only species of which domesticated strains were available. At present the experimentalist desiring to use fishes that will breed regularly in captivity can choose from a number of species belonging to six different families, and if he does not require self-reproducing strains, many more kinds are at his disposal.

The small fresh-water fishes are well suited to the laboratory. They require relatively little space; housing and equipment for their maintenance are not particularly costly. Those who have mastered the technique of aquiculture are convinced that of all vertebrate animals fishes are the easiest and cheapest to breed and to maintain. Compared with mammals or birds, they are the cleanest and least odoriferous of the laboratory animals.

Because these fishes are largely diurnal and non-secretive, their behavior can easily be observed in their aquarium microcosm. Here

* From the Genetics Laboratory of the New York Aquarium at the American Museum of Natural History, New York 24, New York. The work of this laboratory is supported by grants from the National Cancer Institute of the National Institutes of Health, United States Public Health Service, and aided by the facilities of the Department of Birds and of the Animal Behavior Laboratory of the American Museum of Natural History. The author is deeply indebted to Mr. James W. Atz, New York Aquarium, to Dr. Roberts Rugh, Columbia University, to Dr. L. R. Aronson, American Museum of Natural History, and to Dr. Ralph Gordon, College of the City of New York, for their help in the preparation of the manuscript.

their whole life cycle from birth to death can be studied, subject to the experimental conditions the observer wishes to introduce. The use of fishes in a variety of experiments in animal behavior has revealed the fundamental importance of these aquatic vertebrates as an aid in interpreting the behavior of other vertebrate animals, including man.

A survey of the available aquarium fishes will reveal an abundance and variety of types of experimental material. Fertilized eggs of aquarium species may be obtained for study and experimentation every month in the year. Their developmental rate is rapid, usually a few days. The gestation period of the viviparous species is usually less than a month, and for some of them data on graded steps in their embryonic development are available.

The sensitivity of fishes to chemicals in their environment has led to their use as assay animals in both toxicological and endocrinological work. Study of their diseases has proven fruitful, both for successful maintenance in captivity and for the light it sheds on comparative pathology. For example, Nigrelli (1938) and Lucké and Schlumberger (1942) have called attention to the similarity of the neoplasms of fishes and the higher vertebrates, including man.

Basic stocks of fresh-water aquarium fishes may be procured at the neighboring pet shops, the larger fish hatcheries, or well-established biological supply houses and laboratories. Our present-day domesticated fishes have come to us originally from many parts of the temperate and tropic areas of the world. For instance, the danios hail from India, the fightingfishes from Siam, medakas from Japan, mouthbreeders from the Gold Coast of Africa, guppies from Trinidad and northern South America, platyfish and swordtails from Mexico and Guatemala, and so on. Many of these and other species are of particular value in the biology schoolroom for observation and study of animal behavior, embryology, and the relationship of organisms to their environment (see Bullington, 1947, and Rugh, 1948).

Inbred lines are available in some species. The platyfish and swordtails are in their thirteenth inbred (brother-sister) generation in the Genetics Laboratory of the New York Aquarium, and Dr. C. P. Haskins of the Haskins Laboratory, New York City, has inbred the guppy by mass cultures for over twenty generations. Genetically uniform stocks of these three species and a number of their geographically different strains are available. Within these stocks there are representatives of many color patterns, some of which are autosomal in inheritance whereas others are sex-linked. In addition, some cancerous strains can be synthesized genetically at any time while others can be continued generation after generation. In some of these the position on the fish's body where the neoplasm will appear can be predicted (see Gordon, 1948a).

The small aquarium fishes are destined to be as useful to the fish culturist as the laboratory rat and guinea pig have been to the animal husbandryman. Many of the modern methods used in animal husbandry have been developed from the discoveries made with the smaller laboratory mammals. The breeding of rats, brother to sister, generation after generation, has taught us that loss of vigor does not necessarily follow. This closest type of inbreeding often intensifies desirable qualities and makes it possible to establish those good features in our standardized laboratory animals. We have also learned from genetic research that virility does not always result from the breeding of individuals belonging to two unrelated stocks. Indeed, outcrossing is generally practiced for the production of spontaneous cancers in fishes.

As yet fish breeders of our federal, state, and private hatcheries and fish farms have lagged far behind other animal husbandrymen in utilizing the available knowledge of genetics for the improvement of their stocks (see Rice, 1942, and *United States Yearbook of Agriculture* for 1936, 1937). A beginning has been made in the selective breeding of trout by Hayford and Embody (1930), Davis (1931), and Lewis (1944). This meager list of references practically covers the subject, an effort which is far too small compared with the importance and the promise of applying genetic technique to fish culture. The albino brook trout, of all our native cold-water species, is the only fish inbred to any degree.

Although no marine fish has ever been regularly bred in captivity, a few must be considered standard laboratory animals because of their suitability to experimentation and the extensive work done with them. Most noteworthy is the Atlantic killifish, *Fundulus*. Parker (1948), for instance, lists several hundred references to this species, concerning its pigmentation and nerve physiology alone. Such species must be obtained from the wild, from commercial collectors, or from some of the biological stations or supply houses. Their eggs are seasonably available, but work on successive generations has not so far been practicable.

The differences in the life requirements of various laboratory fishes are far greater than those for the dog and cat or the rat and rabbit. Compare the goldfish and the platyfish, for example. The former are relatively large animals (3 to 5 inches) requiring rather cool water, 60 to 70° F., large aquaria, and some running water for stimulating breeding behavior under laboratory conditions; their spawning time is seasonal; when they are adults a large part of their food may be of vegetable origin. The platyfishes are small (1 to 2 inches), require warm water, 70 to 80° F.; they may be kept in relatively small aquaria of a few gallons of quiet but well-conditioned water. In addition, platyfishes are viviparous, giving birth to living young every month in the year; the bulk of their food must be of animal origin.

Owing to the different requirements of the various laboratory fishes, this chapter will be divided into a number of short units. In presenting this material it was found impractical to write this chapter as a continuous narrative. A unit will describe the breeding behavior characteristic of a particular species or group of species. This knowledge is basic to proper management. Other units dealing with such matters as housing, water, temperature, nutrition, and aquatic plants are treated broadly, for the subject matter will apply in part to all species. At the end of each part devoted to the reproductive behavior of laboratory species a short introduction will be given to the work done with that species or group of species. The references to the literature chosen are by no means complete, but they will reveal in part the diversity of the researches made with the use of fishes. In this connection the work of Kosswig (1936b) should be consulted.

HOUSING

Aquarium

The unit container for fishes is the aquarium (fig. 1). The smaller aquaria are generally constructed of glass within a framework of metal bands of iron, aluminum, or stainless steel. Large aquaria, those of fifty or more gallons, require heavy frames of angle iron. The base is usually a slab of smooth, thin slate or plate glass. Thanks to the popular and steadily growing hobby of aquarium fish-keeping, there are many excellent aquarium-manufacturing companies in the United States. Many of them employ modern mass production methods so that aquaria even in these times of high prices are cheaper than they were twenty-five years ago. Aquaria bought from the established manufacturers are generally well made, and their expertness in workmanship cannot be easily duplicated by home or private

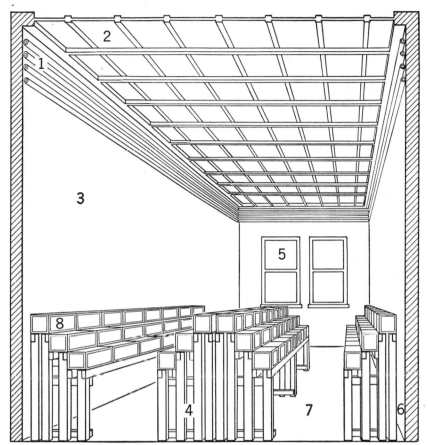

Fig. 1. Diagram representing interior view of the Genetics Laboratory of the New York Aquarium at the American Museum of Natural History. *1.* Steam-heating coils offset the down draft of cold air from skylights. Another set of steam coils controlled by thermostats is located on floor near windows. The floor steam coils are not shown here. *2.* Skylights are constructed of wired glass. Immediately beneath the skylights the laboratory has a system of shades, known as Ventilighter, for controlling the amount of illumination. The Ventilighters are not shown in this diagram. *3.* The sides of the laboratory are concrete. *4.* The stands for the aquaria are constructed from 2 by 4 inch lumber; for detail see figure 2. The central racks are placed back to back. *5.* Windows for ventilation are usually opaqued on the inside to exclude excess light. *6.* The racks near the walls are placed 4 inches away from the wall to avoid the colder concrete. *7.* The aisles between racks are 36 inches wide. The floor is of concrete and has a drain built into it. The laboratory has facilities for artificial illumination, compressed air, and a sink with running hot and cold water. *8.* The aquaria are arranged in three tiers. (Drawn by Donn Eric Rosen.)

laboratory methods (see Sources of Supply). Therefore this chapter will not contain directions for the making of aquaria. Nor will there be a description of the methods used for their repair, except those of a minor nature, since the time and energy spent in this rather specialized job are worth more than the cost of replacement.

All-glass aquaria, such as fish globes, battery jars, large food containers, holding about a gallon or more of water, may be used and are often quite useful in maintaining in isolation a small, warm-water fish for relatively short periods of time. Because of the limited volume of water they hold, small containers should be avoided as permanent or long-time housing units. Larger all-glass aquaria cannot withstand rapid changes in external temperatures and are likely to crack. Metal containers cannot be used because they quickly become poisonous to fishes.

Aquaria are made water-tight by a thin layer of aquarium cement between the glass or slate and the frame. This bond depends upon adhesion to both metal and non-metal and upon the relatively pliable cement between; the latter allows for differential expansion of metal and glass and for changing pressures when water is put into or removed from the tank. As the cement gets older it gradually loses this elasticity and becomes less able to withstand movement. It will also dry out if the tank is allowed to stand without water. No tank should ever be moved without first removing almost all of its water, but special care must be taken with old and large ones. Trying to move a tank heavy with water often distorts the frame or loosens the glass or bottom from the cement, causing it to leak at a later time. A tank should always be grasped by its frame alone; pressure should never be applied to its glass sides or its glass or slate bottom.

Slight leaks along the cemented edges may occasionally be sealed by pouring asphaltum varnish along the entire length of the cement. A crack in the glass in a small aquarium of less than three gallons may be temporarily fixed by applying a piece of commercial adhesive tape across it, but the aquarium must be absolutely dry before fixing (see Gordon, 1946c).

The aquarium in operation should always be covered by glass to keep dust and other foreign material out of the tank. A corner should be cut out of the glass cover for convenience in feeding, but the cut should be made as small as possible, for aquarium fishes are active jumpers. Many valuable specimens are lost in an uncovered aquarium or in one whose cover has too large an opening.

Stands for Aquaria

The stands supporting the aquaria must be sturdy. The true weight of an aquarium complete with water and gravel is seldom appreciated by non-aquarists. For example, consider an aquarium suitable for large goldfish or the large African mouthbreeder. Such a tank is

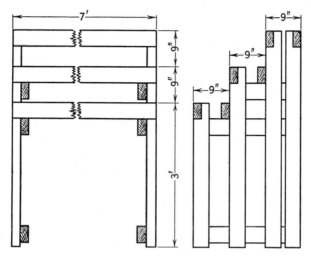

FIG. 2. Diagram of stand for a battery of 15 aquaria, each of which measures 16 inches long, 9 inches high, and 9 inches wide. To the right, the end view shows the arrangement of the 2 by 4 inch uprights and the ends of the 2 by 4 inch planks that serve as runners upon which the aquaria rest. To the left the front view shows the arrangement of the runners, the narrow 2-inch edge is placed uppermost. (Drawn by Donn Eric Rosen.)

about 3 feet long, a foot high and a foot wide, and when complete with water and gravel it weighs over 200 pounds. One gallon of water weighs about 8⅓ pounds; a cubic foot of water contains about 7½ gallons and weighs 62½ pounds.

If properly constructed, wooden stands made of "two by fours" are sufficiently strong to carry the heaviest aquarium or series of aquaria. One of the important points which needs emphasis is arrangement of the boards. The narrower surfaces of a pair of "two by fours" should be uppermost to serve as wooden rails. The aquarium placed upon the 2-inch rails will then straddle an open area, and this is desirable. Placing an aquarium on a flat wood surface should be avoided, for inevitably water will be spilled upon it. This, in turn, will warp the surface and create an uneven base. A plan for a properly constructed aquarium stand is shown in figure 2. This type of stand is designed for small aquaria (16 inches long) arranged in three tiers. Stands can readily be modified to hold larger tanks in any desired arrangement. Any deviation from the model shown in the figure should be made with consideration of the desirability of providing constant visibility and easy means of servicing.

Shape and Size of the Laboratory Aquarium

Shallow aquaria are preferable since they provide the greatest surface area for a given volume of water. Commercial aquaria are not likely to be the most efficient for laboratory purposes because usually they are higher than wide. The vitally necessary exchanges of gases which are dissolved within the water and those which are in the air take place at the critically important surface film. An appreciation of this fact should be one of the guiding principles in aquarium management (see also Aquatic Plants and Their Uses).

Mechanical water aerators are helpful in the maintenance of many species of fishes, but they do not aerate water directly. Little or none of the constant stream of air bubbles produced by mechanical aerators is soluble in water, and only soluble oxygen is of use to most fishes. The constant flow of air bubbles circulates the mass of aquarium water from the lowermost levels so that more of it reaches the surface film, where the exchange of gases takes place. In a way, these aerators substitute for wind and wave action which aid in adjusting the oxygen and carbon dioxide balances in natural ponds and lakes [for a survey of the conditions of existence of aquatic animals, see Shelford (1918)].

In order to attain the greatest and most useful ratio of surface film to volume of water, the aquarium (regardless of its length) should rarely be higher than a foot, unless some unusual species must be accommodated. For rearing platyfishes, guppies, and many of the live-bearing species, an aquarium 16 inches long, 9 inches high, and 9 inches wide is suitable. For swordtails and mollies aquaria somewhat larger are desirable: 24 inches long, 9 inches high, 12 inches wide. Goldfish and large mouthbreeders require aquaria that are about 36 inches long, 12 to 24 inches wide, and 12 inches high. Competent manufacturers may be trusted to choose the proper kind of glass, the suitable gauge of the metal for the frame, and the appropriate thickness of the slate for the various aquaria. If orders are placed for a dozen or more aquaria of a given size, it is likely that they will cost no more than commercial ready-to-use aquaria of an equivalent volume.

The number of fishes that may be reared or maintained in an aquarium varies with the size and habits of the animals and the temperature of the water. The general rule for maintaining fish has been that an inch of fish requires 1 gallon of water. For example, four to five adult platyfish each about 1 to 1½ inches long are usually well managed in a 5-gallon tank at 75° F. An equal number of guppies can get along in a somewhat smaller proportion of water. But swordtails and particularly mollies require wider ratios. Swordtails are a more active fish, and the mollies need greater browsing areas. The ratio of an inch of fish to a gallon of water is especially important during the summer months in handling many goldfish, which must be regarded as cold-water species. The reason for this is that, should the temperature rise unexpectedly, the available amount of dissolved oxygen in the warmer water is lowered, and at the same time, if the fishes are crowded, the carbon dioxide concentration will increase.

Some aquarists have abandoned the rule of the ratio of the volume of aquarium water to the number and size of fish in favor of the ratio of square inch surface area to the number and size of fish. In principle the use of the surface area is better, for it emphasizes the importance and advantages of shallow aquaria over deep ones. In the 5-gallon aquarium recommended here of 16 by 9 by 9 inches, the surface area is 144 square inches. Allowing about 24 square inches of surface for each inch-long fish, one obtains the value of 6.

The surface-breathing fishes like the fightingfishes (*Betta*) or paradise fishes (*Macropodus*) individually can be kept in extremely small containers. Owing to their pugnacity, large aquaria are required when several are maintained together. And in breeding these and other egg-laying species, large volumes of water are needed to provide in abundance the microscopic organisms which the large numbers of tiny fry require. Cannibalism may be avoided among these species by providing plenty of room and the shelter of aquatic plants.

But over and above these forces which determine the size of the population in an aquarium, there are other factors which are not at all clearly understood. It has been observed that only a definite number of fishes are able to live in tanks of a given size regardless of how well aerated the water is or how much water flows through the tank [see Breder (1935)]. Extended experiments with the guppy by Breder and Coates (1932) and Shoemaker (1944) indicated that a standing aquarium of a given size is capable of supporting only a certain mass of fishes.

Light for Aquaria

The beneficial effects of light of the proper kind and in sufficient amount are well known to the professional fish fancier. The *direction* of illumination seems to be almost as important as its source, intensity, and duration. The largest and most successful tropical and goldfish hatcheries in the United States, including some of the sunshine states, such as Florida, are invariably constructed in such a way that most, if not all, of the illumination is admitted by a system of skylights (fig. 1). Top-lighting is, by practical tests, the most desirable. Distinction must be made between an aquarium equipped with skylights and a greenhouse. The differences between the two need consideration because, for keeping fishes, at any rate, the greenhouse admits too much light. Furthermore, the problem of adequate heating and the maintenance of an even temperature is far greater in a greenhouse than in a building that has a limited glass exterior.

Artificial methods of illumination may be substituted for inadequate sources of natural light. The amount required is generally greater than suspected. Working at the National Cancer Institute, Bethesda, Maryland, Dr. Clifford Grobstein maintains and rears swordtails and platyfish entirely under artificial lighting. In a windowless, well-insulated, and air-conditioned room, temperature controlled at 78° F., he installed a strong battery of fluorescent lamps which are turned on about 8:30 A.M. and turned off about 5:30 P.M., giving the fishes an equivalent of eight hours of daylight and sixteen hours of darkness. Under the full intensity of light *Anacharis* grows vigorously, and in water where these plants grow luxuriantly fishes can usually be reared and bred without trouble. With proper general aquarium management, Grobstein obtains maximum growth and maturity in his experimental fishes.

In complete contrast to the use of artificial illumination of great intensity, Mr. Fred Flathman, an amateur tropical fish fancier, keeps his fishes in semi-darkness in the basement of his home. The only light his aquaria receive comes from three small and narrow transom windows; yet, despite this apparent light deficiency, this aquarist is successful because he uses very large, shallow aquaria (36 to 48 inches long). Flathman's method requires considerably more room than most laboratory workers have available to them. It is not recommended, but is mentioned in case basement laboratories have to be used temporarily. It must be emphasized that one is not justified in deviating from conventional methods until one has mastered the techniques of tropical fish management.

With few exceptions, such as the blind cave-characins or the cat-fishes, aquarium fishes are diurnal. Ordinarily 12 daylight hours are desirable for them. In northern latitudes for several months during the summer there are about 15 or 16 hours of light. This does no appreciable harm, except perhaps that of causing an undesirable growth of algae in the aquarium water. In winter, in northern latitudes, natural lighting should be supplemented by artificial sources, so that the animals get about 12 hours of light.

The intensity, duration, and quality of the light have a powerful physiological effect upon the reproduction of some species of fishes. Hoover (1937) and Hoover and Hubbard (1937) have demonstrated experimentally that controlling the light periodicity sequences radically modified the sexual cycle in the brook trout (*Salvelinus fontinalis*). In some species such as the medaka, Robinson and Rugh (1943) have shown that light periodicity has a specific effect on ovulation. Scrimshaw (1944b) found that continuous illumination of guppies reduced the interval between broods from about thirty to twenty-one days. Burger (1939), however, found that changes in the light periodicity had no specific effect on *Fundulus*. For the physiological effects of light on pigmentation of fishes see the many references on this subject by Parker (1948).

Temperature

In the Genetics Laboratory of the New York Aquarium, which contains over 600 aquaria, the thermostat is set for an air temperature of 75° F. In the summer time the temperature occasionally rises to 90° F., and while this is undesirable it cannot be helped. For short periods it does no harm. During the heat of the day the skylight roof is sprinkled with water, and this helps in keeping the temperature down. The water temperature lags considerably behind the air temperature. During the early spring, late fall, and winter the temperature rarely falls below 70° F. The desired range would be between 78° during the day and 72° during the night.

Many aquarists would not agree that the 72 to 78° F. range is the most desirable one and would insist upon a higher average temperature. Zoologists using *Tilapia*, the mouthbreeder, or mollies insist that an average temperature of 80° is preferable, and for these species I believe they are right. But for platyfishes and swordtails the 72 to 78° F. range is better, and there is some evidence from field work that the lower range is preferable.

Gordon (1948b) found a pool in Oaxaca, Mexico, from which thousands of platyfish, swordtails, and their associates were collected. The pool was about 150 feet in its longest part, and for the most part was exposed to strong subtropic sun. The water was kept from evaporating during the dry season in February and March by a tiny spring run at one end. In that part of the pond which was farthest away from the spring the temperature of the water registered 98° F., and here we caught only a few fishes. In the central portion of the pond where the temperature was 84° F., we seined many more fishes, mostly mollies and characins, but we got relatively few platyfish. But within the immediate vicinity of the spring's origin, in an area only 4 feet wide and 10 feet long, in a tangle of aquatic vegetation, and where the water temperature was only 78° F., we made our best collection of over 500 platyfishes and swordtails. There, too, we found gambusias and mollies. There was no question in our minds that the platyfish and the swordtails in this and other areas were attracted to the coolest water available to them.

In another nearby collecting station we caught platyfish and swordtails by the thousands in the early morning hours when the water temperature was 74° F. In the same pool on second try late that afternoon, when the water temperature had risen to over 80° F., we caught only a few. Returning to the same area the following morning when the water was relatively cool, we again collected platyfish and their associated species in large numbers. Apparently the fishes are active in the morning during the coolest period of the day, and then they dig down into the bottom mud during the heat of the day.

Recently information was obtained which indicated that perhaps even the temperatures maintained at the Aquarium's Genetics Laboratory are a bit too high. Wild Rio Jamapa fishes have an average dorsal fin ray count of 9.5, according to Gordon (1947). Yet their aquarium-reared descendants inbred for seven generations had only 8.4 fin rays. Assuming their genetic constitution had not changed, the lower fin ray count can be explained by the fishes being reared, in the laboratory, in water at a higher temperature than is found in nature. This would be in accordance with the general principle that, other conditions being equal, the lower the temperature during early development the larger the number of meristic characters such as the number of vertebrae, or fin rays. For further discussion of this point see Gabriel (1944).

Great fluctuations in water temperatures should be avoided in the aquarium laboratory. Chillings often result in lowering the fishes' resistance to bacterial and protozoan parasites. Sudden rises in temperature are perhaps just as undesirable. According to Sumner and Wells (1935), fishes require time to become acclimatized to the temperature levels imposed upon them, and subsequent adjustments to wide temperature changes tax them greatly.

Owing to the sensitivity of poikilothermal organisms like fishes to various temperature levels, fishes may be utilized in experiments designed to evaluate the specific effects of temperatures on their physiology and longevity. For instance, platyfish may be maintained at 20° C., 25° C., and 30° C., with some assurance that they will survive at each of these temperatures.

Maintaining an even temperature in experimental aquaria is difficult unless a cold room is available. Suitable aquarium heaters and thermostats are available (see Sources of Supply), but the worker must be on his guard to get thermostats which are designed so that they may be regulated for the lower temperatures. Cheap, poorly made thermostats are dangerous. Many valuable specimens have been lost in an overheated aquarium. For most species a water temperature of 105° F. or over is lethal if maintained for a number of hours. As the temperature increases the water's ability to dissolve oxygen is lessened and a point may be reached where the fishes may die of suffocation (see Aquatic Plants and Their Uses).

The temperature requirements of fishes depend upon the species, and some of the details concerning this matter are treated under the various laboratory fishes described. Where large numbers of aquaria are housed, all of which require much the same temperature range, it is far better to regulate the heat of the entire aquarium room rather than attempt to heat each aquarium separately by an individual heating unit. The units are relatively expensive, require heavy electrical current output, and are subject to failure when used constantly over a long period of time.

Use of Siphon

The siphon is the aquarist's handiest tool, for it combines the utility of the broom, shovel, vacuum cleaner, and pipe line. An aquarium, complete with water, should never be moved without first siphoning off most of its water. Flocculent bottom material, dead snails, or plant parts, droppings of aquatic organisms, may be conveniently removed by the siphon without undue disturbance to the aquarium.

A useful siphon consists of a piece of rubber or suitable plastic tubing (½ or ¾ inch diameter) of convenient length—about 3 to 4 feet. To siphon water from an aquarium to another container, the container or pail should be placed about 6 or more inches below the level of the bottom of the aquarium. Place one end of the tube in the aquarium considerably below the water surface and hold it firmly in this position with one hand. (Some aquarists like to place the tube in the water in such a way that the bend in the tube lies at one of the front corners of the aquarium. This enables them to hold the tube with their fingers at a fixed point of the tank.) The opposite end of the tube, held considerably below the water level, is then taken into the mouth and the air is sucked out of it vigorously so that the water will travel up the tube and get over the bend. The trick which should be mastered is to learn just when to quit sucking. If it is done too vigorously or too long, one will get a mouthful of aquarium water as the siphon suddenly goes into operation. Perhaps practicing with tap water may help. Or another method to start a siphon may be used. The entire tube may be filled with tap water, no part of the siphon being permitted to contain a pocket of air. Then, keeping both ends of the water-filled tube tightly closed, insert one end in the aquarium water and the other in the pail below. If no water has been allowed to escape from the tube, the siphon will start working immediately upon opening both ends of the tube.

A glass tube, about a foot long, inserted at one end of the siphon has some advantages for spot cleaning of an aquarium. The rigid glass tube enables the operator to direct the sucking point to the area where the sediment has accumulated. The diameter of the open end of the glass tube should be slightly narrower than that of the main section. This will prevent in some measure the entrance of large, solid particles such as snail shells or big pieces of gravel. These objects often are trapped within the tube and clog the siphon. Should a snail be trapped in the rubber tube it may be crushed; simply place the tube on the floor and step on it.

Sand for the Aquarium

Aquarium sand should be rather coarse, each grain about ⅟₃₂ to ⅟₁₆ inch, commercial grade number 2, white to gray in color, and made from quartz or granite-like minerals. Most aquarium supply houses have the right kind (see Sources of Supply). Newly purchased sand should always be thoroughly washed and rinsed until the water runs off clear even when the sand is vigorously stirred. Avoid the use of broken or whole sea shells or calcareous stones. Under ordinary circumstances use no earth, humus, or any other kind of soil beneath the sand covering. Soil is unnecessary for aquarium plants; sufficient plant food is supplied by the organic fertilizer which constantly accumulates in the aquarium from surplus food and from the fish and the snail wastes. Many aquarists suggest the sand be arranged so that its surface is sloped toward the front, the sand at the back being about 1½ inches thick and at the front about ½ inch. This will enable sediments and detritus of all kinds to work down so that they may be siphoned out without disturbing the aquarium.

At a place where dried fishfood is dropped and uneaten, the sand is likely to turn black in color owing to the metabolic activity of the aquatic bacteria, some of which are anaerobic. This discoloration may extend throughout the bottom sand covering if overfeeding has been practiced over a period of time. Putrefaction at the bottom covering may be stopped by reducing the amount of food. The vigorous growing plants like *Vallisnaria*, *Sagittaria*, *Cryptochoryne* aid in keeping the sand clean and sweet, but the plants, too, will succumb when conditions become intolerably bad (see Aquatic Plants and Their Uses).

The blackened sand should be siphoned out by forcing the end of the siphon, preferably one having a length of glass tubing at the probing end, into the badly affected areas (see Use of Siphon). The black sand should be removed entirely right down to the bottom slate. It may be necessary to lift the glass tube free of the sand and into clear water once in a while to allow the water flow to regain its momentum through the siphon. (The distinctive odor of discolored sand and bottom ooze is characteristic of tidal flats, and for the same reason— bacterial decomposition of organic matter.) When the foul sand has been siphoned off into a pail, allow the sediment to settle and then inspect the water carefully for fish which may have been inadvertently sucked up in the process. Wash the sand in hot water, the hotter the better, stirring the mass constantly to loosen all the accumulated dirt.

It is a good policy not to place the reconditioned sand back in the aquarium immediately. A twenty-four-hour period of drying on paper towelling or on glass or enamel trays in the sun, if possible, will bleach and further purify the sand. Sand to be poured back into the aquarium will be more manageable if it is first wetted.

AQUARIUM WATER

Water, as everybody knows, is composed of hydrogen and oxygen. Its molecules contain two atoms of hydrogen and one of oxygen, and this is represented by the familiar formula H_2O. But as Timm (1932) clearly states, let no one make the mistake of thinking that the combined oxygen in the water molecules is available to the fishes which swim in it.

Fish do not breathe the oxygen of the water; they breathe the *dissolved oxygen* in the water. Fishes can be drowned in pure water as readily as mammals. Aquatic life is made possible because water can and does dissolve or absorb free oxygen from atmosphere. Water is a solvent, and it is the dissolved oxygen in it which fishes and other gilled animals require. Gills are the device which enables them to utilize the available oxygen (and eliminate carbon dioxide). These red-fringed, highly vascularized organs are located behind the mouth cavity above the opening of the gullet. The gill arches, a series of rib-like bones, support the gills, each of which is composed of many delicate filaments. The fish, in breathing, takes water into its mouth and forces it past the barely perceptible spaces between the gill filaments. The water then passes through the gill slits and is ejected in an opening just back of the opercular bones. As the water passes over the gill filaments its dissolved oxygen is absorbed through thin gill membranes and passes into their blood capillaries. At the same time the blood corpuscles within the venules release their burden of carbon dioxide. The blood of a fish and its functions are lucidly explained by Chapman (1948).

Conditioned or "Old" Water

Fishes require specially conditioned water, that is, water at least a proportional part of which has previously supported living fishes. Obviously, in preparing an aquarium for laboratory fishes, if one starts from scratch, obtaining the proper kind of water is quite a problem. However, raw tap water may be conditioned for fishes by allowing it to stand exposed in shallow, non-metallic containers. Conditioning may be speeded by bubbling compressed air through it or by introducing aquatic plants. Water that will support a vigorous growth of alga, *Nitella*, may be regarded as safe and suitable for most of the tropical aquarium fishes. When sprays of *Nitella* are placed in raw water they are likely to die, but the second planting may succeed. If *Nitella* is not available for water conditioning, other types of plants may be used (see Aquatic Plants and Their Uses).

To some degree, fishes create their own protective substances by their excretions. If plants are not available for water conditioning, crowd the fishes into a minimum amount of water in a shallow container for a day or so, but do not feed them much. Then, water which has previously been allowed to stand for several days in suitable containers may be added gradually to build up the supply of conditioned water.

The advantages of "old" or conditioned water for the aquarium, according to Breder (1931), are partly associated with its larger buffer value, but buffering by salts is not the entire cause of the value of "old" water. He indicated that an active bacteriophage exists both in conditioned aquarium water and in the intestinal contents of the smallmouth black bass (*Micropterus dolomieu*). The consequences of a deficiency in this bacteriophage may be demonstrated by placing fishes in an aquarium containing freshly drawn and detoxified tap water. The bacterial count will climb rapidly above that of the tap water. The peak of bacterial activity will usually appear in the first week and will fall rapidly to a point of stability. When many fish are placed in new water and overfed, the water will be in a constant state of cloudiness owing to bacterial activity. Cutting down on the food or eliminating it for a week and adding a quart of "old" water from a well-established aquarium will probably speed the clarification process.

If for any reason it becomes necessary to empty a healthy aquarium, the water should be saved and used again. It is advisable to siphon the water into well-washed enamel-ware pails. Avoid galvanized iron, copper, brass or Monel metal utensils, but if only these are available, paint their insides with asphaltum varnish. Save the aquarium water for refilling the tank in its new location, but do not allow the water to stand in untreated metal pails longer than half an hour.

Chlorine in Water

No laboratory worker would dream of keeping stocks of his experimental mammals in rooms containing chlorine or other noxious gases, yet essentially the same intolerable conditions are created when fishes are confined in water freshly drawn from the tap.

The chlorinated water problem in aquarium management was clarified by Faber (1933). Although the amount of chlorine used by public water-purification plants varies with the season, the usual doses range from 0.2 to 0.8 part of the green gas per million gallons of water. This is equivalent to 1½ to 6½ pounds of chlorine to each million gallons of water. Even in these small doses the gas may be quite harmful, and according to Coventry, Shelford, and Miller (1935), residual chlorine or chloramine as low as 0.05 of a part per million is toxic to some delicate species.

Natural waters contain varying amounts of living and dead organic matter. When chlorine comes in contact with these substances much of its potency is lost. In order to be sure that there is sufficient chlorine at all times to kill harmful organisms, health officers insist that water supplied to the public contain an excess of 0.1 part per million. It is this excess reserve chlorine gas which causes havoc to fishes in the aquarium. Water-purification plants maintain a particularly high excess of chlorine during the hottest summer months when pollution of potential drinking water is at its peak. They keep it high during the months of extreme cold, for then chlorine does not dissolve readily in the water, and again during the spring thaws, for then the water is turbid and contains a high count of soil bacteria.

The aquarist using tap water has several methods for making it fit for sensitive fishes. Suggestions have been made that tap water be allowed to stand exposed to the air until the chlorine has passed off. Some aquarists draw water from the hot-water tap and let it cool before using. Bubbling air through the stored water or adding aquatic plants to the water to hasten the detoxification process have also been suggested. A chemical method utilizes sodium thiosulphate and ferrous iron salts (sulphate or ferrous ammonium sulphate). Chlorine will combine with these substances and thus can be eliminated, but there is a catch to this method. Unless correct amounts of the reagents are used, the new chemical compounds may be just as harmful, or more so, than the objectionable chlorine. When sodium thiosulphate is used, if the water is slightly acid, sulphur dioxide may be liberated to form lethal doses of sulphuric acid. A third method for eliminating chlorine from tap water is recommended by Faber, who suggests that a few ounces of activated carbon be stirred into a container of freshly drawn water. He indicated that one-half pound of powdered, activated carbon will dechlorinate 250 gallons of water containing 0.15 part per million of residual gas, provided that the water is allowed sufficient time (about 2 hours) for complete contact with the carbon. Before being used the treated water must be filtered to remove the carbon particles.

To determine whether water contains chlorine, Faber says to add one cubic centimeter (1 cc.) of a standard solution of orthotolidine to ten cubic centimeters (10 cc.) of water to be tested. If the water appears yellow there is at least 0.02 part of chlorine per million in the water; the yellower the water becomes the more chlorine it contains.

The paper by Coventry, Shelford, and Miller should be consulted if difficulties resulting from chlorine are not to be conveniently remedied by the suggestions given here. Charts to show exactly the amount present and informative pamphlets on the subject are available (see Sources of Supply).

During the winter months tap water is likely to have a temporary excess of dissolved air because cold water can hold more oxygen in solution than warm water. This is easily demonstrated by permitting cold water to stand in a glass container for about ten minutes at room temperature. Tiny bubbles of air will form and cling to the side of the container. A fish from a warm aquarium placed in oversaturated water will attract bubbles of air to its sides, fins, and gills. Under this treatment, if prolonged, some fishes are susceptible to the gas-bubble disease, a gas embolism. This is an additional reason for the recommendation that tap water should be allowed to stand until all excess gases, oxygen as well as chlorine, are eliminated.

The toxic effects of nascent oxygen on fishes were described by Hubbs (1930). Notes on the exposure of several species of pond fishes to sudden changes in pH were prepared by Wiebe (1931), and the effects of water softeners were studied by Miller (1944). Standard methods for the examination of water and sewage were published by the American Public Health Association (1939).

An important review on the determination of water quality with an extensive bibliography was prepared by Ellis, Westfall, and Ellis (1948).

Metals and Water

No metal should come in contact with aquarium water, for any water in which fishes have lived for some time becomes corrosive to a great variety of metals. Atz (1947b) indicated that no metal or metal alloy ever tested by the New York Aquarium has proved capable of resisting the corrosive action of aquarium water. The action may not be extensive, but it is sufficient to poison fishes and invertebrates. It takes amazingly little dissolved metal to injure or kill aquatic organisms. For instance, concentrations of copper and silver of but a few parts per hundred million are quickly fatal to fishes, and much lesser ones can kill if allowed to act over a period of time.

Metal poisoning of aquarium water may inadvertently be caused by the introduction of a gadget of one kind or another. Seeton (1934) reported that an aquarist had constructed a breeding trap which he made of perforated steel plate and had cadmium plated. When it was returned to him from the plating works he first washed it thoroughly with hot, soapy water, flushed it several times, and finally gave it a spray of boiling water. When a number of gravid live-bearing fish were placed in the trap in an aquarium they died within a few days. The trouble was traced to a residual film of cadmium salts used in the plating process. Nickel tubing used in certain types of aerating systems will be attacked by organic acids, over a period of time, and form nickel salts, and these, too, are deleterious.

Marsh (1908) showed that silver nitrate is extremely toxic, reporting that 1 part in 22,500,000 parts of water for 48 hours is fatal to chinook salmon fingerlings. Concentrations of this salt as low as 1 part in 300,000,000 parts of water killed stickelbacks quickly. The deadly effects of copper were indicated by Marsh's experiments; he reported that 20 square inches of metallic copper lying in 6 quarts of Potomac River water killed 8 out of 10 salmon fry within 24 hours; 10 square inches killed 4 out of 6 free-swimming salmon fry within 2 days and 18 hours, and all of them within 3 days and 2 hours. Lead salts are fatal to many fresh-water fishes in concentrations of 0.3 part per million, according to Carpenter (1927). Mercury, chromium, and zinc are also very toxic.

Not only metals can poison fishes, however, and the introduction of anything not chemically inert into a fish tank is to be avoided. Resinous woods, soluble rocks, certain plastics, and most builders' cements are some of the substances that can kill fishes. No paint save a high grade of black asphaltum varnish should be used. Questionable materials can often be made "safe" by completely covering them with this water-proofing, relatively inert, varnish. It should also be remembered that insect sprays, including DDT, are very toxic to fishes.

An extensive review of toxicity experiments with fishes was prepared by Cole (1941), and while he emphasizes toxic substances from pollution, much of his report should be of value to workers dealing with the use of fishes as test animals for various chemical substances, particularly pharmaceutical products. In this connection the work of Lapenta (1932), Macht (1943), and Surber (1948) should be consulted; see also under Goldfish, especially Asher and Allee (1940) on the effects of traces of tin on the rate of growth of the goldfish.

AQUATIC PLANTS AND THEIR USES

"Balance"

Aquarists for over a hundred years have related the carbon dioxide-utilizing activities of plants and their oxygen-releasing properties under strong light, to the utilization of oxygen and the elimination of carbon dioxide by animals, and have claimed that the plants "balanced" the animals in their life requirements. According to Atz (1949) the so-called "balanced aquarium" is a myth. He refers to the work of Breder (1931), who pointed out that the production of O_2 by the photosynthesis of plants under the influence of light in open aquaria contributes little, if any, to the O_2 consumed by the animal life in it. Oxygen that is produced or used up in the water returns to a condition of equilibrium with extreme rapidity. Apparently, carbon dioxide lags considerably behind oxygen in its properties of invading and evading water. It can accumulate in the water.

Since there can be no accumulation of dissolved O_2 in water for any appreciable length of time, and since the respiratory activities of fishes continue as usual at night and on dark days, it follows, stated Breder, that the animal's chief source of O_2 is that which constantly invades water through the surface film from the atmosphere. Contrary to the opinion of many aquarists, water may become dangerous to fish living in it not because its dissolved oxygen is depleted but because its carbon dioxide concentration is high. Carbon dioxide, at certain concentrations, can actually prevent the utilization of the available dissolved oxygen present in water. The probable explanation of this, as Atz (1949) pointed out, is that carbon dioxide at relatively low concentrations increases the efficiency of the fish's blood to deliver oxygen to the tissues, but at the same time decreases the blood's ability to take on oxygen at the fish's gills (the Bohr effect).

The value of plants in reducing the CO_2 content of the aquarium water during their photosynthetic processes is beyond question. In fact, carbon dioxide is one of the limiting factors of growth in plants under the usual aquarium conditions [see Arber (1920)]. On the other hand, plants can and do actually increase the concentration of

CO_2 in the water during dark days and at night, since plants, quite like animals, utilize O_2 and eliminate CO_2 in their respiratory process

Under unusual weather conditions aquarium water may develop an undesirable concentration of carbon dioxide. An aquarium overstocked with overfed fishes and snails may have a good deal of undecomposed organic matter in its bottom muck. At the same time this aquarium may be heavily planted with various aquatics. The color of the water may be deep green because of a strong suspension of unicellular algae (see Green Water). Now when this overstocked, green aquarium receives adequate light and the temperature of the water is moderate (70 to 80° F.), things may keep going in an apparently good condition for some time. But if the temperature should suddenly rise to 90° F., or over, and the intensity of the light should be radically diminished owing to dark, stormy days in the summer, followed by hot nights, the concentration of carbon dioxide in the aquarium water will gradually increase, and it may reach the critical stage not only for the fishes but for the plants as well. Actually, then, the deciding factor in the maintenance of fishes, and other aquatic organisms having somewhat similar requirements, depends upon the quantity of carbon dioxide in the water. Therefore, in doubtful water, it is far more important to determine the concentration of carbon dioxide than of oxygen.

While the usefulness of aquatic plants, as far as their ability to re-oxygenate water is concerned, has been overstated by aquarists, plants do have many important duties in an aquarium. They provide suitable sites for the attachment of the fertilized eggs of those fish which, in spawning, scatter adhesive eggs. They provide convenient attachment areas for a wealth of sessile microorganisms which, in turn, furnish fish fry with needed live food [see Needham and Lloyd (1916) and Shelford (1918)]. Rooted plants utilize some of the organic fertilizer in the bottom ooze [see Pond (1918)]. Finally, plants create shelters for the young fishes and for the less aggressive members of an aquatic community. The specialized uses to which the various species of aquatic plants may be put are given below.

Plants for the Aquarium

Submerged Plants with Centralized Root System. These plants should be rooted in such a way that the crown is fully exposed above the level of the gravel. If parts of the stems or leaves are covered, the plants are likely to decay.

Cryptochoryne.   Many species, all natives of tropical Asia and the
Malay Archipelago, are useful because they do extremely well in
subdued light.   The nomenclature of aquarium species is con-
tradictory.

Cryptochoryne sp. has narrow, lance-shaped green leaves and is
the dwarf of the genus, rarely growing over 6 inches in length,

Fig. 3.   Plants for small aquaria.   On the left the dwarf or microsagittaria (*Sagittaria subulata*), showing runners and three young plants attached.   On the right
the dwarf crypt, *Cryptocoryne* sp.   (Photo by S. C. Dunton, New York Aquarium,
New York Zoological Society.)

and is one of the most desirable of all plants in shallow-water
aquaria.   The leaf stalks, in growth, habitually bend in a gentle
curve (fig. 3).

Cryptochoryne cordata and *Cryptochoryne willisii* resemble each
other.   Both have deep purplish green leaves, the undersides of
which are purple to red.

Cryptochoryne griffithii grows to 12 inches.   It has narrow stalk-like
leaves which broaden out, becoming ovate, spoon-shaped.   These
have a waxy, glistening, deep green surface.

Sagittaria and *Vallisneria* contain the species of aquatic plants which
are the most desirable for aquarium purposes.   The effectiveness of their fine system of bushy roots in utilizing the accumulating fertilizer of the bottom was revealed in an instructive
experiment which was described by Pond (1918).   This work indi-

cated that these plants do take up matter from the substrata and in so doing are effective aids in keeping aquarium gravel clean. There are many anatomical differences between *Sagittaria* and *Vallisneria,* especially in their flowers, but their leaves and roots are superficially similar. However, the outer margins of *Vallisneria* leaves are lighter green than the middle part, whereas the leaves of *Sagittaria* are rather uniform in color. This distinction is made plain if one holds the two kinds of leaves up against a bright light.

Sagittaria natans (common sagittaria) has grass-like leaves about ½ inch wide and grows from 6 to 8 inches tall.

Sagittaria sinensis (giant sagittaria) is suitable in large aquaria where the gravel bottom is 3 to 4 inches thick. Its leaves are fairly thick and brittle compared with others of the genus.

Sagittaria subulata (dwarf or microsagittaria) is particularly suited for the breeding and maintenance of the poeciliid fishes. This plant forms a fine, turf-like mat over the gravel by sending out many runners. The stronger the light, the shorter the plant; some never get larger than 1 to 2 inches. In aquaria of 5 gallons or less, it is preferable to all other species in this group (fig. 3).

Vallisneria spirillis (tape grass) is a tall plant which is better suited to the large aquarium (fig. 4). In buying this or any other aquatic plant for aquarium purposes, it is extremely important to get aquarium-grown plants. There is a tendency for plants collected in the wild to wither away during the dark winter months.

Submerged Plants with Roots Developing from Individual Stalks. These plants should be used unrooted although sometimes the roots will extend to the gravel and become attached. The plants should not be inserted into the gravel, for the stalks decay rapidly when covered (fig. 5).

Cabomba caroliniania (fanwort) has round stalks around which grow hair-like leaves. It is often sold, but in the aquarium it is a rather delicate aquatic plant that is likely to break apart.

Ceratophyllum (hornwort) has a horny feel owing to calcareous deposits upon it. In a vigorously growing state its bushy, horsetail-like arrangement of fine leaves provide a sanctuary for the spawn of goldfish and for their fry.

Anacharis (Elodea, Philotria) canadensis (waterweed, Babbington's curse) has brittle stems around which extend whorls of three or four relatively large spear-shaped, bright green leaves. The useful aquarium plant grows luxuriantly in the smallest bodies of water provided there is sufficient light [see Gordon (1935) for an account of its wild growth in the rivers of England].

FIG. 4. *Vallisneria*, tape grass, showing habit of multiplication by a system of runners. Aquarium-grown varieties are the desirable kind. (Photo from W. T. Innes, 1948a.)

Myriophyllum (millfoil) is so named in allusion to numerous divisions of the foliage which is arranged in fans and whorls. It is a useful plant for spawning of goldfish, but owing to the brittleness of its stalk it is likely to break apart and die out in a small aquarium.

Submerged Plants without Roots

Nitella (stonewort) is, in the opinion of the author, the best plant of its type for providing shelter and food to newborn viviparous fishes.

It is important to obtain aquarium-grown pure cultures, free of filamentous algae and particularly free of the fine bladderwort (*Utricularia minor*). When provided with an abundance of light the plant grows vigorously yet it is easily controlled simply by

Fig. 5. *Cabomba, Myriophyllum, Anacharis,* indicated from left to right. *Cabomba,* best selling but least useful; *Myriophyllum,* useful in goldfish spawning; *Anacharis,* hardiest of all aquarium plants. (Photo from W. T. Innes, 1948a.)

tearing away parts of it. It cannot be destroyed by this crude treatment. *Nitella* is, in a sense, a linear mass of almost naked protoplasm. It is particularly sensitive to chlorine and other toxic substances in the water. As mentioned elsewhere in a discussion of conditioned water, an aquarium which will support a vigorous growth of *Nitella* is quite suitable for fishes [see Gordon (1943) and fig. 23].

Riccia fluitans (crystalwort) has strong, interlacing branches that form a loose tangle of innumerable, interconnected channels (fig. 6). The compact masses provide a sanctuary for the fry of viviparous and egg-laying fishes. Crystalwort is particularly useful in breeding such fishes as *Gambusia* and *Epiplatys* which are extremely predaceous.

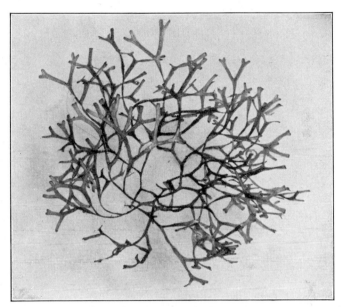

Fig. 6. The interlacing branches of *Riccia;* crystalwort is useful in breeding predaceous killifishes. (Photo from W. T. Innes, 1948b.)

Utricularia minor (lesser bladderwort) is preferred by many fish fanciers for the spawning of many species of fishes. However, this plant has two disadvantages. Its branches are extremely thin and tough, and as the plant grows in thick mats, the entangling threads may gill-net the adult fishes. The traps may engulf the tiny fry of certain egg-laying fishes.

Utricularia vulgaris (common bladderwort) is dangerous owing to its definitely carnivorous properties—its bladders are quite large and fully capable of trapping young fishes. A comprehensive monograph on these carnivorous plants has been written by Lloyd (1942).

Surface Plants Which Grow Partly above the Water-level. Some of these plants are useful for reducing the amount of top light. Some, because of their elaborate root system, are useful for spawning purposes.

Ceratopteris thalicroides (water fern) is a true fern. It has sharply cut leaves, much like parsley. Indeed, it is eaten as a vegetable in the Indian Archipelago, where it is native. The water ferns are

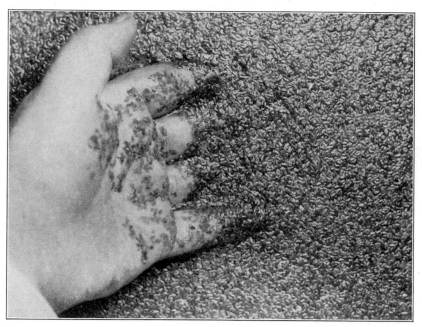

Fig. 7. Duckweed, *Lemna minor,* a tiny plant but one which multiplies rapidly in the presence of strong light. Thick surface mats of this plant are dangerous, for they may suffocate fishes. (Photo by Lilo Hess, from Myron Gordon, 1945, courtesy of *Nature Magazine.*)

viviparous and reproduce themselves by means of young plants which appear in large numbers along the edges of the older leaves. Under strong light their growth is rapid.

Eichhornia speciosa (water hyacinth) is one of the showiest of all flowering aquatic plants. The petioles of the leaves are inflated to an extraordinary degree. These bulbous structures are filled with spongy aerenchymal tissue which enables the plant to become extremely buoyant. The light purple flowers are beautiful but last only a few hours. In its native, semi-tropical habitat and adjacent area the plant grows luxuriantly, covering many acres of water surface, but it is regarded as a grotesque novelty in countries foreign to it. Water hyacinths are a serious hindrance to navigation in the rivers and canals in Florida and Louisiana. The feature which makes these plants important to the fish breeder is their long, purplish roots. Their fine, bushy roots and rootlets which hang down 8 or 10 inches below the water surface provide one of the best places for the adhesive eggs of goldfish.

Lemna minor (common duckweed) is one of the world's tiniest flowering plants. Fish fanciers like it because this duckweed is eaten by goldfish. *Lemna* also harbors many microorganisms. Duckweeds

FIG. 8. A cluster of *Salvinia,* a surface-growing aquatic plant useful in providing shade. (Photo from W. T. Innes, 1948b.)

grow at a prodigious rate, and for that reason they should be used in the aquarium with caution [see Gordon (1945) and fig. 7]. They are difficult to eliminate once they are well established. Some aquarists use them to cut down the amount of top light, but other larger and more manageable plants should be used for this purpose.

Pistia stratiotes (water lettuce) is a decorative plant of limited utility for the fish breeder. Its ribbed, pale green leaves are fan-shaped and form rosettes. The plant's usefulness lies in its white, long, feathery root system which provides a sanctuary for developing fish eggs and fry.

Salvinia natans (cat's tongue) has small, swollen floating leaflets about ¼ to ½ inches (fig. 8). They are bright green above, while the undersides are matted with clusters of brown root hairs. Where shade is required, this plant should be used. It does not grow as rapidly as *Lemna* and can be kept under control.

In addition to Lloyd, the subject of aquatic plants has been covered by Arber (1920), Perry (1938), Fassett (1940), and Muenscher (1944).

Green Water

When the submerged, bright green plant microorganisms burst into rapid reproductive activity they create a deep green fog in the aquarium water. Spores of the microscopic green algal plants are carried with dust particles everywhere. They arise from the drying edges of fresh-water ponds, swamps, and river banks, their natural habitat, and are blown away in all directions. Many algal species are highly specialized and cannot withstand dryness, but others like *Pleurococcus* normally live exposed to the air. This form rarely is found alive in the water, but according to Snow (1918) species like *Chlorella*, *Stichococcus*, and *Hormidium* may adapt themselves to a land or water environment. These algae are very widely distributed under many different conditions. Thus it would be quite easy for a few algal cells to get to any new body of water, large or small. It is unlikely that any aquarium is free of them for very long.

The immediate conditions in and around an aquarium will determine whether the algal spores will develop into plants which then will multiply seemingly without restraint. The greater the amount of light, other conditions being equal, the more rapid the plants will grow. This factor usually can be controlled. Often the simple expedient of laying a piece of cardboard on top of the aquarium, or along the side of the tank which receives the sun's rays, will cut down the critical amount of light needed by algal plants for rapid multiplication.

If this does not work, then the chances are that the aquarium water contains an overabundance of plant food, a condition which may have developed in a number of ways. Feeding too much food to the fishes provides the fertilizer for the water plants. Keeping too many fishes in a relatively small tank transforms the food eaten into fertilizer rapidly. At the same time the crowded fishes raise the concentration of carbon dioxide in the water by their respiratory activity, and, of course, the plants utilize carbon dioxide in the presence of light for their photosynthetic processes. If it is impractical to alleviate crowded conditions, artificially aerate the water; feed live food as much as possible and for as long a period as necessary and at the same time cut down on the light.

Some aquarists claim that when a tank is in a process of turning green with algae, a few quarts of water from a long-set-up clean aquarium will often clear the water overnight. Some recommend copper sulphate to clean the water, but this is extremely dangerous. The safest principle upon which to act is to starve the microscopic algal plants. This may be done by reducing the algae's available food supply. Remove the plant fertilizer in the bottom sediment by siphoning. Make it impossible for the plants to utilize the available food by cutting down on the light source by artificial methods, or by adding some surface-loving plants like *Salvinia* or *Riccia* which will produce shade and at the same time utilize some of the plant food in the water. In addition, one may introduce a number of plant competitors within the water, preferably those which are easily controlled, such as *Anacharis* or *Nitella*.

Some aquarists believe green water has a beneficial effect upon aquarium fishes, and they often recommend that sick ones be placed in such water. The usual reason given for the alleged beneficial effects of green water is that it provides a super-abundance of oxygen. Perhaps all it provides is protection by isolation. It must be emphasized that green water may be dangerous, for on dark, hot days and hot nights the algae use up oxygen and at the same time liberate carbon dioxide. Green algae are particularly dangerous when dying out rapidly, for they are decomposed by aquatic bacteria, and this increases the CO_2 content of the water.

NUTRITION

Prepared Foods

Any of the standard brands of commercial, dried aquarium fish foods may be used equally well, but as much attention should be given to the physical shape of the food particles as to their chemical composition. No matter how well formulated and carefully prepared the food may be, unless the fish eat it, the food will rot and foul the water. As such it is not a food but an undesirable fertilizer. Moreover, it must be remembered that *overfeeding is responsible for more failures in aquarium management than any other single error.*

A review of the subject of nutrition in fishes is not feasible here, but a number of references may serve if this subject needs further elaboration. Cultivated fishes do suffer from dietary deficiency diseases [see Wolf (1945)]. Embody and Gordon (1924) compared the composition of the live food which trout eat in their natural waters with the artificial foods used by fish hatcherymen. Introductions to fish nutrition were presented by McCay and Phillips (1940) and Phillips and Brockway (1948); and Tunison (1945) prepared a useful summary on fish feeds and feeding. Kenyon (1925) studied the digestive enzymes in poikilothermal vertebrates.

A successful formula for the preparation of a dried food is given later in this section, but it may be varied to suit the needs of the experimenter. White fish meal, salmon egg meal, clam meal, dried egg yolk, dried liver, brewers' dried yeast, dried and ground insects have been used successfully. Dried shrimp, which aquarists regard as the best single food, is the most expensive. However, when pure shrimp is bought in bulk, the cost is less than that of commercially prepared foods. On the whole, fishes being some of the smallest of laboratory animals, the food bill, on the basis of each experimental individual, is relatively low.

Liver-cereal Wet Food. The liver-cereal food described below is designed for feeding as a thick, wet mash. The following is a modification of an earlier formula of Gordon (1943b).

INGREDIENTS:

> 1 pound of fresh beef liver
> 20 tablespoonfuls of Pablum or Ceravim
> 2 teaspoonfuls of table salt

APPARATUS:

> Waring blender or K-M liquidizer (fig. 9).

1. Skin the liver of its connective tissue covering; remove the larger blood vessels and other tough or fibrous tissues.

2. Cut liver into ½-inch cubes.

3. Measure 2 ounces each of cubed liver and cold water. Place in blender and liquidize. Pour liquidized liver through the strainer into a 2-quart bowl. Repeat until all the liver is liquidized and strained. Add salt.

4. Add Pablum, Ceravim, or any other similar dry, precooked cereal, stirring thoroughly all the time. Add as much cereal as the liquidized liver will take or until a thick peanut-butter-like consistency is attained.

5. Fill 1-, 2-, or 3-ounce glass containers, the size depending upon the amount used for a single day's feeding of the liver paste.

6. Place filled glass jars in water. Heat until water begins to boil, turn off heat, and allow jars to stand in hot water for about ½ hour.

7. Cool and cover the glass containers and place in the coldest part of refrigerator. Some may be frozen.

Heating the liver mixture is necessary to coagulate the liquid elements of the gland. If heat were not applied, the liver fluids would separate from the paste, and when placed in the aquarium water the fluids would foul the water just as quickly as ordinary liver particles do.

For ordinary routine in feeding, the fishes are given a portion of the liver paste, the size varying from that of a rice grain to that of a coffee bean or larger, depending upon the number and size of the fishes and the temperature of the water. The fishes are usually fed once a day, early in the morning so that they may have an entire day to finish the portion given them. If any food remains uneaten on the following day, no further feeding is made until all is gone, or the food removed by some other method. The uneaten food, if properly prepared, will remain intact and will not dissolve for a day or so.

Very young live-bearing fishes will learn to peck at and eat pieces of the liver paste since it crumbles into extremely tiny particles. In these tanks, be sure that there are not too many snails because these scavengers will be attracted to the liver, crawl over, and cover it entirely, making it difficult if not impossible for the very small fish to get near the food.

Fig. 9. The electrical blender liquidizes liver, which is then combined with precooked cereals for a standardized fish food. (Photo by S. C. Dunton, courtesy of *Animal Kingdom*, New York Zoological Society.)

In the Genetics Laboratory of the New York Aquarium this food is fed every other day. On alternate days, the fishes are given pure, shredded, dried shrimp or live food such as tubificids, whiteworms, or *Daphnia*. In the aquarium of the Department of Animal Behavior at the American Museum of Natural

History, the dried shrimp is combined with the liver paste at the time that the dried cereal is added to the liquidized liver.

Yudkin (1949) warns that raw fish, particularly that of cyprinids, contains the Chastek-paralysis factor, an enzyme called thiaminase, which destroys the vitamin thiamine. Raw fish should be used with caution in the diet of other fishes, and if used, should be given separately, not mixed with other ingredients of the food.

A Standard Dried Food. The liver-cereal food has been made the base for the standard dried fish food for *Tilapia*, the African mouth-breeder, and *Anoptichthys*, the blind cave-characin, both of which have been maintained and bred for many years for experimental work at the American Museum of Natural History. The ingredients and proportions are as follows:

	Pounds
Liver of beef	5
Pablum or Ceravim	14
Shrimp shell meal	6
Shrimp meat shredded	6
Lettuce	3
Spinach	3

The raw beef liver is cut into 2-inch pieces and placed in a kettle, covered with water, and boiled for 15 minutes. The liver is then removed from the kettle, but the liver liquid is used to boil the shrimp shell meal, the shredded shrimp, the vegetables, and cereal. The boiled liver pieces are cut finely by passing them through a meat grinder, and then the chopped liver is added to the shrimp-vegetable-cereal combination and everything is cooked together for an additional 15 minutes. The warm paste is spread over large trays and cut across to form patties about 2 inches square. It is dried in the sun or over heated steam coils like those of a radiator. Then the food is scraped from the drying pans and ground in a mill. The broken pieces are passed through several screens and sorted for fine, medium, and coarse grades. Dr. Lester R. Aronson has found that a combination of the crude shrimp shell meal and pure shredded shrimp is better for fishes than the pure shrimp meat itself. Apparently the crude shell material adds bulk and carotin pigment elements to the diet, both of which are highly desirable.

Live Foods

Daphnia and Their Culture. Any aquarium laboratory having fifty or more tanks devoted to rearing small fishes should have facilities

for the home cultivation of *Daphnia*, the prime live food of all fishes
(fig. 10). In the Genetics Laboratory of the New York Aquarium

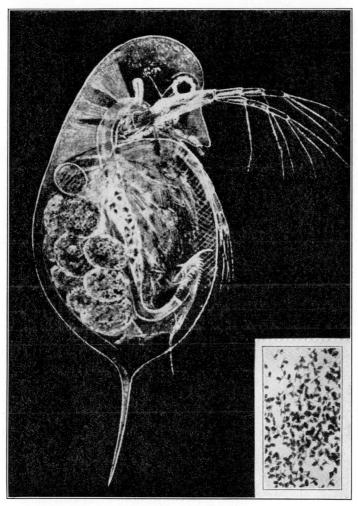

FIG. 10. *Daphnia pulex*, the waterflea of prime importance as food for fishes.
The oval objects are the ova which develop parthenogenetically into females.
Insert shows natural size of the *Daphnia*. (Photo by D. J. Scourfield, courtesy
of W. T. Innes, 1948b.)

a circular, six-foot tub made of cedar staves, holding water to the
depth of a foot, has yielded an adequate supply of water-fleas the year
round for fry up to a month of age. And it has been found that when

Daphnia are fed to the fishes just after they are born, they subsequently grow better on other foods.

There have been any number of suggestions for the cultivation of *Daphnia*. The fertilizing substances for use in outdoor ponds are unsuited for the cultivation of *Daphnia* in the laboratory. Such water fertilizers as the fresh manures of cattle, horses, or chickens are impractical. These substances have adequate, less odoriferous, substitutes in dried skim milk, brewers' dried yeast, bone meal, soybean meal, and the outer leaves of lettuce plants (those which are generally discarded by retail grocers). Small amounts of sugar or molasses and vitamin B tablets have been used to advantage. A strong current of air bubbles is necessary to bring carbon dioxide and other gases to the surface of the water. No soil is required or desirable in the indoor *Daphnia* pond.

It is difficult to prescribe any definite rules for the maintenance of *Daphnia* culture, for the variables as regards the temperature of the water, the amount of available *Daphnia* food, the amount of decomposing substances in the water, and the number and condition of *Daphnia* are too great. One should avoid a rapid rise in the *Daphnia* population because it is likely to be succeeded by a dearth. If a swarm of *Daphnia* appears net them out, reducing the population, and feed them liberally to the fishes. Try to stabilize their increase by regulation of the amount of their food.

The appearance of ephippia or winter eggs in the brood chamber of females is indicative of adverse conditions. This may call for the partial drainage of the pool and filling again with fresh, well-aerated water.

A mixed culture of daphnioid species is preferable to one containing a single species. *Moina* is smaller and will tolerate higher temperatures than *Daphnia pulex*. *Daphnia magna* is one of the largest and is the species generally cultivated at the larger fish hatcheries. [Holm-Jensen (1948) has used this species as a test animal in studying the physiological effects of heavy metals.]

In general, most species of water-fleas do best in cool water, 70° F. being about right. Cultures have been maintained continuously for over six years, but conditions will arise when draining and refilling with fresh water are desirable. Often partial drainage, and at the same time the removal of the bottom products of decomposition, will restore the *Daphnia* pool to normal productivity. If complete drainage is necessary, a few *Daphnia* should be saved to seed the water for the new culture, but they should not be placed in fresh water until the organic fertilizers have had a chance to condition the water for them. Obviously there is a saving in effort if adverse conditions may be anticipated.

The cultural methods suitable for *Daphnia* are also useful for raising the small tubificid worm *Aulophorus,* but perhaps the worms require more lettuce leaves than *Daphnia* ordinarily need. These worms are especially desirable because they are small enough to be eaten by day-old poeciliid fishes. Hyman (1937) furnishes more details on their culture [see Tubificid (*Tubifex*) Worms].

For the cultivation of *Daphnia* and other useful invertebrates the grand compendium gathered by Needham et al. (1937) must be consulted. For a popular account of their life history and species, see Gordon (1941). For information on invertebrates often found in the aquarium, see Ward and Whipple (1918) and Boardman (1939).

Whiteworms and Their Culture. For the culture of whiteworms light, loamy, porous soil containing some disintegrating wood may be gathered from a woodland area, sometimes from within the trunk of a dead tree. Humus, sterilized and neatly bagged, may be purchased from any of the larger garden supply houses. The soil is placed in wooden boxes about a foot square and six inches deep. A variety of foods has been given with equal success. Some whiteworm culturists take a handful of raw, rolled oats and mix it in the soil, combing it through a few times. Advocates of rolled oats claim that the soil remains sweeter longer. The soil at the beginning may be slightly moist and crumbly, but after the worms and food are placed together in the box, the cultures are wetted down. The soil should be well moistened at all times but not soggy. Some fanciers place a pane of glass directly upon the soil to maintain an even degree of moisture; others just stack one box of worms upon another. The experienced aquarist Stoye (1935) suggests covering the cultures with a wet cloth or a dozen thicknesses of well-moistened newspaper.

White bread soaked in milk is the aquarist's favorite food for whiteworms, but many kinds of cooked cereals and vegetables have also been used. The food is placed about two inches below the surface of the soil and covered over with it. Food should not be placed on the surface, for it will be consumed by molds, nor should too much food be fed at one time.

The temperature level at which worm cultures should be kept is just as important as the kind of food they get. Indeed, in many instances the temperature factor decides the issue. Worms do not thrive well above 70° F. (22° C.). They do best between 60 and 70° F. Next to the proper temperature and food, a proper degree of moisture is of greatest importance.

Whiteworms (*Enchytareus albidus*) are quite similar to earthworms in their mating behavior. They lie apposing the ventral surfaces of their anterior ends, the heads pointing in opposite directions. Mucus is secreted until each worm becomes enclosed in a tubular "slime tube." They are hermaphroditic, and after an interchange of sperms the worms separate. Fertilization of the eggs takes place within the mucous ring, which finally slips past the anterior tip of the worm and becomes closed at both ends to form a sealed capsule or "cocoon." The fertilized eggs develop directly into young worms and escape from this capsule. Cultures of whiteworms are invaded by red mites and by the larvae of fungus gnats, but these cause little trouble. However, ants and particularly mice do considerable damage and these must be excluded.

Tubificid (Tubifex) Worms. No matter how highly *Tubifex* and related aquatic worms may be recommended to the aquarist, he will do well to remember that these worms are organisms of aquatic filth. To the aquatic biologist inspecting watercourses for sources of pollution, the discovery of living tubificid worms is his best positive clue to the presence of organic sewage. In nature these worms have a vital role in transforming the raw organic wastes into living organisms suitable and acceptable as food for fishes, amphibia, and other higher forms of aquatic life.

Tubificids are usually most plentiful about one-half mile downstream from the spot where raw pollution is dumped. On its passage down the river courses it is aerated and acted upon by bacteria. The suspended matter settles in low water areas of the flat lands and forms the characteristic black, evil-smelling muck soil. Continuing bacterial action eventually transforms the organic sewage to a substance which can be utilized by tubificid worms.

The methods of a tubificid collector were described by Armbruster (1937), who declared that the best collecting grounds were in muddy flats of tidewater streams and ponds. In the water the worms form quivering masses which resemble animated rust-colored fibrous mats. When disturbed the worms disappear into their slimy tubular homes. Armbruster scoops up the tubificid-worm-infested mud by cutting off the upper 2 inches of the soil. He uses a stout metal-rimmed net fitted with strong curtain scrim of $\frac{1}{8}$-inch mesh. The mass is washed in a sluiceway for a rough separation of the worms from the soft muck.

Armbruster's method of separating worms from the muck is hard on the worms but it is effective. He places the crude worm mass in cans to the depth of about 3 inches. The cans are tightly closed and placed in a very warm room at about 80 to 90° F. This causes some suffocation, and the worms, after a 2-hour treatment, are forced to come to the surface. They gather in dense masses and are picked up

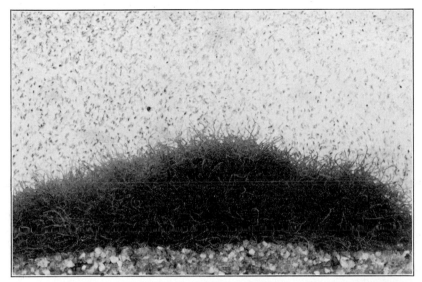

Fig. 11. Live food. The worms are tubificids, the common genera of which are *Tubifex* and *Limnodrilus*. The waterfleas above the worm mass are *Daphnia pulex*. (Photo by S. C. Dunton, New York Aquarium, New York Zoological Society.)

by hand and placed in cool running water. The colder they are kept, the better they will last; a few degrees above freezing will do them no harm. As the worms lose the protection of mud they cling to each other, forming red-brown balls and mats (fig. 11). In this condition the worms are sold and arrive at the laboratory. They are still far from clean, and it is dangerous to feed them immediately. The worm mass should be placed in running water in a shallow container like a large white-enameled photographic tray about 2 inches deep. One end of the tray should be tipped slightly by placing a wedge underneath it; then let a stream of water play upon the surface so that it overflows. It is far better to lose a few worms, when some of them become disengaged from the main mass and float over the side of the tray, than to pollute the water of aquaria by feeding the worms inadequately washed. Not only must all the residual filth surrounding the worms be washed away, but they must be washed until the contents of their alimentary canals are entirely evacuated. In the laboratory adequate washing requires about 48 hours, but the mass of worms should be checked for cleanliness regularly by being turned completely over in its water basin. The presence of a gray, foul-smelling slimy mass indicates that the washings should be continued. The putrid waste may be removed conveniently with a small, square-shaped, fine-meshed fish net.

When the worms are completely purged they are fed to the fishes by taking an appropriate pinch of them, nipped away from the mass with a pair of tweezers. The fish should be given as many as they can eat. Tubificids as fish food are regarded as a rich diet and should not be given continuously day after day. In our laboratory, worms are fed, when available, every third day.

Many aquarists suspect tubificids as carriers of diseases, and this may be so, in consideration of the polluted areas in which they live. Dr. Clifford Grobstein of the National Cancer Institute has obtained evidence that these worms carry the organism—or a related organism—that is probably responsible for the red-leg disease of frogs, and a somewhat similar disease was found in fishes fed on commercial "Tubifex."

According to Hyman (1937), the worms sold under the name of "Tubifex" generally consist of mixtures of genera and are seldom pure *Tubifex*. Often there are no *Tubifex* at all present, but instead other genera of the family Tubificidae, notably *Limnodrilus*, which in many localities are much more common than *Tubifex*. It is therefore better, Hyman suggests, to refer to these worms as *tubificids* than *Tubifex*.

Tubifex and *Limnodrilus* are aquatic annelid earthworms belonging to the Oligochaeta. In nature they dig their heads down an inch or so into the bottom mud, and build themselves tiny protective tubes. Their tails possess breathing organs which stick up above the surface of the surrounding muck, waving back and forth, gaining as much oxygen as they can in an environment which does not contain much oxygen. When tubificids are kept in small containers in a laboratory they form compact masses, and they need all the oxygen that can be provided. If experiments with fishes are planned which call for a continuous supply of tubificid worms, thought must be given to the possibility that these worms cannot be collected in large numbers when the collecting grounds are swept over by flood waters or when the mud flats are covered with unbreakable ice.

A related oligochaet, *Aulophorus*, may be raised in small quantities together with *Daphnia* in the laboratory, and the methods are presented in the section under *Daphnia* and Their Culture.

Snails as Scavengers

A limited number of aquarium snails are useful in an aquarium, for they aid in converting uneaten food into comparatively innocuous organic substances. Snails eat and remove some of the algae that grow on the surface of the aquarium glass and plants. Having snails in an aquarium, however, should not encourage carelessness in overfeeding fishes, for snails cannot recondition a poorly managed tank.

Snails should be eliminated from the breeding tank, or they will devour newly fertilized or developing fish eggs. Nor should many of them be permitted in the rearing tanks holding small fishes, for then they will find the food and eat it before the fishes can get at it.

Two species of snails may be recommended, the ramshorn (*Planorbis*) and the pond snail (*Physa gyrina*). The red or black ramshorn is preferable under good aquarium conditions when plants are growing vigorously. The pond snails' gluttony and great reproductive capacity often make them a nuisance, for they overdo the job that aquarists wish them to do.

If it should be necessary to propagate snails rapidly, feed them on lettuce leaves, but watch the pH of the water, for they are likely to acidify it. The use of *Nitella* in their breeding tanks is recommended.

REPRODUCTIVE BEHAVIOR PATTERNS IN SOME LABORATORY FISHES

The equipment needed and the methods employed for the breeding of the various species of laboratory fishes are determined by their reproductive behavior patterns. Aquarists usually classify the types as follows.

1. Egg Scatterers. This category is further subdivided into those fishes which scatter adhesive eggs and those which scatter non-adhesive ones. The goldfish (*Carassius*) scatter their eggs in the brush-like leaves of submerged plants or on the bushy roots of floating plants and the eggs adhere to them. The danios (*Brachydanio*) scatter their non-adhesive eggs. Those that fall in crevices of rock work or within a tangle of water plants on the bottom may escape being eaten by the fish that spawned them.

2. Egg Carriers. The medaka (*Oryzias*) male is capable of inseminating the female so that fertilization is internal. The eggs are extruded but held fast to the female by a membranous sac. The female eventually brushes itself against branches of aquatic plants, and the eggs become detached.

3. Bubble Nest Builders. The Siamese fightingfish (*Betta*) and the paradisefish (*Macropodus*) rise to the surface, engulf mouthfuls of air, and blow them out in bubbles with films of mucus. The fertilized eggs are placed in the surface nest by the male who guards them.

4. Mouthbreeders. After spawning non-adhesive eggs, parent fish of the Egyptian mouthbreeder (*Haplochromis*) and the African mouthbreeder (*Tilapia*) pick them up in their mouths and hold them there throughout the incubation period.

5. Live Bearers. This group may be subdivided into those which give birth to all their young within a few hours or minutes like the platyfishes, swordtails, and guppies, and those like the mosquitofish (*Heterandria*) and some species of *Gambusia* which produce a small number of living young every day for several days.

In the following section only the outstanding laboratory species will be treated. For general information on the aquarium and its fishes consult the books by Innes (1948a, 1948b), Stoye (1935), and Coates (1933). The following books on fishes and fisheries biology will be helpful, too: Curtis (1938), Norman (1948), and Schultz and Stern (1948). Hubbs and Lagler (1947), Eddy and Surber (1947), and Swingle and Smith (1947) will provide information and references on general ichthyology, on the life histories, and on the culture of our native fishes.

Fishes Which Scatter Their Adhesive Eggs

Goldfish. A large aquarium 36 inches or more long and 12 to 24 inches wide is highly desirable for breeding goldfish. Fanciers take plants like *Myriophyllum* with its many tiny leaflets, tie them together at the base of their stems, and allow them to float just below the surface of the water in a circular pattern (fig. 12), or they may use water hyacinths with their extensive root system. Goldfish in spawning cast their adhesive eggs upon the root hairs of surface plants or on the leaflets of the submerged aquatics.

The sexes of goldfish may be recognized most easily during the early spring months just before breeding. Like most minnows (family Cyprinidae), to which they belong, the males develop small, white, pearl-like excrescences, in this case on their opercula. The ripe, ova-

FIG. 12. Goldfish spawning ring designed by Franklin Barrett, made by tieing bunches of *Myriophyllum* together. The ring floats just beneath the surface. Two male telescope goldfish follow the female and fertilize the eggs as they are emitted. The eggs adhere to the plant filaments. (Drawing by Franklin Barrett, courtesy of W. T. Innes, 1948b.)

carrying females may be distinguished by their rotundity. As the spawning period approaches in the early spring months, the goldfish may be stimulated by the addition of some fresh water and by raising the temperature of the water a few degrees. Ordinarily these fish require cool water, between 60 to 70° F.

Fanciers generally use two or three males to every female for breeding purposes, saying that this results in a higher rate of fertilization of the eggs. Innes (1948b) reports that in the initial stages of reproductive behavior the males drive the females for several days. When both sexes are ready for spawning the males butt the female vigorously with their snouts. She swims over the cluster of tiny branches of the bushy *Myriophyllum* or the roots of the water hyacinth and releases a dozen or more eggs. The males follow immediately behind the female and release their sperm over the eggs. The fertilized eggs are adhesive and are locked in place in the vegetation. The spawning activities may continue from early morning till early afternoon.

The plants containing the adhering eggs may be removed from the breeding tank and placed in a rearing tank or pond. If the breeding tank is suitable for hatching the eggs, the parent goldfish may be taken out. The eggs, if left with the goldfish, may be eaten by them.

A large 6-inch female may produce 2000 eggs, but in general, smaller ones yield about 300 to 600. The maximum reported of a 5-year-old female was 70,000 eggs. They hatch in 3 to 6 days at 60 to 70° F. The unfertilized eggs are likely to be attacked by fungus (*Saprolegnia*), the natural history of which was presented by Gordon (1936). Some fanciers prefer to transfer the eggs to large, shallow enameled trays, where the early stages of development may be watched carefully. The eggs that die are removed immediately to prevent the spread of the invading fungus.

The very young fry should be placed in tanks with no more than 6 inches of water. They require protozoa, small *Daphnia*, brine shrimp naupuli, and fine, dust-like powdered fish food. As goldfish reach 6 weeks of age, an increasing amount of various cereals may be fed to them. For this purpose strained oatmeal is recommended as part of the diet, but it should not be used exclusively. As fishes increase in size, the kind of food indicated under Nutrition is adequate. Allee and Frank (1949) have pointed out that goldfish utilize minute particles of food in the water, and in the process goldfish ingest water.

The care, breeding, and history of the goldfish have been described in detail by Innes (1948b), Smith (1909), Wolf (1908), Matsubara (1908), and Mitsukuri (1904).

As early as 1894 William Bateson suggested that goldfish would be useful to the experimental zoologist. Berndt (1925, 1928) studied the relation of variation to heredity, and similar studies were made by Chen (1925, 1926 to 1927, 1928), Fukui (1927, 1930), Goodrich (1929), Goodrich and Anderson (1939), Koh (1934), and Wolf (1908). The genetic analyses of the many color and fin varieties (fig. 13) were carried out by Berndt (1925), Chen (1926 to 1927, 1928, 1934), Goodrich and Hansen (1931), Hance (1924), and Matsui (1934). Kasansky (1930) studied goldfish-tench hybrids. The inheritance and histology of the warty growth on the head of the lionhead goldfish were dealt with by Matsui (1925). Bell (1938) studied the pituitary,

FIG. 13. Goldfish, *Carassius auratus*, shown here are variously colored; some have comet tails while others are fantailed. (Photo courtesy of New York Aquarium, New York Zoological Society.)

and Levenstein (1939) related the cytology of this gland to the morphology of different strains. A sex-ratio analysis was made by Sasaki (1926).

The embryological and larval development of the goldfish was studied by Battle (1940) and by Goodrich and Hansen (1931). Tung, Chang, and Tung (1945) studied potencies of blastoderms. Rhythmic movements in the developing egg were revealed by Yamamoto (1934). Artificial fertilization was attempted by Berkhouse (1909). In the field of behavior and physiology, Allee, Finkel, and Hoskins (1940) and Allee, Frank, and Berman (1946) may be consulted. Fry, Black, and Black (1947) reviewed the influence of temperature and other factors on asphyxiation. In a related form, Gerbilsky (1937) traced the development of ovocytes in *Carassius carassius* under different temperatures.

Much work on the non-sexual social behavior of goldfish has been done. Schuett (1934), Escobar, Minahan, and Shaw (1936), Breder and Nigrelli (1938), and Shlaifer (1938, 1939, 1940) demonstrated and analyzed the effects of mass physiology on the individual fish, while Breder and Halpern (1946) and Breder and Roemhild (1947) discussed goldfish aggregations.

McVay and Kaan (1940) described the goldfish's digestive tract. Evans (1936) reported the relationship between vitamins and growth. Walker and Bennett (1945) reported the size relationship in the optic system of the telescope-eyed fish. A study of their lateral line system of canals was made by G. M. Smith (1930).

The effects of thyroid feeding and oxygen consumption of the goldfish were studied by Etkin, Root, and Mofshin (1940); the effects of testicular substance on their weight were worked out by Stanley and Tescher (1931, 1932). The goldfish was also used for assay of the thyrotropic hormone by Gorbman (1940). Radiation studies were made by Smith (1932a, 1932b), Goodrich and Trinkaus (1939), Ellinger (1940), and Davison and Ellinger (1942). Rasquin (1946) studied the relationship of blindness to pigmentation. Castration experiments were performed by Hansen (1931). Goodrich and Nichols (1933) succeeded in transplanting scales of one variety in another.

When maintained in cold water, 60 to 70° F., and not crowded, the goldfish is extremely hardy. It has been used extensively in studies of toxicity [see Gersdorff (1930), Lapenta (1932), Asher and Allee (1940), Macht (1943), and Surber (1948)].

A fibrosarcoma of the skin in a goldfish was described by Schamberg and Lucké (1922).

Blind Characin. In the underground streams and pools of La Cueva Chica, the blind characins (*Anoptichthys jordani*) were discovered near Tampaon, Mexico [see Jordan (1937), Innes (1937, 1948a), Bridges (1940), and Osorio-Tafall (1943)]. Breder and his associates have demonstrated that this and related species are extremely useful as laboratory animals (fig. 14).

The Everglades Aquatic Nurseries of Tampa, Fla., among others, have specialized in the breeding of the blind characin, which fortunately is in demand as an aquarium fish (Greenberg, 1938). The fish is hardy, omnivorous, and tolerant of temperature changes, and despite

its lack of eyesight manages to thrive as well as, and in many instances
better than, those with normal vision. In an interview at Tampa, Roy
Bast, in charge of these fishes, gave me the following notes on their
reproductive behavior. To secure breeders he maintains about twenty
males and an equal number of females in separate aquaria. He prepares the breeding tank in the following manner: A 50-gallon tank
is filled with clear, artesian-well water which, in Florida, has a slightly
sulphurous taste and odor. The water temperature is 75° F., and it

Fig. 14. The blind characin, *Anoptichthys jordani,* an egg layer from La Cueva
Chica, San Luis Potosi, Mexico, reaching 4 inches. (Photo by S. C. Dunton,
New York Aquarium, New York Zoological Society.)

is slightly alkaline, registering a pH of 7.5. Roy Bast uses no sand
or gravel in the breeding tank. Since the blind characin's eggs are
adhesive, he prepares an appropriate attachment structure for them
by introducing large clusters of *Anacharis* which completely cover the
surface layer of water.

Two large ripe males and one female are chosen from the breeders
and placed in the 50-gallon tank sometime during an afternoon. The
lighting conditions are seemingly unimportant, yet spawning, if it takes
place, occurs usually between dawn and 8 A.M., sometimes extended to
10 A.M. Roy Bast has never observed spawning during the afternoon.
If spawning does not take place within 3 days, the breeders are
changed.

If all conditions are right, the males sense the presence of the
female, become more active, and swim faster. A pair meets at a
corner at the bottom of the aquarium; they swim side by side and
dash upward together. As they hit the surface layer of plants they
violently break loose, and the female's eggs are scattered in all direc-

tions and most of them become attached to the leaves and stems of the *Anacharis*. As the breeding pair breaks apart, a bloody red patch appears around the base of the male's anal fin. Bast thinks that the tiny hook on the male's anal fin is somehow engaged and holds the female as they rise together in the spawning act. He also observed that while two males were available only one takes part in spawning.

According to this fancier, female blind characins may deposit 5000 eggs each, and he has raised 2000 to 5000 young from a pair. The female seems emaciated after spawning, but if well fed, particularly with mosquito larvae, it regains its rotundity and is capable of spawning again in about 2 months. After spawning the parents are removed from the breeding tanks. The fertilized eggs are tiny, amber-colored, glass-like spheres. After 48 hours most of the plants are shaken to free them of the eggs, and then they are removed from the breeding tank. The transparent fry struggle towards the darker section of the aquarium and attach themselves to the glass at the backs of their heads. When 3 to 4 days old they creep up the side of the glass, looking for food, releasing and regaining their hold. When born their eyes seem to be normal and appear to function, for the fry were observed to catch tiny organisms by sudden sideswipes of their heads. Later, in 3 weeks, their eyes are covered over by the growth of the adjacent tissue.

As soon as they hatch they are fed a pulverized dried food, tiny *Daphnia* and mosquito larvae sifted through fine organdy cloth. The size of food particles is increased later, and larger food organisms are fed. When an inch long, the fish will accept any kind of food, seeking it out by their constant wandering movements.

In an effort to elucidate broadly the factors involved in the evolution of cave forms, Breder *et al.* have studied the several kinds of Mexican blind characins (*Anoptichthys*), the closely related, eyed *Astyanax mexicanus,* and various tank-bred hybrids and naturally occurring intermediates between them. The structure of these fishes' eyes was correlated with their behavior by Gresser and Breder (1940), Breder and Gresser (1941a, 1941b), Breder (1943b, 1944), and Breder and Rasquin (1947b). That eye remnants had no influence on growth was demonstrated by Breder and Rasquin (1947a). The ecology of *A. jordani* was described by Breder (1942, 1943a) and Osorio-Tafall (1943). The taxonomy of related forms was reviewed by Alvarez (1946, 1947). Chemical sensory reactions were analyzed by Breder and Rasquin (1943), and pigmentary changes associated with blindness and life in total darkness by Rasquin (1947). Social behavior and oxygen consumption in light and darkness were measured by Breder and Roemhild (1947) and Schlagel and Breder (1947). Breder (1945) described an anomalous specimen, living without a lower jaw. For further information concerning the structure of the characin eye and the phylogenetic history of the teleost eye see Walls (1942). For a description of a remarkable case of lymphosarcoma in the eyed *Astyanax,* see Nigrelli (1947).

Fɪɢ. 15. The tiger killifish, *Epiplatys chaperi,* an egg-laying cyprinodont species from the Gold Coast streams of Africa, reaching 2 inches. (Photo by S. C. Dunton, the New York Aquarium, New York Zoological Society.)

Tiger Killifish. The tiger killifish (*Epiplatys chaperi*) are egg-laying cyprinodonts from the west coast streams of equatorial Africa (fig. 15). They thrive at temperatures between 70 and 80° F. Their eggs are deposited on submerged, floating plants and upon the roots and root hairs of surface aquatics. In the process of spawning the male drives the female, then they swim side by side, pressed closely together. At the instant of oviposition the pair bend their bodies into an S-shaped curve and hold this position rigidly for 5 to 15 seconds. Then, perfectly synchronized, they straighten out quickly and at that instant the eggs, apparently fertilized, are extruded into the water. Every egg has several long, sticky threads which act as holdfasts and anchor it to a plant leaf or stem. Occasionally an egg falls free and lands on the bottom of an aquarium. Roberts (1935) noticed that one spawning lasted from 7:30 A.M. to midafternoon, but the usual egg-laying period is generally over by late morning. Eggs hatch in about 6 days when the temperatures run from 80 to 85° F., or in about 12 days at 70 to 75° F. The fry move along the surface and at times browse on the protozoan life attached to plants. When well fed and cared for, they reach adult size in about 7 months, but the males attain their nuptial colors at about 1 year. Sexually mature males have brilliant red throats and lower lips. Their sex may be recognized early by a gradually developing pointed extension of the lower rays of the tail fins. Those of the female retain their well-rounded contours.

Epiplatys has been used successfully by Rasmussen (1948) and others in experimental embryology.

Fig. 16. The zebrafish, *Brachydanio rerio,* an egg-scattering minnow from India and Bengal, reaching 1½ inches. (Photo from the New York Aquarium, New York Zoological Society.)

Fishes Which Scatter Their Non-adhesive Eggs

Zebrafish. The zebra danio (*Brachydanio rerio*) is a common aquarium fish about 1½ inches long (fig. 16). It and a related species, *B. albolineatus* (fig. 17) are readily available and have been used in experimental work, particularly in the fields of embryology. In spawning, these cyprinid fish scatter their eggs, which are not adhesive. They fall to the bottom of their spawning grounds where they are in great danger of being eaten by other fishes and sometimes by those that spawned them. Should they fall in bacterial ooze, they probably would suffocate. For this reason clean spawning aquaria are desirable.

The breeding tank for zebrafish should be not less than 16 inches long and the water no deeper than 6 inches. The bottom covering should consist of fairly coarse pebbles or large marbles between which the eggs will fall during spawning activities. To further protect the eggs, sprays of *Nitella* or *Riccia* should be anchored in the pebbles to discourage the adults' tendencies toward ovacide.

When a female is ripe with eggs the distension of her body is quite perceptible. The male is slightly smaller, slimmer, and more intensely colored. The general practice is to use two males and one female in a breeding aquarium containing clear, conditioned water. The female

FIG. 17. The pearl danio, *Brachydanio albolineatus,* an egg-laying species from Burma and India, reaching 2½ inches. (Photo by Myron Gordon.)

is first placed in the tank toward evening, and 24 hours later during the evening the two males are introduced.

Spawning is influenced by proper light conditions. If the morning light is adequate, the fish will most likely spawn in the early hours following daybreak. In spawning the males pursue and butt the female, and as the eggs are expelled they are fertilized. One to four hundred slightly tinted, glass-like eggs are shed and lodge safely in between the pebbles and marbles. When eggs are detected among the pebbles, the parent fish should be removed from the spawning tank and placed in separate tanks for another spawning at a later time. When the fishes are properly maintained, ten spawnings a year may be expected.

Danio eggs hatch in 48 hours at 80° F. The young fry hold fast to the glass, plants, or other objects. Within 2 days more the fry begin eating vigorously the various microorganisms or pulverized fish food with which they must be supplied. Breeders usually sprinkle the food in dust-like particles almost as soon as the parents are removed, so that an infusion of protozoa may develop within the next 48 hours. As the fry grow they should be transferred to larger aquaria, but the depth of the water should not be greater than 6 inches.

Creaser (1934) appreciated early the suitability of the zebrafish for embryological research and provided information and special techniques for handling them so that the developmental stages of the eggs might be available at any specified time throughout the year. Roosen-Runge (1938, 1939) presented the original data on the bipolar differentiation cleavage in the eggs of the zebrafish, and Rugh (1948) has summarized the developmental stages of its eggs at approximately 15-minute intervals to the age of 146 minutes at 27° C. Lewis and Roosen-Runge (1943) described the formation of its blastodisk, which was recorded by motion pictures. Lewis (1943) emphasized the role of the superficial gel layer in gastrulation. In the field of pigment-cell growth and regeneration, Goodrich and Nichols (1937) found the zebrafish a useful experimental animal.

The aggregating behavior of the danio was studied by Noble (1939), Breder and Halpern (1946), and Breder and Roemhild (1947).

Egg Carriers

Medaka. The medakas (*Oryzias latipes*) are small fish about 1 to 2 inches long, hardy, omnivorous, and tolerant of wide ranges of temperature, 50 to 90° F. (fig. 18). According to Robinson and Rugh (1943), spawning is determined by certain conditions of light periodicity. Oka (1931b, 1931c) pointed out that the males have longer dorsal and broader anal fins than the females and other different secondary sex characters. In spawning the male approaches the female in semi-circular movements. During the actual act the pair are side by side, and at this time the female is inseminated. Instead of being released, a complement of seven to eighteen fertilized eggs is held together within a membranous sac, and this sac is attached to the abdomen of the female. She swims carrying the grape-like cluster of crystal-clear eggs for varying periods of time up to several hours. They are ultimately scraped from her body as she passes through the tangles of aquatic plants. When freed from their sac the eggs have thread-like holdfasts which adhere to the filaments of such plants as *Nitella* and *Riccia*. In an exceptional medaka embryos were found within the body of a female by Amemiya and Murayama (1931).

Two female medakas spawned fourteen times in May, and one of them produced about 450 eggs, according to Lee (1936). The eggs hatched out in 12 days at 70° F. Medaka fry grow rapidly on protozoa cultures, brine shrimp (*Artemia*) naupuli, tiny *Daphnia*, fine grades of dried fish foods, and small amounts of the yolk of a hard-boiled egg squeezed through a fine mesh cloth. The number of fry reared depends to a great measure on the availability of the right kind, and upon the abundance, of protozoa, copepods, and other appropriate microscopic organisms during the first few days of their life. The developmental stages are described by Solberg (1938b) and by Rugh (1948).

FIG. 18. The medaka, *Oryzias latipes*, an egg-carrying cyprinodont from Japan. The female is above, the male below. (Photo by S. C. Dunton, the New York Aquarium, New York Zoological Society.)

The medaka is native to Japan, China, and Korea, but it is a common aquarium species all over the world. There are several color varieties, the genes for which are sex linked. They have been studied by Aida (1921, 1930, 1936), who has advanced our knowledge of sex determination in the lower vertebrates. Oka (1931a) discussed allelism of the genes for color, and Winge (1937) reviewed the work on sex determination. Preliminary studies had been begun by Ishikawa (1913), Toyama (1916), and Ishiwara (1917). Goodrich (1927, 1933) used the genetic strains of the medaka for a study of the development of Mendelian characters.

The Japanese biologists have utilized their species in a number of experiments. Kamito (1928) observed their habits and early development. Amemiya and Murayama (1931) pointed out the usefulness of the *Oryzias* eggs for student embryological demonstrations. Yamamoto (1931a, 1939c) in a long series of nine studies described the rhythmic movements within the egg protoplasm and in early embryos. Ikeda (1934a, 1934b, 1937a, 1937b) studied the chemical constitution of *Oryzias* ova and the effects of various agents on their development. The egg membrane was studied by Kamada (1935) and Yamamoto (1936, 1939a, 1939b), and the latter extended his investigations (1940) to the changes in volume of the eggs at fertilization. Waterman (1940a) repeated some of Yamamoto's work on the contractile activity of the extra-embryonic cells. Waterman (1939, 1940b) also studied the effects of colchicine and 2,4-dinitrophenol on the early development of *Oryzias*. In other experiments with the medaka, Yamamoto (1931b) and Matsuura (1934) studied the influence of drugs on the heart beat. Ono (1927) used tissue culture techniques and later studied the problem of orientation (1937a). Solberg (1938a) applied x-rays to their germ cells, and Yamamoto (1938) subjected the fish to variations of temperature. For purposes of comparative study, information concerning the spawning habits and developmental stages of a related species, *Oryzias melastigma,* is supplied by Mookerjee and Basu (1946).

Bubble Nest Builders

Fightingfish and Paradisefish. The male fightingfish, *Betta,* constructs a nest by rising to the surface, taking a gulp of air, and blowing it out in a form of a bubble which is surrounded by a mucous secretion. The fish continues to do this until hundreds of bubbles are formed. The bubbles stick to each other, and all cluster about a floating plant or in the corner of the tank. As more bubbles are formed and are inserted beneath the mass, some of them are lifted above the surface of the water and form a frothy mound.

Soon upon completion of the nest the male pays increasing attention to the female, displaying by spreading its fins and special movements. Eventually the receptive female approaches the nest and assumes a somewhat vertical position, and the male curves his body around hers in a semi-circular embrace (fig. 19). At this moment the eggs are shed and at the same time fertilized. Embraced, the pair slowly sinks toward the bottom of the tank, while the thin stream of eggs floats down after them. The male breaks away and gathers the slowly descending eggs in his mouth, and spews them into the bubbly surface nest. Subsequently, should the eggs drop from the nest, the male gathers them in his mouth and places them back among the bubbles.

FIG. 19. The Siamese fightingfish, *Betta splendens,* a bubble nest builder, shown in a mating embrace with the bubble nest immediately above, the fish reaching 3 inches. (Photo by courtesy of the Department of Animal Behavior, American Museum of Natural History.)

The Siamese fightingfish has many color varieties and is used in genetic, physiological, and animal behavior studies. The most interesting and informative account of its ecology in its native Siam is given by an eye witness, Smith (1945). As an experimental animal for analysis of behavior, the fightingfish is one of the best, as shown by the fundamental work of Lissman (1932). The breeding reactions in *Betta splendens* were also described by Breder (1934a) and Good-

MIDDLEBURY COLLEGE LIBRARY

rich and Taylor (1934). Ono (1937b) produced conditioned orientation in the *Betta,* and Noble (1939) described dominance behavior in the species. Noble and Borne (1941) reported the effect of forebrain lesions on sex and fighting behavior; the removal of one lobe did not inhibit courtship, but with most of the forebrain removed, all sex behavior was eliminated. A case of sex reversal was described by Svärdson and Wickbom (1942), who reviewed other instances of this phenomenon and also described the chromosomes in both *Betta* and *Macropodus.* The genetics of color varieties of the fightingfish was

FIG. 20. The paradisefish, *Macropodus opercularis,* a bubble nest builder from China and Formosa, reaching 3½ inches. (Photo by Myron Gordon.)

described by Umrath (1939) and Wallbrunn (1948) and germ-cell origin and spermatogenesis were worked out by Bennington (1936).

In the similarly behaving and phylogenetically related paradisefish, *Macropodus* (fig. 20), Dalton and Goodrich (1937) studied the chromatophore reactions in the normal and albino, and Goodrich and Smith (1937) described the genetics and histology of the pigment cells in the two-color varieties. Tomita (1936) studied the reaction of melanophores to light during the early stages of development. Schwier (1943) reviewed his previous work on the genetics of *M. opercularis, M. concolor,* and *chinensis.*

Tilapia the Mouthbreeder

The African mouthbreeder (*Tilapia macrocephala*) requires well-conditioned aquarium water. It thrives at 75 to 85° F., breeding best at 80 to 85° F. This species seems to do better when it has a considerable amount of sunlight. Since the mouthbreeders are heavy eaters, the accumulated feces and other débris should be removed

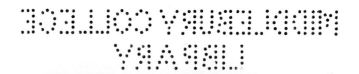

regularly, or, better, an efficient filtering mechanism should be installed. They prefer clear rather than turbid water.

The breeding males and females should be segregated in large stock tanks (48 inches long, 24 inches wide, and 12 inches high). About seventy-five adults may be kept together without fighting, but two or more fish in a smaller aquarium (24 by 12 by 12 inches) often fight to the death even when numerous hideouts are furnished.

When fish are about 3 inches long their sex may be distinguished. The males develop bright yellow opercula. In the female the central part of the operculum becomes partly transparent, and the blood-red gills beneath the "window" of the gill cover may be seen as a deep red spot. Immature, castrated male and castrated female *Tilapia* have silvery gill covers.

The breeding tank should be about 15 gallons in capacity (24 by 12 by 12 inches), filled with conditioned water, and the bottom covered with about 3 inches of well-washed gravel (Commercial Grade no. 2). The tank should have an automatic filter.

When placed together, a pair of mouthbreeders may start their breeding activities within 24 hours. The male uses his mouth to scoop out in the sand a nest area which is about 12 inches in diameter at the rim. The male is joined in the activity by the female, and when this occurs, spawning is not far off. However, preparations for the spawning act may extend from several hours to several days. Since the actual spawning act may last for only a minute or two, the experimenter wishing to observe the activity should watch for the following types of behavior: first, the presence of the nest; second, nest-building, particularly on the part of the female; third, extrusion of the female genital tube. Finally, probably the last prespawning action occurs when the male and female swim over their nest and rub their extended genital tubes on the substratum. This is called "nest-passing" and is usually followed by the deposition of the fertilized eggs (fig. 21).

After each spawning the eggs are picked up in the mouth of the male, although occasionally the female will do the same. (In the Egyptian mouthbreeder, *Haplochromis multicolor*, it is the female which generally takes over the mouth-incubation duty.)

During the first few days of oral incubation the mouth of the parent African mouthbreeder is distinctly distended. This is the sign for which the fish fancier must watch once or twice a day. By this sign the presence of embryos may be told even though the actual spawning act was not observed. Five or six days after the date of spawning, the number of eggs may diminish so much that it may be somewhat difficult to detect the presence of the developing eggs in the parent's mouth.

These notes were kindly given to me by Dr. Lester R. Aronson, who tells me that in his Animal Behavior Laboratory at the American

Fig. 21. The African mouthbreeder, *Tilapia macrocephala,* depositing its eggs in an excavated nest; later these eggs are picked up by the male in its mouth and there incubated until they hatch. (Photo by courtesy of the Department of Animal Behavior, American Museum of Natural History.)

Museum of Natural History about 60 per cent of the spawnings are ineffectual, but those which produce viable young yield a surprisingly high number of offspring.

The eggs hatch within the parent fish's mouth and the fry are released between the 7th and 20th day (usually the 14th). The danger of the parents' eating their newly liberated fry may be minimized by heavy plantings of *Nitella* just before the release of the young. When known ages of the developing embryos are required for experimental work, Dr. Aronson recommends that the parent be caught with a large net on the 8th to the 10th day after oviposition, and the young may be forcibly ejected by scooping the fry out of the parental mouth by means of a wire loop.

The fry, from the moment of their independent existence, are fed on the finest grade of powdered food (see Nutrition). As they grow older larger particles are given of the dried food and the wet mash, too. Live food is always acceptable at every age.

Tilapia have a great tolerance for common salt. This knowledge may be used to advantage in curing fish which have become fungused after injury or after surgical operations. They may be placed, until recovery is complete, in a solution of 30 grams of rock or evaporated sea salt dissolved in 2 gallons of aquarium water.

Tilapia about 3 to 4 inches long are especially useful in experimental work involving several types of surgical operations. Castration in both males and females, forebrain and cerebellar lesions and ablations, and partial hepatectomy have been performed successfully. Dr. Aronson claims that with a little practice the mortality following these operations can be reduced almost to zero.

Aronson (1945) analyzed the influence of the stimuli provided by the male cichlid on the spawning frequency of the female and (1948) described their reproductive behavior. Aronson and Holz-Tucker (1947) studied the effects of hormones on pigmentation. Tavolga and Nigrelli (1947) described one of this fish's protozoan parasites, *Costia*.

Undoubtedly cichlids of several different species will become regularly employed in experimental laboratories. Their elaborate courtship and egg and brood care make them especially interesting to the behaviorist. Notable experimental studies on reproductive activities in *Aequidens* were made by Breder (1934b), in *Hemichromis* by Noble, Kumpf, and Billings (1938) and Noble and Curtis (1939), and in *Astatotilapia* by Seitz (1940). For a general review of behavior in cichlid fishes see Aronson (1948). For a general account of the relationship of hormones to behavior, see Beach (1948).

Livebearers

Within this group are the following species widely used as laboratory animals: the platyfishes (*Platypoecilus*), the swordtails (*Xiphophorus*), the guppy (*Lebistes*), the mollies (*Mollienisia*), the gambusias (*Gambusia*), and several others. All these give birth to living young which when born are capable immediately of fending for themselves. At the time of birth they may retain a bit of their embryonic yolk sac, but this is usually absorbed within a few hours, so that an additional source of food must be made available to them very soon. That these fishes are truly viviparous and not simply ovoviviparous has been demonstrated by Turner (1939, 1940a) and Tavolga and Rugh (1947), who described the extensive extra-embryonic membranes developed by poeciliid embryos, and by Scrimshaw (1945), who showed that no *dry* weight at all was lost by embryos during the month-long period of development inside the mother. Materials equal in weight to those lost through metabolism must therefore be supplied from the mother. In *Heterandria*, Scrimshaw (1944a) determined that the embryos increased considerably in dry weight during development.

In preparation for the birth of a brood of living young, a single gravid mother is allowed an aquarium which is heavily stocked with aquatic plants. If plenty of light is available, the most useful plant in providing sanctuary for young fishes in aquaria is *Nitella*, although many fanciers prefer the lesser bladderwort, *Utricularia minor*. At this time a generous diet, including some live food, is highly desirable in order to allay the female's desire to prey upon the newborn fishes. Some gravid females, particularly the mollies, are quite sensitive to handling. Thus, if a pair of fish have been together for purposes of breeding, remove the male rather than the female from the mating tank.

Platyfishes, swordtails, guppies, and mollies release all their young within a relatively short period of possibly several hours. In that time as many as 80 to 120 fishes may emerge. But in a number of species of *Gambusia* and in *Heterandria*, the mosquitofish, a few young are born every day for 3 or 4 days. This type of viviparity is superfetation—a method of parturition which has been studied in detail by Turner (1937, 1940b) and by Scrimshaw (1944b, 1945).

After the young are born, the spent female is removed to another aquarium, usually to the container of its mating partner. In genetic work a given female of a viviparous type is generally never used again with a different male, since successive broods can result from a single mating, but a single male may be tested with many females.

The young fish are fed on finely powdered food. A small portion of live *Daphnia* is most desirable at this time. While the small fish will be unable to eat the adult *Daphnia*, they can and do eat the newborn water-fleas. The large *Daphnia* will continue to produce young crustaceans until the fish reach a size which enables them to destroy the parent crustaceans. Remarkable differences in rate of growth and maturity may be noticed between broods of fishes which have been started on *Daphnia* and those which have had only dried food for the first month or two.

Basic studies of anatomy, including secondary sexual characters in viviparous species, were made by Phillipi (1909) and by Langer (1913). A valuable recent species survey was presented by Stoye (1947 to 1948) who, in a series of papers, has collected the classifica-

FIG. 22. The platyfish, *Platypoecilus maculatus,* a livebearer from the Rio Jamapa, Mexico. The upper platy is the male with the spotted-dorsal-fin gene (*Sd*), the stripe-sided gene (*Sr*), and the comet-tailed gene (*Co*). The lower left platy is a female with the spot-sided gene (*Sp*) and the crescent-tailed gene (*Cc*). The female on the lower right has the one-spot gene (*O*) and the comet-tailed gene (*Co*.). (Photo by S. C. Dunton, New York Aquarium, New York Zoological Society.)

tion data on the cyprinodont fishes and arranged them in their systematic order.

Platyfishes and Swordtails. The poeciliids have been favorite experimental animals, especially for the geneticists, endocrinologists, and investigators studying normal and atypical pigment-cell physiology and growth. Summaries of the work in the genetics of the platyfish, *Platypoecilus* (fig. 22), may be found in Gordon's (1948a) paper. Details may be found in papers by Bellamy (1922 to 1936), by Kosswig (1927 to 1948), and by Breider (1935 to 1942, bibliography in 1938) and Rust (1941). The genetics of the swordtail, *Xiphophorus* (fig. 23), parallels that of the platyfish, and similar references may be used for the most part. However, in addition, the papers of Kerrigan (1934), Breider (1938), and Gordon (1937b, 1938, 1946a) should be consulted.

The work on the genetics of melanomas in platyfish-swordtail hybrids (fig. 24) has been reviewed by Gordon (1948a), while the histological details were given by Reed and Gordon (1931), Gordon and Smith (1938a), and Levine (1948). From tissue culture tech-

FIG. 23. The swordtail, *Xiphophorus hellerii,* from the Arroyo Zacatispan, Oaxaca, Mexico. The male is above, the female below. This pair represents the seventh inbred generation of the wild strain. (Photo by S. C. Dunton, New York Aquarium, New York Zoological Society.)

niques Grand, Gordon, and Cameron (1941) described the definitive cells of the fish melanomas. These tumors, writes Heston (1948), are of particular interest in that they appear to offer an approach to the study of malignant change and open the way for the identification of the specific genes affecting malignant transformations. Gordon (1948a) has emphasized that a clearly identified giant pigment cell, the macromelanophore (fig. 25), is probably the cell of origin of the melanoma, but the problem is still under investigation. Heston suggests that normal macromelanophores and their apparent malignant cell counterparts in melanomas offer the cytologist an excellent opportunity for a comparative study of the stages in the transformation of normal to atypical cell growth. Gordon (1947c) has recently indicated that certain macromelanophores of *Xiphophorus montezumae* may be transformed in hybrids of *X. montezumae* and *X. hellerii*. The resulting tumors usually originate on the posterior part of the caudal peduncle of the hybrid and present to the experimental oncologist a convenient tool for a variety of operations, including chemical and physical methods of treatment. Li and Baldwin (1944) claimed that sesame oil induced testicular tumors in swordtails.

Fig. 24. Melanomas in hybrids of the platyfish and the swordtail. The upper fish has a melanoma at the base of the gonopodium and at the tail fin. The one below had one at the base of the tail which was amputated; the fin has regenerated and so has the melanoma. The lowermost one has a melanoma at the base of the caudal peduncle. These three fish are siblings, and the tumors on the midventral line develop in response to the spotted-belly gene (Sb) of the platyfish. (Photo by S. C. Dunton, New York Aquarium, New York Zoological Society.)

Gordon (1950) has pointed out the importance of the spatial relationship of genes, or linkage, in the production of a new pigmented tumor, an erythromelanoma in platyfish-swordtail hybrids.

The early embryology of the platyfish was described by Hopper (1943) and in greater detail by Tavolga and Rugh (1947) and Tavolga (1949). The chromosomes and chromosome numbers of the xiphophorin fishes were described and counted by Oguma and Makino (1932), Ralston (1934), Friedman and Gordon (1934), and later by Wickbom (1941, 1943). There are twenty-four pairs, except in *P. variatus*, which has twenty-five. The number in *X. pygmaeus* is unknown. Wolf (1931) worked out the history of the germ-cell origin. A discussion of sex chromosomes in this and other groups was presented by Svärdson (1945). Behavior studies with platyfish and swordtails were made by Noble (1938, 1939), Noble and Borne (1940),

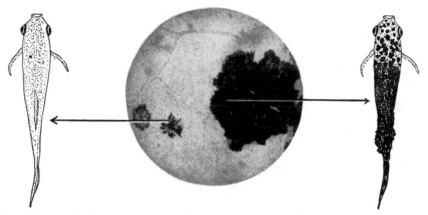

FIG. 25. Small black pigmented cells, when they appear in a hybrid fish, behave in a normal manner and the fish is not diseased. Large black pigmented cells, when they appear in a hybrid fish, grow out of bounds and eventually produce a black tumor. (From Myron Gordon, 1940.)

Braddock (1945), Gordon (1947b), and Clark, Aronson, and Gordon (1948). The development of secondary sex characters in *Xiphophorus* was described by Van Oordt (1925) and Essenberg (1923). Gordon and Benzer (1945) and Gordon and Rosen (in press) have presented the history of our knowledge of the development and transformation of the sexually modifiable anal fin into the male's gonopodium. The effects of hormones on these species were studied by Regnier (1938), Grobstein (1940, 1942a, 1942b, 1947), Baldwin and Goldin (1939, 1941), Baldwin and Li (1942), Cohen (1946), and Tavolga (1948). Grobstein (1948) recently studied the effects of extremely low doses of hormones on fishes by using well-aerated distilled water 25 per cent of which was made up of amphibian Ringer's solution (without bicarbonates). The water was changed every 48 hours to avoid contamination because of the buffering action of conditioned water. For summaries of the work in the field of sex differentiation, see Bailey

(1933) and Regnier (1938). Studies on the longevity of the sperm were made by Van Oordt (1928). Sex reversals were described by Essenberg (1926) and Gordon (1947a).

The structure and chemistry of pigment cells of swordtails, chiefly of domesticated strains (which may be influenced by platyfish genes), were carefully studied by Breider and Seeliger (1938) and by Goodrich, Hill, and Avrick (1941). Baker and Ferguson (1942) developed methods of rearing platyfish free from bacteria and other organisms (axenically). This should open up a series of experiments where the bacteria-free animals are required. Crozier and Wolf (1939) showed in a series of papers that these species and others can be tested for their visual acuity by their responses to flicker.

GENETIC METHOD OF DETERMINING THE SEX OF NEWBORN PLATYFISH. Poultrymen have used their knowledge of sex-linked inheritance in chickens to advantage in determining the sex of newborn chicks. The barring of the barred Plymouth rock is dominant to non-barring and is sex linked. When a Rhode Island red male is mated to a barred Plymouth rock female, the chicks at hatching time can be separated according to sex by their coloration. The male chicks have down feathers which are black except for a white spot on the head, and their beaks and legs are yellow. The down feathers of female chicks are solid black, and their beaks and shanks are very dark or almost black. Poultrymen have several sex-linked genes (sometimes called autosexing genes) which they employ to aid them in determining the sex of chicks. This knowledge is used in a practical way, for producers of broilers may purchase and raise the cockerel chicks, while egg producers may take the pullet chicks.

In experimental work with fishes, particularly in the field of endocrinology, it becomes extremely important to know definitely the sex of newborn fishes. In young platyfishes, swordtails, and guppies there are no reliable criteria for sex recognition by external features. [In *Gambusia*, Turner (1942a) uses early dimorphic features in the pectoral fins.] In the platyfishes the use of sex-linked genes makes it possible for the biologist to separate the sexes at the time of birth. This technique has been employed by Gordon, Cohen, and Nigrelli (1943), by Cohen (1946), and by M. C. Tavolga (1949); see figure 26.

The two opposing genetic mechanisms for sex determination have both been described in the platyfish (figs. 27, 28). The wild-caught stocks have the male heterogametic type (Gordon, 1946b, 1947a), while in domesticated stocks the females are heterogametic (Gordon, 1927, 1937a). By a series of matings between wild and domesticated individuals a special stock was developed by Gordon (1946b) in which the gene for heavy spotting (Sp) was carried by the Y-chromosomes in males, indicated as follows: $(X)+/(Y)Sp$. When this type of male was mated to the females of the wild stocks whose genetic constitution was recessive, $(X)+/(X)+$, their male offspring were heavily spotted like their fathers, $(X)+/(Y)Sp$, while the female offspring, like their

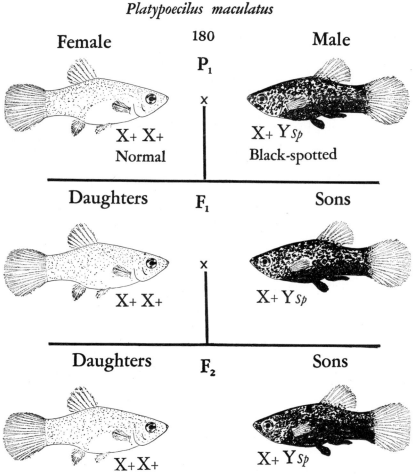

Platypoecilus maculatus

Fig. 26. Father to son sex-linked inheritance. A strain (180) of *P. maculatus* has been developed in which the dominant black-spotted gene (Sp) is carried on the Y-chromosome. When spotted males are mated to recessive females, the spotted pattern is transmitted regularly from father to son. The spotted fish may be recognized on the day of birth, and so the brood may be accurately sexed early for endocrine and other studies.

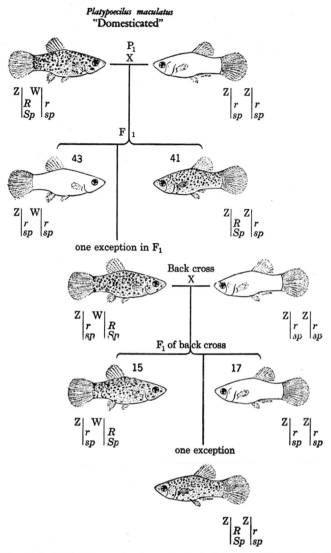

FIG. 27. Sex-linked inheritance in domesticated platyfish. This chart shows the breeding behavior of a red, black-spotted female when mated to a double recessive, gold-colored male (*shown on top row*). Criss-cross inheritance was evident: 41 sons were red and black spotted, while 43 females were golden. There was 1 exceptional female which was red and black spotted. This exceptional female (*shown to the left on row three*), when mated to a golden male, produced 15 daughters colored like herself and 17 golden sons with 1 exceptional red, black-spotted son. The (Z)r sp/(W)R Sp stock has been continued. This stock exhibits mother to daughter inheritance with an occasional exception. (Figure and explanation from Fraser and Gordon, 1929.)

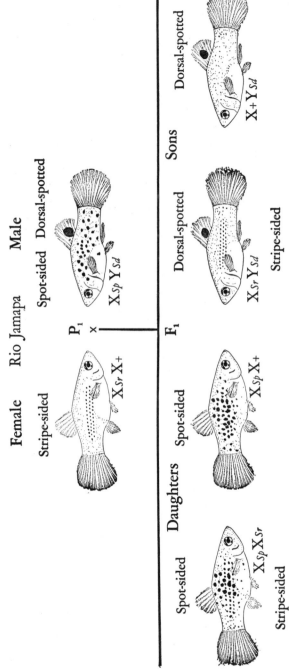

FIG. 28. Sex-linked inheritance in wild platyfish. Father to son inheritance: the dorsal-spotted gene, *Sd*; criss-cross inheritance: from father to daughters, the spot-sided gene, *Sp*; from mother to sons and daughters, the stripe-sided gene, *Sr*. Based on the XY sex-determination mechanism in the male. (From Myron Gordon, 1946b.)

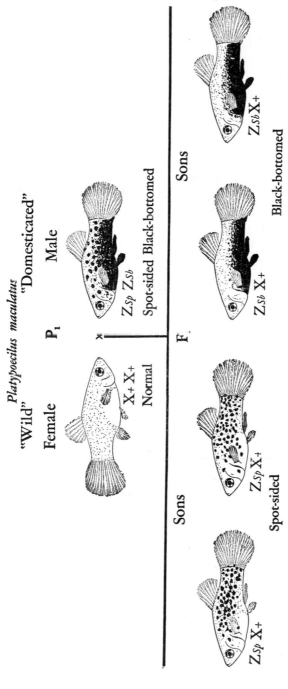

FIG. 29. One hundred per cent males in platyfish crosses. Unsexed broods may be produced by crossing a platyfish female from the "wild" strain with a male from the "domesticated" stock. The genetic constitution of the male may be written more conveniently $(Y)Sp/(Y)Sb$, because the Z- and Y-chromosomes are apparently homologous. (From Myron Gordon, 1946b.)

mothers, lacked the spotting, $(X)+/(X)+$. Thus when the fish were born those that were black spotted could be recognized by their markings as males, and those that were unspotted were females. In eight generations of brother to sister matings in this stock, only one exceptional individual appeared. This was less than one might expect. According to Fraser and Gordon (1929), crossing over of the sex chromosomes has been recorded as occurring at the rate of 1 per cent.

In studies of the influence of genetic factors on the normal and atypical growth of pigmented cells, the various stocks containing the sex-linked genes are particularly valuable. For example, one may demonstrate that in some matings the gene for the macromelanophores, the large pigment cells, is transmitted from father to son (fig. 26). In other matings, the same gene may be transmitted from mother to daughter (fig. 27). In still another mating the gene may be transmitted in criss-cross fashion (fig. 28) from a mother to her sons or from a father to his daughters. By the use of homozygous stocks the gene may be transmitted from a mother to both her sons and daughters or from the father to both his sons and daughters. By the use of heterozygous stocks the gene may be transmitted from the mother to half of her sons and to half of her daughters, or from the father to half of his sons and to half of his daughters. When a "wild" female platyfish recessive for two sex-linked genes, $(X)+/(X)+$, is mated with a "domesticated" male dominant, $(Z)Sp/(Z)Sb$, the offspring will all be males, figure 29.

Guppy. A general discussion of the variability and behavior of *Lebistes reticulatus*, the guppy (fig. 30), was presented by Fraser-Brunner (1946). The basic work on the genetic mechanism for sex determination was done by Winge (1922 to 1938), and he and Ditlevsen (1947) reviewed the status of the genetics of this species. Haskins and Haskins (1948) described the interaction of the albino gene with other color factors in the guppy. Earlier work was done by Schmidt (1919), Blacher (1928), Kirpichnikov (1935), and Haskins and Druzba (1938). Winge and Ditlevsen (1938) studied a lethal gene in the Y-chromosome of the guppy. Svärdson (1944) discussed polygenic inheritance. Goodrich, Dee, Flynn, and Mercer (1934) related the development of germ cells and sexual differentiations, and Goodrich, Josephson, Trinkaus, and Slate (1944) related the cellular expressions of various pigment cells in response to two autosomal genes. Winge (1922a) and Iriki (1932) found thirty-six diploid chromosomes. Ginsburg (1929) described the origin of the pigment cells.

Breder and Coates (1932) and Purser (1937, 1938) studied reproduction in the guppy. Dildine (1936a, 1936b) studied its embryology
and the effect of light and temperature on the gonadal development.
Vaupel (1929) described the spermatogenic process. Its follicular cells
were studied by Nichita (1928), and related studies were made by
Tuchmann (1936) on the effects of the hypophysis on morphogenesis.

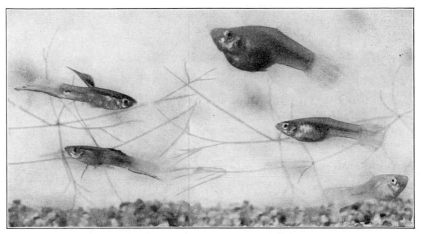

Fig. 30. The guppy, *Lebistes reticulatus,* a livebearing cyprinodont from northern South America and some islands of the Lesser Antilles The males are to
the left and show the spotted dorsal fin carrying the *maculatus* gene (*Ma*); the
lower lobe of the tail is prolonged into a sword. The uppermost female on the
right is an albino; the middle female is a wild type; and the lowermost one is
a golden. (Composite photo by S. C. Dunton, New York Aquarium, New York
Zoological Society.)

Sexual dimorphism and its development were described by Blacher
(1926) and Gallien (1946a, 1946b). In the field of endocrinology
Lebistes has been used in studying the effect of synthetic steroids by
Berkowitz (1937, 1941), Eversole (1939, 1940), Svärdson (1943), and
Scott (1944). Samokhvalova (1933, 1935) correlated the effects of
x-ray dosage upon the gonadal structure and secondary characters.
Gallien (1946a), employing endocrinological methods, used very dilute
doses of androgens on immature *Lebistes* as an aid in sex determination.
Under his treatment the very young males were recognized by their
color patterns.

Hooker (1932) showed that this species may be used for studies
of spinal cord regeneration, and Mookerjee, Mitra, and Mazumdar
(1940) used it for their analysis of the development of the vertebral

column with special reference to the centrum, notochord, hemal arch, ribs, and caudal skeleton. The effects of the thyroid hormone on the growth rate, time of sexual differentiation, and oxygen consumption in *Lebistes* were studied by Smith and Everett (1943). Observations and experimental studies on behavior and sex recognition were made by Breder and Coates (1935), Breder (1941), Fraser-Brunner (1946), and Breder and Roemhild (1947). Zahl (1933) performed a successful

Fig. 31. The black mollie, the melanistic phase of *Mollienisia sphenops* from Louisiana and inland streams through Mexico. This color variety is a cultivated production. The plant in the figure is *Chara*, but just in front of the fish's snout is a sprig of the large bladderworth, *Utricularia*. (Photo by Myron Gordon.)

Caesarean operation on a gravid female guppy. Chambers and Harvey (1931) showed that ultrasonic waves produced by the piezoelectric oscillator induced pathological changes in the tissues of the guppy.

Mollies. Several *Mollienisia* species have featured prominently in studies devoted to the analysis of apparent parthenogenesis in fishes. Hubbs and Hubbs (1932) found that when *M. sphenops* was outcrossed to *M. latipinna* the hybrids resembled the species previously known as *M. formosa.* Female hybrids (*M. formosa*) mated to males of either parental species, or to male *Lebistes*, *Limia*, or *Gambusia*, produced female offspring only, all of which were similar to their mother; none were intermediates. Meyer (1938) believed that the chromosomes of the male entered the egg at fertilization but were later inactivated.

FIG. 32. The sailfin mollie, *Mollienisia latipinna,* a livebearing cyprinodont from the coastal areas of southern United States through Mexico. (Photo from New York Aquarium, New York Zoological Society.)

FIG. 33. *Mollienisia latipinna,* the sailfin mollie. This is a rare black-spotted variety found in nature in the brackish waters of the west coast of Florida by Charles M. Breder, Jr. (Photo by S. C. Dunton, New York Aquarium, New York Zoological Society.)

The common aquarium species of mollies and their hybrids were described by Hubbs (1933). Gordon (1943a) indicated that the albino strain is recessive to the normal *M. sphenops* and that when a black mollie (fig. 31) is mated to a normal one they produce spotted intermediates. The effects of x-rays on the development of secondary sexual characters were studied by Cummings (1943a, 1943b). Scrimshaw (1946) measured the ovarian eggs. Nigrelli and Breder (1935) described a specimen that lived for three months with a prolapsed and occluded gut. *Mollienisia latipinna* are shown in figures 32 and 33.

Breider (1935a, 1935b, and 1939) has made a thorough study of the geographical distribution, sex determination, hybridization effects, and general evolution of the related species, *Limia nigrofasciata*, *L. caudofasciata*, and *L. vittata*.

Dixie Mosquitofish. The Dixie mosquitofish (*Gambusia affinis affinis* and *G. affinis holbrookii*) are not generally cultivated by aquarists owing to their cannibalistic habits. However, Dr. C. P. Haskins in his broad evolutionary and genetic work with the Poeciliini has succeeded in breeding and rearing them in numbers by the use of large containers and dense masses of *Riccia*. This plant provides cover in which the young hide from their extremely voracious parents. Owing to their pugnacity, *Gambusia* should not be placed with other cyprinodonts or with aquarium fishes in general.

These viviparous species, native of our southeastern and south central states, have been introduced into many foreign countries as aids in the campaign to control mosquito larvae. *Gambusia's* small size, hardiness, great reproductive abilities, and voracity have made them exceptionally useful in destroying the potential insect vectors of malaria and yellow fever organisms. An early study of their habits and anatomy was presented by Kuntz (1914). Hildebrand (1917) observed that a female *Gambusia* ate 165 mosquito larvae in 12 hours, and Self (1940) discussed the reproductive cycle and habits of the species in relation to mosquito control. In this country, too, *Gambusia* have been planted for mosquito control in the fairly northern latitudes of Chicago and the Great Lakes (see Krumholz, 1948a, 1948b). They may be collected during the summer months in great numbers for experimental work. In Italy, where they have been introduced, they have become useful as laboratory animals in the hands of Dulzetto (1931, 1932, 1933), Leopori (1941, 1942), Artom (1924), and other Italian biologists.

The reproductive habits of *Gambusia* were described by Seal (1911), and the ovarian egg size by Scrimshaw (1946). A detailed account of the organography of the closely related *Gambusia patruelis* was given by Potter and Medlin (1935).

Barney and Anson (1921) and Geiser (1924) studied the sex ratios in nature. Turner (1941a, 1941b) demonstrated morphological changes in the normal and regenerating anal fin of the male. He studied the effects of testosterone on the development of anal and pectoral fins in females (1941c, 1942a, 1942b, 1942c). Castration and hormone treatments were made by St. Amant (1941). The work of Sumner (1943) and his associates on *Gambusia* should be consulted for an introduction to the experimental study of protective coloration in fishes, but on this subject the suggestive paper of Breder (1947) should also be read.

LABORATORY FISHES NOT BRED IN AQUARIA

Many important contributions to biology and medicine have been made by research workers using fishes which are available from federal, state, and private hatcheries and farms. Other kinds of fishes are collected when needed from their natural habitats. The artificially propagated species are the salmons, graylings, trouts, whitefishes, and basses. Those collected from the wild include, among many others, the marine killifishes, the sticklebacks, the bitterling, the elritze, the electric eel, and the lungfishes. Methods for rearing marine fishes in the laboratory are given by McHugh and Walker (1948).

Wylie (1948) has described the opening, the facilities, and the possibilities of the Lerner Marine Laboratory of the American Museum of Natural History at the subtropic island of Bimini in the British West Indies. Aquaria with running water as well as large enclosed areas for big fishes and sharks will make it possible within the near future to add many species to the list of "laboratory" fishes. Many experiments may be conducted in the fishes' native environment.

Much of the work in fish nutrition was done with trout. It was begun at the Connecticut State Hatcheries and Cornell University, and is being continued at the Cortland, New York, hatchery of the U. S. Fish and Wildlife Service in cooperation with the New York State Department of Conservation. Most of the published results appear in the *Transactions of the American Fisheries Society,* in the *Progressive Fish Culturists,* and in the official state and federal bulletins (see Nutrition).

In embryology Eakin (1939) used the trout, Pelluet (1944) and Battle (1944) used the salmon, and Price (1934 to 1939) used the whitefish. The classical embryological study of a teleost was done on the sea bass (*Serranus*) by Wilson (1891). Reighard (1890, 1906) has also contributed detailed embryological studies of the wall-eyed pike and the fresh-water black bass. The secondary sexual characters in the salmon were studied by Tchernavin (1943), whose bibliography is extensive. Back (1932) studied the mesenchymal cells of the salmon by means of tissue cultures.

Research on bioelectricity, nerve physiology, and enzymatic agents such as cholinesterase has been performed by a team composed, in part, of an aquarist, a physiologist, and a physicist, C. W. Coates, D. Nachmansohn, and R. T. Cox, who used the South American electric eel, *Electrophorus;* see Cox, Coates, and Brown (1946) and Nachmansohn (1946). Wells (1935a, 1935b) and Sumner and Wells (1935) used the Pacific killifish, *Fundulus parvipinnis*, for studies in respiratory metabolism. The lungfishes have aided research workers in elucidating metabolism and kidney function [see H. M. Smith (1930)].

While most sticklebacks and their eggs are collected from the wild for study, some have been bred in aquaria. They are good experimental animals in work dealing with the subjects of evolution [see Heuts (1947)] and embryology and animal behavior [see Wunder (1930), Craig-Bennett (1931), Leiner (1934), Tinbergen and Van Iersel (1947)].

The bitterling, *Rhodeus amarus* (fig. 34), has been used as a test animal in assaying gonadotropins in human pregnancy urine by Kleiner, Weisman, Mishkind, and Coates (1936), Weisman and Rothburd (1938), and Kleiner (1940). A related form, the European elritze, *Phoxinus laevis* (fig. 35), has been imported into the United States for somewhat similar purposes, and Astwood and Geschickter (1936) used this species in testing for the presence of pituitary hormone in the blood of a patient suffering with a melanoma.

Methods for tagging small fishes are given by Hasler and Faber (1941).

The Atlantic coastal killifish (*Fundulus heteroclitus*), an egg-laying cyprinodont, has probably been used by more American experimental zoologists than any other species of fish. This is due, most likely, to its great abundance, convenient size, and availability, particularly at the marine biological laboratories, and especially at Woods Hole, Mass. *Fundulus heteroclitus* is not usually bred in the aquarium for the purpose of maintaining the stock, but ripe adults are caught from tidal pools and basins, and their ova and sperm taken by stripping or dissection. Methods for artificial insemination are described by various workers mentioned below.

Experimentalists unfamiliar with the taxonomy of the species of *Fundulus* along the Atlantic coast should be on their guard against using *F. majalis*, which looks superficially much like *F. heteroclitus*. The latter is a tough, hardy species that will take a great deal of

Fig. 34. The bitterling, *Rhodeus amarus,* female showing ovipositor which enables it to deposit its eggs in the mantle of a freshwater clam. It is used for test for presence of gonadotropins. (Photo by S. C. Dunton, New York Aquarium, New York Zoological Society.)

manipulation. On the other hand, *F. majalis* is more delicate and will not stand up as well under aquarium conditions. One biologist using killifish in testing the potency of a drug complained bitterly that his *Fundulus* were unreliable, for he was getting just about as many deaths among the control fish as the group being given the drug. One glance at his stock revealed that he was using two species, both of which he obtained in the same traps in a tide pool. The discrepancies in his experiments ceased after he was instructed to differentiate the species and eliminate *F. majalis* from his tests.

The reproductive behavior and secondary sexual characters of *F. heteroclitus* were observed by Newman (1907). Armstrong (1936) described the mechanisms of the hatching of the eggs. Nicholas (1927) showed how experimental methods may be applied to the study of the developing embryos. Oppenheimer (1937) delineated the normal developmental stages of the eggs and in a series of papers described (1936a) isolated blastoderms (1936b), transplantation (1936c), localization. In another paper (1936b), she presented a historical introduction to the study of teleost development. Later (1947), she re-

Fig. 35. The elritze, *Phoxinus laevis,* from Europe used for test of gonadotropins. (Photo by S. C. Dunton, the New York Aquarium, New York Zoological Society.)

viewed the subject of organization in the teleost blastoderm. The effects of external agents on developing *Fundulus* embryos were studied by Stockard (1907, 1921) and Hinrichs (1925, 1938) by means of radiation and in the formation of monsters. Solberg (1938c) studied the effects of x-rays on the embryos.

Child (1941) summarizes his views on gradients in developing embryos. Hyman (1921) indicated a secondary metabolic gradient in the posterior region. Gabriel (1944) described the critical effects of temperature on young embryos on the definitive number of vertebrae in the adults. Gordon and Benzer (1945) related Gabriel's conclusions to those of Hyman's, pointing out that there appears to be a critical developmental stage at which the number of vertebrae is probably determined. Hoadley, in two papers (1928a, 1928b), studied the potencies and viscosity changes in embryos, in developing eggs, and Coghill (1934) was concerned with myogenic activity. Richards

(1917) studied the chromosome activity in *Fundulus,* and Richards and Potter (1935) developed the mitotic index determination; this study was followed up by Jones (1939). Localization in the developing killifish was studied by Lewis (1912), regulation by Nicholas and Oppenheimer (1942), oxygen consumption by Philips (1940). Pickford and Thompson (1948) studied the effects of mammalian growth hormone in this species.

In the field of endocrinology, Matthews (1937) described the pituitary and (1938) the gonadal cycle; then he (1939a, 1939b) studied the relationship between the gonad and the pituitary under the influence of light. Marza, Marza, and Guthrie (1937) showed histochemical changes in the *Fundulus* ovary. Spermatogenesis was studied by Burger (1939). Later he experimented in hypophysectomy (1941) and the effects of androgens (1942).

Work in regeneration, transplantation, and tissue culture with *Fundulus* was done by Stone (1940) and Alexander (1942) with the eyes, Matthews (1933) with optic nerves. Tail-fin regeneration was studied by Birnie (1934). Bobkin and Bowie (1928) described the digestive system and function. Abramowitz and Fevold (1937) produced exophthalmia by use of anterior pituitary extract. Brinley and Jenkins (1939) described some of the antagonisms in regeneration processes. The ectodermal cells of *Fundulus* were studied with tissue culture methods by Dederer (1921). Warmke (1944) used the adults as test animals to estimate the potencies of the marijuana content of various genetic strains of hemp plants.

For additional references to this important species see Parker (1948), especially for pigment-cell physiology, and see Morgan (1927) for work on experimental embryology.

DISEASES

In a useful outline of the causes of diseases and death in captive fresh-water fishes at the New York Aquarium, Nigrelli (1943) discusses the following:

I. Diseases of the skin and gills
   1. *Mycobacterium* (tuberculosis)
   2. *Bacillus columaris*
   3. *Saprolegnia* (fungus)
   4. *Ichthyophthirus* (ciliate)
   5. *Childon* (ciliate)

   6. *Cyclochaeta* (ciliate)
   7. *Costia* (flagellate)
   8. *Myxosporidia* (cnidosporodial)
   9. *Lymphocystis*
   10. *Gyrodactylus* (trematode)
 II. Diseases of internal organs
   11. Gordius worms in sinus venosus
   12. Strigeid worms in kidney, liver, gonad
   13. Nematode worms in pericardium
III. Neoplastic diseases and many degenerative diseases of the
   ovary, kidney, liver, gall bladder, bone, cardiovascular system,
   etc.

For information on neoplasms in fishes see Schmey (1911), Thomas (1931), Takahashi (1934), Nigrelli (1938), and Schlumberger and Lucké (1948). For a review of melanomas in fishes see Gordon (1948). Breed, Murray, and Hitchens (1948) reviewed the bacterial diseases of fishes.

The incidence of disease may be minimized by preventing crowding, avoiding rapid temperature changes, seeing that there are no metals in contact with aquarium water, and using only conditioned water, especially free of chlorine. A balanced diet containing live food is essential, but unclean tubificid worms should not be fed.

Common rock salt is generally the best medicine for externally parasitized fish. Thirty grams of rock or evaporated sea salt (avoid table salt) dissolved in 2 gallons of aquarium water is a standard remedy. Keep the fish in the solution until they show signs of distress, then remove to fresh, conditioned aquarium water. Repeat treatment if necessary. A teaspoonful of salt may occasionally be added to a 5-gallon tank in operation, but avoid the accumulative effects of constant additions of salt.

Formaldehyde, acetic acid, sulfa drugs, and other substances have been used by fish pathologists, but their original work must be consulted [see Davis (1937), Fish (1947), Wolf (1935), and the standard works on fish diseases, Hofer (1904), Plehn (1924), and Schäperclaus (1941)].

Mention has previously been made of the work of Baker and Ferguson (1942), who reared platyfish under axenic conditions and who suggested that these fish be inoculated with specific pathogens for the purpose of studying their unique effects. Axenic fish would be useful also in the study of the effects of intestinal bacteria on the utilization of specific foods.

SOURCES OF SUPPLY

In its many years of operation, the New York Aquarium has had occasion to test a wide variety of materials, aquaria, and aquarium devices. From this experience a few fundamental pieces of equipment and material have emerged which meet its rigid specifications. The New York Aquarium does not claim that the limited numbers of items listed below are the only satisfactory ones of their type, nor does it necessarily endorse or recommend all the products of companies mentioned in this section. It is also entirely possible that companies other than those listed here manufacture equally satisfactory products.

Small aquaria, up to 48-inch size, may be obtained in quantity from the Metal Frame Aquarium Company, Pine Brook, N. J. Frames for larger aquaria requiring special types of construction and made up of angle iron may be obtained from the Pan-American Iron Works, 447–553 East 120 Street, New York 35, N. Y. To lighten the tremendous weight of very large aquaria, C. W. Coates of the New York Aquarium has designed one of heavy angle iron and panel boards, replacing the usual concrete box-like type. Atz (1947a) described the advantages of this newer type of construction. The boards are called Diestocks (Benaloid or Benalite) and are manufactured by Masonite Corporation, 630 Fifth Avenue, New York 20, N. Y. Plate glass may be obtained from either the Pittsburgh Plate Glass Company, 30 Rockefeller Plaza, New York 20, N. Y., or the Libbey-Owens Ford Glass Company, Toledo 3, Ohio.

Cements for making concrete tanks or lining wooden ones suitable for aquarium purposes without extensive and time-consuming "curing" delays are desirable. The New York Aquarium uses Medusa Waterproof Cement and Saylor Cement, which are available throughout the United States.

Aquarium cement must be tough, flexible, and non-toxic. A high grade is manufactured by J. W. Fiske Iron Works, 78 Park Place, New York 7, N. Y. Like all aquarium cements, it must be thoroughly hand-worked and hand-mixed with whiting to the proper consistency just before using.

Non-metallic, chemically inert water-conducting apparatus is essential in a closed circulation. The New York Aquarium uses hard rubber pumps and plumbing material almost exclusively. These were obtained from The American Hard Rubber Company, 11 Mercer

Street, New York 13, N. Y. Glass-lined pumps may be obtained from the Nash Engineering Company, 203 Wilson Road, South Norwalk, Conn. Ceramic-lined pumps may be purchased from the General Ceramics Company, Keasbey, N. J. Transite, a chemically inert pipe, is manufactured by Johns-Manville Company, 22 East 40 Street, New York 17, N. Y. Plastic tubing has many advantages over soft rubber tubing in conveying compressed air which is used to operate aerators, filters, and air-lifting devices. Tubing and sheeting for filters made of Saran, which is a plastic manufactured by Dow Chemical Company, 30 Rockefeller Plaza, New York 20, N. Y., is recommended. Saran tubing is fabricated in many sizes by the Elmer E. Mills Corporation, 812 West Van Buren Street, Chicago, Ill. Information concerning pH and chlorine is available in a useful pamphlet "Modern pH and Chlorine Control," issued by W. A. Taylor & Company, 7300 York Road, Baltimore 4, Md.

Water filters, aerator-stones, air-line conveyors, air valves and pumps, and a variety of connecting units may be obtained from companies which advertise their products in *The Aquarium* and *The Aquarium Journal*. The following is a partial list:

Aquarium Stock Company, 66 Warren Street, New York 4, N. Y., and 8070 Beverly Boulevard, Los Angeles 36, Calif.
Beldt's Aquarium, 2222 Crescent Avenue, St. Louis, Mo.
Crescent Fish Farm, 1624 Mandeville Street, New Orleans, La.
Eastern Gardens, Kissena Boulevard and Rose Avenue, Flushing, N. Y.
Everglades Aquatic Nurseries, 706 Plaza Place, Tampa, Fla.
General Biological Supply House, 761 East 69th Street, Chicago, Ill.
Henry's Tropical Fish Hatchery, 103-31 111th Street, Richmond Hill, Long Island, N. Y.
Nippon Goldfish Company, 1919 Bush Street, San Francisco, Calif.
Paramount Aquarium, Inc., 61 Whitehall Street, New York 4, N. Y.
William Tricker, Saddle River, N. J., or Independence, Ohio.
San Francisco Aquarium Society, Inc., Golden Gate Park, San Francisco 18, Calif. [for brine shrimp (*Artemia*) eggs].

Some of these companies sell aquaria, fishes, plants, fish food, gravel, and general aquarium supplies. Send for their catalogues. The leading journals devoted to the aquaria, fish breeding, freshwater biology, limnology, and aquiculture in general are as follows:

The Aquarist and Pondkeeper, The Butts, Half Acre, Brentford, Middlesex, England.
The Aquarium, Innes Publishing Company, 129 North 12th Street, Philadelphia 7, Pa.
The Aquarium Journal, San Francisco Aquarium Society, Golden Gate Park, San Francisco 18, Calif.

Blätter für Aquarien und Terrarienkunde, Verlag Julius E. G. Wegner, Winnenden-Stuttgart, Germany.

Deutsche Aquarien- und Terrarien-Zeitschrift. Alfred Kernen Verlag, 14a Stuttgart-W, Schloss-Strasse 80, Germany.

The Progressive Fish Culturist, United States Department of the Interior, Fish and Wildlife Service, Washington 25, D. C.

Transactions of the American Fisheries Society, P. O. Box 1495, Hartford, Conn.

Water Life, Dorset House, Stamford Street, London S.E. 1, England.

Wochenschrift für Aquarien und Terrarienkunde, Verlag Gustav Wenzel & Sohn, Braunschweig, Germany.

REFERENCES

Abramowitz, A. A., and H. L. Fevold. 1937. Experimental production of exophthalmos in *Fundulus.* Abst., *Anat. Rec.,* vol. 70, p. 123.

Aida, Tatsuo. 1921. On the inheritance of color in a fresh-water fish: *Aplocheilus latipes,* Temminck & Schlegel, with special reference to sex-linked inheritance. *Genetics,* vol. 6, pp. 554–573.

——. 1930. Further genetical studies of *Aplocheilus latipes. Genetics,* vol. 15, pp. 1–6.

——. 1936. Sex reversal in *Aplocheilus latipes* and a new explanation of sex differentiation. *Genetics,* vol. 21, pp. 136–153.

Alexander, L. E. 1942. The capacity of the eyecup of *Fundulus heteroclitus* for induction and regeneration of lenses as studied in lensless eyes. *Jour. Exp. Zool.,* vol. 91, pp. 111–117.

Allee, W. C., A. J. Finkel, and W. H. Hoskins. 1940. The growth of goldfish in homotypically conditioned water; a population study in mass physiology. *Jour. Exp. Zool.,* vol. 84, pp. 417–443.

Allee, W. C., and P. Frank. 1949. The utilization of minute food particles by goldfish. *Physiol. Zool.,* vol. 22(4), pp. 346–358.

Allee, W. C., P. Frank, and M. Berman. 1946. Homotypic and heterotypic conditioning in relation to survival and growth in certain fishes. *Physiol. Zool.,* vol. 19(3), pp. 243–258.

Alvarez, J. 1946. Revision del genero Anoptichthys con descripcion de uno especie nueva. *Anales esc. nacl. cien. biol.,* vol. 4(2–3), pp. 263–282.

——. 1947. Descripcion do *Anoptichthys hubbsi* caracinido ciego de la Cueva de los Sabinos, S.L.P., *Rev. soc. mex. hist. nat.,* vol. 8, pp. 215–219.

Amemiya, I., and S. Murayama. 1931. Some remarks on the existence of developing embryos in the body of an oviparous Cyprinodont, *Oryzias (Aplocheilus) latipes* (Temminck et Schlegel). *Proc. Imp. Acad. Japan,* vol. 7, pp. 176–178.

American Public Health Association. 1939. *Standard methods for the examination of water and sewage.* 8th ed., 4th printing. New York. Pp. 1–309.

Arber, Agnes. 1920. *Water plants.* The Cambridge University Press. Pp. xvi + 436.

Armbruster, Gustav. 1937. Truths about *Tubifex. Aquarium,* vol. 6, pp. 122–123.

Armstrong, P. B. 1936. Mechanism of hatching in *Fundulus heteroclitus.* Abst., *Biol. Bull.,* vol. 71, p. 407.

Aronson, L. R. 1945. Influence of the stimuli provided by the male cichlid fish, *Tilapia macrocephala,* on the spawning frequency of the female. *Physiol. Zool.,* Vol. 18(4), pp. 403–415.

Aronson, L. R. 1949. An analysis of the reproductive behavior of the mouth-breeding cichlid fish, *Tilapia macrocephala* (Bleeker). *Zoologica,* vol. 34(3), pp. 133–158.

Aronson, L. R., and A. M. Holz-Tucker. 1947. Morphological effects of castration and treatment with gonadal hormones on the female cichlid fish, *Tilapia macrocephala*. Abst., *Anat. Rec.,* vol. 99(4), pp. 16–17.

Artom, Cesare. 1924. La specie di Gambusia acclimata in Italia (*G. holbrookii*) in relazione colla stabilità del carattere del gonopodio. *Atti reale accad. Lincei,* Rendiconti, serie V, vol. 33, pp. 278–282.

Asher, J. F., and W. C. Allee. 1940. The effect of traces of tin on the rate of growth of goldfish. *Amer. Jour. Physiol.,* vol. 130, pp. 665–670.

Astwood, E. B., and C. F. Geschickter. 1936. The pigmentary response in *Phoxinus laevis*. The effect of blood from a patient with melanosarcoma. *Amer. Jour. Cancer,* vol. 27(3), pp. 493–499.

Atz, James W. 1947a. New tanks for old. *Animal Kingdom,* vol. 50(3), pp. 84–88.

——. 1947b. Water is much more than H$_2$O. *Animal Kingdom,* vol. 50(4), pp. 114–119.

——. 1949a. The myth of the balanced aquarium. *Nat. Hist.,* vol. 58(2), pp. 72–77, 96.

——. 1949b. The balanced aquarium myth. *Aquarist,* vol. 14, pp. 159–160, 179–182.

Back, A. 1932. Origine et développement du mesenchyme chez les salmonides. *Arch. anat. microsc.,* vol. 28, pp. 223–267.

Bailey, R. J. 1933. The ovarian cycle in the viviparous teleost *Xiphophorus helleri*. *Biol. Bull.,* vol. 64, pp. 206–225.

Baker, J. A., and M. S. Ferguson. 1942. Growth of Platyfish (*Platypoecilus maculatus*) free from bacteria and other organisms. *Proc. Soc. Exp. Biol. Med.,* vol. 51, pp. 116–119.

Baldwin, F. M., and H. S. Goldin. 1939. Effects of testosterone propionate on the female viviparous teleost, *Xiphophorus helleri* Heckel. *Proc. Soc. Exp. Biol. Med.,* vol. 42, pp. 813–819.

Baldwin, F. M., and H. S. Goldin. 1941. Effect of testosterone propionate on anal fin transformation of female viviparous teleost, *Xiphophorus helleri* Heckel. *Proc. Soc. Exp. Biol. Med.,* vol. 46, pp. 283–284.

Baldwin, F. M., and Min-Hsin Li. 1942. Effects of gonadotropic hormone in the fish, *Xiphophorus helleri* Heckel. *Proc. Soc. Exp. Biol. Med.,* vol. 49, pp. 601–604.

Barney, R. L., and B. J. Anson, Jr. 1921. Seasonal abundance of the mosquito destroying top-minnow, *Gambusia affinis,* especially in relation to male frequency. *Ecology,* vol. 2, pp. 53–69.

Bateson, Wm. 1894. *Materials for the study of variation.* Macmillan, London. Pp. 1–598.

Battle, Helen I. 1940. The embryology and larval developments of the goldfish (*Carassius auratus* L.) from Lake Erie. *Ohio Jour. Science,* vol. 40, pp. 82–93.

——. 1944. The embryology of the Atlantic salmon (*Salmo salar* Linnaeus). *Can. Jour. Res.,* Sec. D, Zool. Sci., vol. 22(5), pp. 105–125.

Beach, F. A. 1948. *Hormones and behavior.* Paul B. Hoeber, New York. Pp. xiv + 368.

Bell, W. R. 1938. Morphology of the hypophysis of the common goldfish (*Carassius auratus* L.). *Zoologica,* vol. 23(3), pp. 219–234.

Bellamy, A. W. 1922. Sex-linked inheritance in the teleost, *Platypoecilus maculatus*. *Anat. Rec.*, vol. 24, pp. 419–420.

——. 1924. Bionomic studies of certain teleosts, Poeciliinae. I. Statement of problems, description of material, and general notes on life history and breeding behavior under laboratory conditions. *Genetics*, vol. 9, pp. 513–529.

——. 1936. Interspecific hybrids in Platypoecilus, one species ZZ-ZW, the other XY-XX. *Proc. Nat. Acad. Science*, vol. 22, pp. 531–535.

Bennington, N. L. 1936. Germ cell origin and spermatogenesis in the Siamese fighting fish, *Betta splendens*. *Jour. Morph.*, vol. 60, pp. 103–125.

Berkhouse, Jerry R. 1909. Some experiments on the artificial expression and fertilization of goldfish. *Trans. Amer. Fish. Soc.*, 1908, pp. 142–144.

Berkowitz, Philip. 1937. Effect of estrogenic substances in *Lebistes reticulatus* (Guppy). *Proc. Soc. Exp. Biol. Med.*, vol. 36, pp. 416–418.

——. 1938. The effects of estrogenic substances in *Lebistes reticulatus* (Guppy). *Anat. Rec.*, vol. 71, pp. 161–176.

——. 1941. The response of fish (*Lebistes reticulatus*) to mammalian gonadotropins. *Jour. Exp. Zool.*, vol. 86, pp. 247–254.

Berndt, Wilhelm. 1925. Vererbungsstudien an Goldfischrassen. *Zeit. Induk. Abstamm. u. Vererb.*, vol. 36, pp. 161–349.

——. 1928. Wildform und Zierrassen bei der Karausche. *Zool. Jahrb. Abteil. Zool. u. Physiol.*, vol. 45, pp. 841–972.

Birnie, J. H. 1934. Regeneration of the tail fins of *Fundulus* embryos. *Biol. Bull.*, vol. 66, pp. 316–325.

Blacher, L. J. 1926. Fall von Hermaphrochitismus bei *Lebistes*. *Trans. Lab. Exp. Biol. Moscow*, vol. 1, pp. 90–95.

——. 1928. Materials on the genetics of *Lebistes reticulatus*. II. *Trans. Exp. Biol. Zoopark Moscow*, vol. 4, pp. 245–253.

Boardman, E. T. 1939. Field guide to lower aquarium animals. *Cranbrook Inst. Science Bull.* 21, pp. 1–86.

Bobkin, B. P., and D. J. Bowie. 1928. The digestive system and its function in *Fundulus heteroclitus*. *Biol. Bull.*, vol. 54, pp. 254–278.

Braddock, J. C. 1945. Some aspects of the dominance–subordination relationship in the fish *Platypoecilus maculatus*. *Physiol. Zool.*, vol. 18(2), pp. 176–195.

Breder, C. M., Jr. 1931. On the organic equilibria in aquaria. Abst., *Copeia*, (2), p. 60.

——. 1934a. The reproductive habits of the painted betta. *Bull. N. Y. Zool. Soc.*, vol. 37(5), pp. 126–133.

——. 1934b. An experimental study of the reproductive habits and life history of the cichlid fish, *Aequidens latifrons* (Steindachner). *Zoologica*, vol. 18(1), pp. 1–42.

——. 1935. The aquarium and research. *Bull. N. Y. Zool. Soc.*, vol. 38(4), pp. 110–119.

——. 1941. Respiratory behavior in fishes not especially modified for breathing air under conditions of depleted oxygen. *Zoologica*, vol. 26(25), pp. 243–244.

——. 1942. Descriptive ecology of La Cueva Chica, with especial reference to the blind fish, *Anoptichthys*. *Zoologica*, vol. 27(3), pp. 7–15.

——. 1943a. Apparent changes in phenotypic ratios of the characins at the type locality of *Anoptichthys jordani*. *Copeia*, (1), pp. 26–30.

——. 1943b. Problems in the behavior and evolution of a species of blind cave fish. *Trans. N. Y. Acad. Science*, ser. II, vol. 5(7), pp. 168–176.

Breder, C. M., Jr. 1944. Ocular anatomy and light sensitivity studies on the blind fish from Cueva de los Sabinos, Mexico. *Zoologica*, vol. 29(13), pp. 131–144.

——. 1945. Compensating reactions to the loss of the lower jaw in a cave fish. *Zoologica*, vol. 30(10), pp. 95–100.

——. 1947. A note on protective behavior in *Gambusia*. *Copeia*, (4), pp. 223–227.

Breder, C. M., Jr., and C. W. Coates. 1932. A preliminary study of population stability and sex ratios of *Lebistes*. *Copeia*, (3), pp. 147–155.

Breder, C. M., Jr., and C. W. Coates. 1935. Sex recognition in the guppy, *Lebistes reticulatus* Peters. *Zoologica*, vol. 19(5), pp. 187–207.

Breder, C. M., Jr., and E. B. Gresser. 1941a. Correlations between structural eye defects and behavior in the Mexican blind characin. *Zoologica*, vol. 26(16), pp. 123–132.

Breder, C. M., Jr., and E. B. Gresser. 1941b. Further studies on the light sensitivity and behavior of the Mexican blind characin. *Zoologica*, vol. 26(28), pp. 289–296.

Breder, C. M., Jr., and F. Halpern. 1946. Innate and acquired behavior affecting the aggregation of fishes. *Physiol. Zool.*, vol. 19(2), pp. 154–190.

Breder, C. M., Jr., and R. F. Nigrelli. 1938. The significance of differential locomotor activity as an index to the mass physiology of fishes. *Zoologica*, vol. 23(1), pp. 1–29.

Breder, C. M., Jr., and P. Rasquin. 1943. Chemical sensory reactions in the Mexican blind characins. *Zoologica*, vol. 28(20), pp. 169–200.

Breder, C. M., Jr., and P. Rasquin. 1947a. Evidence for the lack of a growth principle in the optic cyst of Mexican cave fish. *Zoologica*, vol. 32(3), pp. 29–33.

Breder, C. M., Jr., and P. Rasquin. 1947b. Comparative studies in the light sensitivity of blind characins from a series of Mexican caves. *Bull. Amer. Mus. Nat. Hist.*, vol. 89(5), pp. 321–351.

Breder, C. M., Jr., and J. Roemhild. 1947. Comparative behavior of various fishes under differing conditions of aggregation. *Copeia*, (1), pp. 29–40.

Breed, R. S., E. G. D. Murray, and A. P. Hitchens. 1948. *Bergey's manual of determinative bacteriology*. 6th ed. Williams and Wilkins, Baltimore. Pp. xvi + 1529.

Breider, Hans. 1935a. Geschlechtsbestimmung und Differenzierung bei *Limia nigrofasciata, caudofasciata* und *vittata* und deren Artbastarden. *Zeit. f. Induk. Abstamm. u. Vererb.*, vol. 68(2), pp. 265–299.

——. 1935b. Eine Mendel-Analyse von Artmerkmalen geographisch vik ariieran-der Arten der Gattung *Limia*. *Zeit. f. Induk. Abstamm. u. Vererb.*, vol. 71, pp. 441–499.

——. 1938. *Die Gesetz der Vererbung und Züchtung*. Gustav Wenzel & Sohn, Braunschweig. Pp. 1–188.

——. 1939. Untersuchungen zur Rassen-, Art-, und Gattungsdifferenzierung lebendgebärender Zahnkarpfenarten. *Zool. Anz. Suppl. (Verh. deutsch. Zool. Ges.,* 41), vol. 12, pp. 221–237.

——. 1942. ZW Maennchen und WW Weibchen bei *Platypoecilus maculatus*. *Biol. Zentralbl.*, vol. 62, pp. 187–195.

Breider, H., and R. Seeliger. 1938. Die Farbzellen der Gattung *Xiphophorus* und *Platypoecilus* und deren Bastarde. *Zeit. wiss. Zool.*, vol. 151(2), pp. 243–285.

Bridges, W. 1940. The blind fish of La Cueva Chica. *Bull. N. Y. Zool. Soc.*, vol. 43(3), pp. 74–97.

Brinley, F. J., and G. B. Jenkins. 1939. Studies of the antagonism between implant and host in fish embryos. *Physiol. Zool.*, vol. 12, pp. 31–38.

Bullington, Robt. A. 1947. Teaching biology with *Gambusia*. *Amer. Biol. Teacher,* vol. 9(9), pp. 261–264.

Burger, J. W. 1939. Some experiments on the relation of the external environment to the spermatogenetic cycle of *Fundulus heteroclitus* (L). *Biol. Bull.*, vol. 77, pp. 96–103.

——. 1941. Some experiments on the effects of hypophysectomy and pituitary implantations on the male *Fundulus heteroclitus*. *Biol. Bull.*, vol. 80(1), pp. 31–36.

——. 1942. Some effects of androgens on the adult male *Fundulus*. *Biol. Bull.*, vol. 82, pp. 233–242.

Carpenter, Kathleen E. 1927. The lethal action of soluble metallic salts on fishes. *Jour. Exp. Biol.*, vol. 4(4), pp. 378–390.

Chambers, L. A., and E. N. Harvey. 1931. Some histological effects of ultrasonic waves on cells and tissues of the fish *Lebistes reticulatus* and on the larva of *Rana sylvatica*. *Jour. Morph.*, vol. 52, pp. 155–164.

Chapman, W. McL. 1948. The blood of a fish. *Aquarium Jour.*, vol. 19(6), pp. 11–14.

Chen, Shisan C. 1925. Variation in external characters of the goldfish, *Carassius auratus. Contrib. Biol. Lab. Science Soc. China,* vol. 1, pp. 1–84.

——. 1926–1927. Variation, evolution, and heredity of goldfish, *Carassius auratus. Peking Soc. Nat. Hist. Bull.,* vol. 1(4), pp. 52–50 (in Appendix).

——. 1928. Transparency and mottling, a case of Mendelian inheritance in the goldfish, *Carassius auratus. Genetics,* vol. 13, pp. 434–452.

——. 1934. The inheritance of blue and brown colours in the goldfish, *Carassius auratus. Jour. Genetics,* vol. 29, pp. 61–74.

Child, C. M. 1941. *Patterns and problems of development.* University of Chicago Press. 811 pp.

Clark, E., L. R. Aronson, and M. Gordon. 1948. An analysis of the sexual behavior of 2 sympatric species of poeciliid fishes and their laboratory induced hybrids. Abst., *Anat. Rec.*, vol. 101(4), p. 42.

Coates, C. W. 1933. *Tropical fishes for a private aquarium.* Liveright, New York. Pp. xi + 226.

Coghill, G. E. 1934. Somatic myogenic action in embryos of *Fundulus heteroclitus. Proc. Soc. Exp. Biol. Med.*, vol. 31, pp. 62–64.

Cohen, H. 1946. Effects of sex hormones on the development of the platyfish, *Platypoecilus maculatus. Zoologica,* vol. 31(9), pp. 121–128.

Cole, A. E. 1941. The effects of pollutional wastes on fish life. Pp. 241–259 in *A Symposium on Hydrobiology.* University of Wisconsin Press.

Coventry, F. L., V. E. Shelford, and L. F. Miller. 1935. The conditioning of a chloramine-treated water supply for biological purposes. *Ecology,* vol. 16, pp. 60–66.

Cox, R. T., C. W. Coates, and M. V. Brown. 1946. Electrical characteristics of electric tissue. *Annals N. Y. Acad. Science,* vol. 47(4), pp. 487–500.

Craig-Bennett, A. 1931. The reproductive cycle of the three-spined stickleback (*G. aculeatus* L.). *Phil. Trans. Royal Soc. London,* ser. B, vol. 219, pp. 197–279.

Creaser, Chas. W. 1934. The technique of handling the zebra fish (*Brachydanio rerio*) for the production of eggs which are favorable for embryological research and are available at any specified time throughout the year. *Copeia*, (4), pp. 159–161.

Crozier, W. J., and Ernst Wolf. 1939. Temperature and critical illumination for reaction to flickering light. V. *Xiphophorus, Platypoecilus,* and their hybrids. *Jour. Gen. Physiol.*, vol. 23, pp. 143–163.

Cummings, Jean Bruce. 1943a. Morphogenesis of the gonopodium in *Mollienisia latipinna. Jour. Morph.*, vol. 73, pp. 1–18.

——. 1943b. Quantitative studies on the induction of gonopodia in females of *Mollienisia latipinna. Jour. Exp. Zool.*, vol. 94, pp. 351–386.

Curtis, Brian. 1938. *The life story of the fish.* D. Appleton-Century, New York. Pp. xiv + 260.

Dalton, H. C., and H. B. Goodrich. 1937. Chromatophore reactions in the normal and albino paradise fish *Macropodus opercularis* L. *Biol. Bull.*, vol. 73, pp. 535–541.

Daniel, R. J. 1947. Distribution of glycogen in the developing salmon (*Salmo salar* L.). *Jour. Exp. Biol. Med.*, vol. 24, pp. 123–144.

Davis, H. S. 1931. The influence of heredity on the spawning season of trout. *Trans. Amer. Fish. Soc.*, vol. 61, pp. 43–46.

——. 1937. Care and disease of trout. *U. S. Dept. Comm. Bur. Fish. Invest. Rept.*, vol. 35, pp. 1–76.

Davison, C., and F. Ellinger. 1942. Radiation effects on nervous system and roentgen-pigmentation of goldfish (*Carassius auratus*). *Proc. Soc. Exp. Biol. Med.*, vol. 49, pp. 491–495.

Dederer, P. H. 1921. The behavior of cells in tissue cultures of *Fundulus heteroclitus* with special reference to the ectoderm. *Biol. Bull.*, vol. 41, pp. 221–240.

Dildine, G. C. 1936a. Studies in teleostean reproduction. I. Embryonic hermaphroditism in *Lebistes reticulatus. Jour. Morph.*, vol. 60(1), pp. 261–277.

——. 1936b. The effect of light and temperature on the gonads of *Lebistes.* Abst., *Anat. Rec.*, vol. 67, suppl., pp. 61–66.

Dulzetto, F. 1931. Sviluppo e struttura del gonopodio di *Gambusia holbrookii* Grd. *Pubbl. staz. zool. Napoli,* vol. 11, pp. 62–85.

——. 1932. Sulla struttura del testicolo di *Gambusia holbrookii* Grd. *Boll. zool. nat. Napoli,* vol. 3, pp. 57–61.

——. 1933. La struttura del testicolo di *Gambusia holbrookii* Grd. e la sua evoluzione in rapporto con lo sviluppo del gonopodio. *Arch. zool. Torino,* vol. 19, pp. 405–437.

Eakin, R. M. 1939. Regional determination in the development of the trout. *Roux's Arch. f. Entwick. der Organismen,* vol. 139, pp. 274–281.

Eddy, S., and T. Surber. 1947. Northern fishes. Rev. ed. University of Minnesota Press. Pp. xii + 276.

Ellinger, F. 1940. Roentgen-pigmentation in the goldfish. *Proc. Soc. Exp. Biol. Med.*, vol. 45, pp. 148–150.

Ellis, M. M., B. A. Westfall, and M. D. Ellis. 1948. Determination of water quality. *Res. Rept.* 9, U. S. Fish and Wildlife Service, pp. 1–122.

Embody, G. C., and M. Gordon. 1924. A comparative study of natural and artificial foods of brook trout. *Trans. Amer. Fish. Soc.*, vol. 54, pp. 185–200.

Escobar, R. A., R. P. Minahan, and R. J. Shaw. 1936. Motility factor in mass physiology: locomotor activity of fishes under conditions of isolation, homotypic grouping, and heterotypic grouping. *Physiol. Zool.*, vol. 9(1), pp. 66–78.

Essenberg, J. M. 1923. Sex differentiation in the viviparous teleost *Xiphophorus helleri* Heckel. *Biol. Bull.*, vol. 45, pp. 46–97.

———. 1926. Complete sex-reversal in the viviparous teleost *Xiphophorus helleri*. *Biol. Bull.*, vol. 51, pp. 98–111.

Etkin, Wm., R. W. Root, and B. P. Mofshin. 1940. The effect of thyroid feeding on oxygen consumption of the goldfish. *Physiol. Zool.*, vol. 13, pp. 415–429.

Evans, G. 1936. The relation between vitamins and the growth and survival of goldfishes in homotypically conditioned water. *Jour. Exp. Zool.*, vol. 74, pp. 449–476.

Eversole, W. J. 1939. The effects of androgens upon the fish (*Lebistes reticulatus*). *Endocrinology*, vol. 25, pp. 328–330.

———. 1940. The effects of pregneninolone and related steroids on sexual development of fish (*Lebistes reticulatus*). *Endocrinology*, vol. 28, pp. 603–610.

Faber, H. A. 1933. The chlorine problem. *Aquarium*, vol. 1, pp. 332–333.

Fassett, N. C. 1940. *A manual of aquatic plants.* McGraw-Hill, New York. Pp. vii + 382.

Fish, F. F. 1947. A technique for controlling infectious diseases in hatchery fish. *Trans. Amer. Fish. Soc.*, vol. 74, pp. 209–222 (1944).

Fraser, A. C., and M. Gordon. 1929. The genetics of *Platypoecilus*. II. The linkage of two sex-linked characters. *Genetics*, vol. 14(2), pp. 160–179.

Fraser-Brunner, A. 1946. *The guppy.* Buckley Press, Brentford, Middlesex. Pp. 1–13.

Friedman, B., and M. Gordon. 1934. Chromosome numbers in xiphophorin fishes. *Amer. Naturalist*, vol. 68, pp. 446–455.

Fry, F. E. J., V. S. Black, and E. C. Black. 1947. Influence of temperature on the asphyxiation of young goldfish (*Carassius auratus* L.) under various tensions of oxygen and carbon dioxide. *Biol. Bull.*, vol. 92(3), pp. 217–224.

Fukui, Ken'ichi. 1927. On the color pattern produced by various agents in goldfish. *Folia Anat. Japon.*, vol. 5, pp. 257–302.

——— . 1930. The definite localization of the color pattern in the goldfish. *Folia Anat. Japon.*, vol. 8, pp. 283–312.

Gabriel, M. L. 1944. Factors affecting the number and form of vertebrae in *Fundulus heteroclitus*. *Jour. Exp. Zool.*, vol. 95(1), pp. 105–147.

Gallien, M. Louis. 1946a. Masculinisation des caractères sexuals somatiques et analyse genetique chez *Lebistes reticulatus*. *C. r. acad. sci.*, vol. 222, pp. 1522–1524.

———. 1946b. Sur un cas d'arrhenoidie chez *Lebistes reticulatus*. *Bull. soc. zool. France,* vol. 70(3), pp. 104–106.

Geiser, S. W. 1924. Sex ratio and spermatogenesis in the top-minnow *Gambusia*. *Biol. Bull.*, vol. 47, pp. 175–213.

Gerbilsky, N. L. 1937. Development of ovocytes in *Carassius carassius* and its dependence upon the temperatures. *Bull. Biol. Med. USSR*, vol. 3, pp. 160–161.

Gersdorff, W. A. 1930. A method for the study of toxicity using goldfish. *Jour. Amer. Chem. Soc.*, vol. 52, pp. 3440–3445.

Ginsburg, J. 1929. Beiträge zur Kenntnis der Guaninophoren und Melanophoren. *Zool. Jour.* (*Abt. Anat.*), vol. 51, pp. 227–260.

Goodrich, H. B. 1927. A study of the development of Mendelian characters in *Oryzias latipes*. *Jour. Exp. Zool.*, vol. 49, pp. 261–287.

——. 1929. Mendelian inheritance in fish. *Quart. Rev. Biol.*, vol. 4, pp. 83–99.

——. 1933. One step in the development of hereditary pigmentation in the fish *Oryzias latipes*. *Biol. Bull.*, vol. 65, pp. 249–252.

Goodrich, H. B., and P. L. Anderson. 1939. Variation of color pattern in hybrids of the goldfish, *Carassius auratus*. *Biol. Bull.*, vol. 77, pp. 184–191.

Goodrich, H. B., J. E. Dee, C. M. Flynn, and R. N. Mercer. 1934. Germ cells and sex differentiation in *Lebistes reticulatus*. *Biol. Bull.*, vol. 67, pp. 83–96.

Goodrich, H. B., and I. B. Hansen. 1931. The post-embryonic development of Mendelian characters in the goldfish, *Carassius auratus*. *Jour. Exp. Zool.*, vol. 59, pp. 337–358.

Goodrich, H. B., G. A. Hill, and N. S. Arrick. 1941. The chemical identification of gene-controlled pigments in *Platypoecilus* and *Xiphophorus* and comparisons with other tropical fishes. *Genetics*, vol. 26, pp. 573–586.

Goodrich, H. B., R. L. Hine, and H. M. Lesher. 1947. The interaction of genes in *Lebistes reticulatus*. *Genetics*, vol. 32, pp. 535–540.

Goodrich, H. B., N. D. Josephson, J. P. Trinkaus, and J. M. Slate. 1944. The cellular expression and genetics of two new genes in *Lebistes reticulatus*. *Genetics*, vol. 29, pp. 584–592.

Goodrich, H. B., and R. N. Mercer. 1934. Genetics and colors of the Siamese fighting fish, *Betta splendens*. *Science*, vol. 79, pp. 318–319.

Goodrich, H. B., and R. Nichols. 1933. Scale transplantation in the goldfish *Carassius auratus*. *Biol. Bull.*, vol. 65, pp. 253–265.

Goodrich, H. B., and R. Nichols. 1937. The development and the regeneration of the color pattern in *Brachydanio rerio*. *Jour. Morph.*, vol. 52, pp. 513–523.

Goodrich, H. B., and M. A. Smith. 1937. Genetics and histology of the color pattern in the normal and albino paradise fish. *Biol. Bull.*, vol. 73, pp. 527–534.

Goodrich, H. B., and H. C. Taylor. 1934. Breeding reactions in *Betta splendens*. *Copeia*, (4), pp. 165–166.

Goodrich, H. B., and J. P. Trinkaus. 1939. The differential effect of radiation on Mendelian phenotypes of the goldfish, *Carassius auratus*. *Biol. Bull.*, vol. 77, pp. 188–195.

Gorbman, Aubrey. 1940. Suitability of the common goldfish for assay of thyrotropic hormone. *Proc. Soc. Exp. Biol. Med.*, vol. 45, pp. 772–773.

Gordon, Myron. 1927. The genetics of a viviparous top minnow *Platypoecilus;* the inheritance of two kinds of melanophores. *Genetics*, vol. 12(3), pp. 253–283.

——. 1931a. Morphology of the heritable color patterns in the Mexican killifish, *Platypoecilus*. *Amer. Jour. Cancer*, vol. 15(2), pp. 732–787.

——. 1931b. Hereditary basis of melanosis in hybrid fishes. *Amer. Jour. Cancer*, vol. 15(3), pp. 1495–1523.

——. 1935. The strange history of Babington's curse. *Amer. Mercury*, June, 1935, and *Aquarium*, vol. 4, pp. 216–218 (1936).

——. 1936. Fishes beware the fungus. *Nature Mag.*, vol. 27, pp. 145–146.

——. 1937a. Genetics of *Platypoecilus*. III. Inheritance of sex and crossing over of the sex chromosomes in the platyfish. *Genetics*, vol. 22(3), pp. 376–392.

Gordon, Myron. 1937b. Heritable color variations in the Mexican swordtail-fish. *Jour. Heredity,* vol. 28, pp. 222–230.

——. 1938. The genetics of *Xiphophorus hellerii* heredity in Montezuma, a Mexican swordtail fish. *Copeia,* (1), pp. 19–29.

——. 1940. The fish that was jungle born and city bred. *Nat. Hist. Mag.,* vol. 45, pp. 96–106.

——. 1941. Water flea circus. *Nat. Hist. Mag.,* vol. 46, pp. 77–82.

——. 1943a. A study of mollies in black and white. *Aquarium,* vol. 12, pp. 8–9.

——. 1943b. Feeding platyfishes and swordtails. *Aquarium,* vol. 12, pp. 86–88.

——. 1945. The world's smallest flowering plants. *Nature Mag.,* vol. 38, pp. 202–204, 218.

——. 1946a. Introgressive hybridization in domesticated fishes. I. The behavior of comet—a *Platypoecilus maculatus* gene in *Xiphophorus hellerii. Zoologica,* vol. 31(6), pp. 77–88.

——. 1946b. Interchanging genetic mechanisms for sex determination in fishes under domestication. *Jour. Heredity,* vol. 37(10), pp. 307–320.

——. 1946c. Tape up broken aquarium glass. *Aquarium,* vol. 15(4), p. 72.

——. 1947a. Genetics of *Platypoecilus maculatus.* IV. The sex-determining mechanism in two wild populations of the Mexican platyfish. *Genetics,* vol. 32(1), pp. 8–17.

—— 1947b. Speciation in fishes. Distribution in time and space of seven dominant multiple alleles in *Platypoecilus maculatus. Adv. Genetics,* vol. 1, pp. 95–132.

——. 1947c. Melanomas in the hybrid offspring of two species of swordtails, *Xiphophorus montezumae* and *Xiphophorus hellerii.* Abst., *Anat. Rec.,* vol. 99(4), p. 57.

——. 1948a. Effects of five primary genes on the site of melanomas in fishes and the influence of two color genes on their pigmentation. *Biology of Melanomas, Spec. Pub. N. Y. Acad. Science,* 4, pp. 216–268.

——. 1948b. Get your fishes out of hot water. *Aquarium,* vol. 17, pp. 33–34.

——. 1950. Heredity of pigmented tumours in fish. *Endeavour,* vol. 9(33), pp. 26–34.

Gordon, Myron, and P. Benzer. 1945. Sexual dimorphism in the skeletal elements of the gonopodial suspensorium in xiphophorin fishes. *Zoologica,* vol. 30(6), pp. 57–72.

Gordon, Myron, Herman Cohen, and R. F. Nigrelli. 1943. A hormone-produced character in *Platypoecilus maculatus* diagnostic of wild *P. xiphidium. Amer. Naturalist,* vol. 77, pp. 569–572.

Gordon, Myron, and Allan C. Fraser. 1931. Pattern genes in the platyfish. *Jour. Heredity,* vol. 22, pp. 168–185.

Gordon, Myron, and D. E. Rosen. Genetics of species differences in the morphology of the male genitalia of xiphophorin fishes. *Bull. Amer. Mus. Nat. Hist.* (in press).

Gordon, Myron, and G. M. Smith. 1938a. Progressive growth stages of a heritable melanotic neoplastic disease in fishes from the day of birth. *Amer. Jour. Cancer,* vol. 34(2), pp. 255–272.

Gordon, Myron, and G. M. Smith. 1938b. The production of a melanotic neoplastic disease in fishes by selective matings. IV. Genetics of geographical species hybrids. *Amer. Jour. Cancer,* vol. 34(4), pp. 543–565.

Grand, C. G., M. Gordon, and G. Cameron. 1941. Cell types in tissue culture of fish melanotic tumors compared with mammalian melanomas. *Cancer Res.*, vol. 1(8), pp. 660–666.

Greenberg, Albert. 1938. Strange spawning of the blind characin. *Aquarium*, vol. 7, p. 80.

Gresser, E. B., and C. M. Breder, Jr. 1940. The histology of the eye of the cave characin. *Zoologica*, vol. 25(10), pp. 113–116.

Grobstein, Clifford. 1940. Endocrine and developmental studies of gonopod differentiation in certain poeciliid fishes. I. The structure and development of the gonopod in *Platypoecilus maculatus*. *Univ. Calif. Pub. Zool.* 47, pp. 1–22.

——. 1942a. Endocrine and developmental studies of gonopod differentiation in certain teleosts. II. Effect of testosterone propionate on the normal and regenerating anal fin of adult *Platypoecilus maculatus* females. *Jour. Exp. Zool.*, vol. 89(2), pp. 305–328.

——. 1942b. Effect of various androgens on regenerating anal fin of adult *Platypoecilus maculatus* females. *Proc. Soc. Exp. Biol. Med.*, vol. 49, pp. 477–478.

——. 1947. Decline in regenerative capacity of the *Platypoecilus maculatus* gonopodium during its morphogenesis. *Jour. Morph.*, vol. 80(2), pp. 145–160.

——. 1948. Optimum gonopodial morphogenesis in *Platypoecilus maculatus* with constant dosage of methyl testosterone. *Jour. Exp. Zool.*, vol. 109, pp. 215–237.

Hance, R. T. 1924. Heredity in goldfish. *Jour. Heredity*, vol. 15, pp. 177–182.

Hansen, J. 1931. Gonadectomy in the fish *Carassius auratus*. *Science*, vol. 73, pp. 293–295.

Haskins, C. P. 1948. Albinism, a semi-lethal autosomal recessive in *Lebistes reticulatus*. *Heredity*, vol. 2, pp. 251–262.

Haskins, C. P., and J. P. Druzba. 1938. Note on anomalous inheritance of sex-linked color factors in the guppy. *Amer. Naturalist*, vol. 72, pp. 571–574.

Haskins, C. P., and E. L. Haskins. 1948. Albinism, a semi-lethal autosomal mutation in *Lebistes reticulatus*. *Heredity*, vol. 2, pp. 251–262.

Hasler, A. D., and W. M. Faber. 1941. A tagging method for small fish. *Copeia*, (3), pp. 162–165.

Hayford, C. O., and G. C. Embody. 1930. Further progress in the selective breeding of brook trout at the New Jersey State Hatchery. *Trans. Amer. Fish. Soc.*, vol. 60, pp. 109–113.

Heston, W. E. 1948. Genetics of cancer. *Adv. Genetics*, vol. 2, pp. 99–125.

Heuts, M. J. 1947. Experimental studies on adaptive evolution in *Gasterosteus aculeatus* L. *Evolution*, vol. 1(½), pp. 89–102.

Hildebrand, S. F. 1917. Notes on the life history of the minnow *Gambusia affinis* and *Cyprinodon variagatus*. Appendix VI, Rept. U. S. Comm. Fish. for 1917. *Bur. Fish. Doc.* 857, pp. 3–15.

Hinrichs, M. A. 1925. Modification of development on the basis of differential susceptibility to radiation. I. *Fundulus heteroclitus* and ultra-violet radiation. *Jour. Morph. Phys.*, vol. 41, pp. 239–263.

——. 1938. The microscopic anatomy of turns and double monsters of *Fundulus heteroclitus*. *Physiol. Zool.*, vol. 11, pp. 155–157.

Hoadley, L. 1928a. Viscosity changes during early cleavage stages of *Fundulus* eggs. *Science*, vol. 68, pp. 409–410.

Hoadley, L. 1928b. On the localization of developmental potencies in the embryo of *Fundulus heteroclitus*. *Jour. Exp. Zool.*, vol. 52, pp. 7–44.

Hofer, B. 1904. *Handbuch der Fischkrankheiten.* Munich. Pp. xv + 359.

Holm-Jensen, I. 1948. Osmotic regulation in *Daphnia magna* under physiological conditions and in the presence of heavy metals. *Det. Kgl. Danske Vidensk. Selsk. Biologiske Meddelelser*, vol. 20(11), pp. 1–64.

Hooker, D. 1932. Spinal cord regeneration in the young rainbow fish, *Lebistes reticulatus*. *Jour. Comp. Neur.*, vol. 56, pp. 277–297.

Hoover, E. E. 1937. Experimental modification of the sexual cycle in trout by control of light. *Science*, vol. 86, pp. 425–426.

Hoover, E. E., and H. E. Hubbard. 1937. Modification of the sexual cycle in trout by control of light. *Copeia*, (4), pp. 206–210.

Hopper, Jr., A. F. 1943. The early embryology of *Platypoecilus maculatus*. *Copeia*, (4), pp. 218–224.

Hubbs, C. L. 1926. The structural consequences of modifications of the developmental rate in fishes, considered in reference to certain problems of evolution. *Amer. Naturalist*, vol. 60, pp. 57–81.

———. 1930. The high toxicity of nascent oxygen. *Physiol. Zool.*, vol. 3(4), pp. 441–460.

———. 1933. Species and hybrids of *Mollienisia*. *Aquarium*, vol. 1(10), pp. 263–268, 277.

Hubbs, C. L., and L. C. Hubbs. 1932. Apparent parthogenesis in nature, in a form of fish of hybrid origin. *Science*, vol. 76(1983), pp. 628–630.

Hubbs, C. L., and K. F. Lagler. 1947. Fishes of the Great Lakes region. *Cranbrook Inst. Science Bull.*, vol. 26, pp. xi + 186.

Hyman, L. H. 1921. The metabolic gradients of vertebrate embryos. I. Teleost embryos. *Biol. Bull.*, vol. 40, pp. 32–73.

———. 1937. Species of small worms suitable as fish food. *Aquarium*, vol. 5, pp. 247–249.

Ikeda, Yuitiro. 1934a. Permeability of the egg membrane of *Oryzias latipes*. *Jour. Fac. Science Tokyo Imp. Univ.*, vol. 3, pp. 499–504.

———. 1934b. Change of body weight of *Oryzias latipes* in an isotonic media. *Jour. Fac. Science Tokyo Imp. Univ.*, vol. 3, pp. 505–507.

———. 1937a. Effect of sodium and potassium salts on the rate of development of *Oryzias latipes*. *Jour. Fac. Science Tokyo Imp. Univ.*, vol. 4, pp. 307–312.

———. 1937b. Potassium accumulation in the egg of *Oryzias latipes*. *Jour. Fac. Science Tokyo Imp. Univ.*, vol. 4, pp. 313–323.

Innes, W. T. 1937. A cavern characin. *Aquarium*, vol. 5, pp. 200–202.

———. 1948a. *Exotic aquarium fishes.* 9th ed. Innes Pub. Co., Philadelphia. Pp. 1–520.

———. 1948b. *Goldfish varieties and water gardens.* Innes Pub. Co., Philadelphia. Pp. 1–381.

Iriki, S. 1932. Preliminary note on the chromosomes of fishes (1) *Aplocheilus latipes* and *Lebistes reticulatus*. *Proc. Imp. Acad., Tokyo*, vol. 8, pp. 262–263.

Ishikawa, Chiyomatsu. 1913. *Dobutsu-gagu* (*Lectures on zoology*). Vol. 1, p. 372. (Cited by Ishiwara.)

Ishiwara, Makoto. 1917. On the inheritance of body color in *Oryzias latipes*. *Mitteil. Med. Fakul. Kaiser. Univ. Kyushu Fuknoka*, vol. 4, pp. 43–51.

Jones, R. W. 1939. Analysis of the development of fish embryos by means of the mitotic index. V. The processes of early differentiation of organs in *Fundulus heteroclitus*. *Trans. Amer. Micro. Soc.*, vol. 58, pp. 1–23.

Jordan, C. Basil. 1937. Bringing in the new cave fish *Anoptichthys jordani* Hubbs and Innes. *Aquarium*, vol. 5, pp. 203–204.

Kamada, T. 1935. Membrane potential of the egg of *Oryzias latipes*. *Jour. Fac. Science Tokyo Imp. Univ.*, vol. 4, pp. 203–213.

Kamito, A. 1928. Early development of the Japanese killifish (*Oryzias latipes*) with notes on its habits. *Jour. Coll. Agr. Imp. Univ. Tokyo*, vol. 10, pp. 21–38.

Kasansky, W. J. 1930. Zur Morphogie der Hybriden von *Carassius carassius* L. × *Tinca Tinca* L. *Zool. Anz.*, vol. 90, pp. 168–175.

Kenyon, W. A. 1925. Digestive enzymes in poikilothermal vertebrates. *Bull. Bur. Fish.*, vol. 16, pp. 181–200.

Kerrigan, Alice M. 1934. The inheritance of the crescent and twin-spot markings in *Xiphophorus hellerii*. *Genetics*, vol. 19, pp. 581–599.

Kirpichnikov, W. 1935. Autosomal genes in *Lebistes reticulatus* and the problem of the arising genetic sex determination. *Zeit. f. Biol. Moscow*, vol. 4, pp. 343–354. (English summary.)

Kleiner, I. S. 1940. The relative effects of progesterone and testosterone propionate upon ovipositor lengthening of the female bitterling. *Endocrinology*, vol. 26(3), pp. 534–535.

Kleiner, I. S., A. I. Weisman, D. I. Mishkind, and C. W. Coates. 1936. The female bitterling as a biologic test animal for male hormone. *Zoologica*, vol. 21(21), pp. 241–250.

Koh, T. P. 1934. Notes on the evolution of goldfish. *China Jour.*, vol. 20, pp. 101–107.

Kosswig, Curt. 1927. Über Bastarde der Teleostier *Platypoecilus* und *Xiphophorus*. *Zeit. f. Induk. Abstamm. u. Vererb.*, vol. 44, p. 253.

———. 1932. Hermaphroditismus im Tierreich vom Genetischen Standpunkt. *Der Züchter*, vol. 4, pp. 22–32.

———. 1936a. Homogametische ZZ- und WW-Weibchen entstehen nach Artkreuzung mit dem in weiblichen Geschlecht heterogametischen *Platypoecilus maculatus*. *Biol. Zentralbl.*, vol. 56, pp. 409–414.

———. 1936b. Methoden der Zierfischhaltung und zucht fur wissenschaftliche (besonders genetische) Zwecke. In E. Abderhalden's *Handbuch der biologischen Arbeitmethoden*. Abt. IX, Teil 7, Heft 4, pp. 653–710.

———. 1948. Genetische Beiträge zur Präadaptationstheorie. *Rev. fac. sci. univ. d'Istanbul*, ser. B, vol. 13(3), pp. 176–209.

Krumholz, L. A. 1948a. Reproduction in the western mosquitofish, *Gambusia affinis affinis* (Baird & Girard), and its use in mosquito control. *Ecol. Monogr.*, vol. 18(1), pp. 1–43.

———. 1948b. The mosquitofish, *Gambusia*, established in the Great Lakes region. *Copeia*, (2), p. 144.

Kuntz, A. 1914. Notes on the habits, morphology of the reproductive organs, and embryology of the viviparous fishes: *Gambusia affinis*. *Bull. U. S. Bur. Fish.*, vol. 33, pp. 177–190.

Langer, W. F. 1913. Beitrage zur Morphology der viviparen Cyprinodontiden. *Gegenbauer's Morph. Jahrb.*, vol. 47, pp. 193–307.

Lapenta, V. A. 1932. Physiological assay of glucosides, toxins, and poisons on goldfish *Carassius auratus*. *Proc. Indiana Acad. Science*, vol. 41, pp. 445–448.

Lee, H. F. 1936. On behalf of the medaka. *Aquarium*, vol. 5, pp. 139–142.

Leiner, M. 1934a. Die drei europäische Stichlinge (*Gasterosteus aculeatus* L., *Gasterosteus pregnitius* L., und *Gasterosteus spinachia* L.) und ihre Kreuzungsprodukte. *Zeit. Morph. Okel. Tiere*, vol. 28, pp. 107–154.

——. 1934b. Beiträge zur ontogenetischen Entwichlung der drei europäischen Stichlingsarten und ihrer Kreuzungsprodukte. *Zeit. wiss. Zool.*, vol. 145, pp. 366–388.

Leopori, N. G. 1941. Differenziamento sessuale delle gonadi intersessulalità transitoria in *Gambusia Holbrookii* Grd. *Arch. ital. Anat. e embriol.*, vol. 46(2), pp. 170–204.

——. 1942. Inversione sperimentale delle gonadi femminili di *Gambusia Holbrookii* Grd. trattata con testosterone. *Monitore Zool. Ital.*, vol. 53, pp. 117–131.

Levenstein, I. 1939. The cytology of the pituitary gland of two varieties of goldfish *Carassius auratus* L., with some reference to variable factors in the gland which may possibly be related to the different morphological types. *Zoologica*, vol. 24, pp. 47–60.

Levine, Michael. 1948. The cytology of the typical and the amelanotic melanomas. *Spec. Pub. N. Y. Acad. Science* 4, pp. 177–215.

Lewis, R. C. 1944. Selective breeding of rainbow trout at Hot Creek Hatchery. *Calif. Fish and Game*, vol. 30(2), pp. 95–97.

Lewis, Warren H. 1912. Experiments on localization and regeneration in the embryonic shield and germ ring of a teleost fish (*Fundulus heteroclitus*). *Anat. Rec.*, vol. 6, pp. 325–333.

——. 1943. The role of the superficial gel layer in gastrulation of the zebra fish egg. *Anat. Rec.*, vol. 85, p. 326.

Lewis, Warren H., and Edward C. Roosen-Runge. 1943. The formation of the blastodisc in the egg of the zebra fish, *Brachydanio rerio*. *Anat. Rec.*, vol. 85, p. 326.

Li, Min Hsin, and F. M. Baldwin. 1944. Testicular tumors in the teleost (*Xiphophorus helleri*) receiving sesame oil. *Proc. Soc. Exp. Biol. Med.*, vol. 57(1), pp. 165–167.

Lissman, H. W. 1932. Die Umwelt des Kampffisches (*Betta splendens* Regan). *Zeit. vergl. Physiol.*, vol. 18, pp. 65–112.

Lloyd, F. E. 1942. *The carnivorous plants*. Chronica Botanica Co., Waltham, Mass. Pp. xv + 352.

Lucké, B., and H. Schlumberger. 1942. Common neoplasms in fish, amphibians, and reptiles. *Jour. Tech. Methods Bull. Internat. Assoc. Med. Mus.*, vol. 22, pp. 4–16.

McCay, C. M., and A. M. Phillips, Jr. 1940. Feeds for the fish hatcheries. *Progr. Fish Cult.*, no. 52, pp. 18–21.

McHugh, J. L., and B. W. Walker. 1948. Rearing marine fishes in the laboratory. *Calif. Fish and Game*, vol. 34, pp. 37–38.

McVay, J. A., and Helen W. Kaan. 1940. The digestive tract of *Carassius auratus*. *Biol. Bull.*, vol. 78, pp. 53–67.

Macht, David L. 1943. Ichthyometric comparison of cobra venom, morphine, and saponins. *Proc. Soc. Exp. Biol. Med.*, vol. 52(2), pp. 111–113.

Marsh, M. C. 1908. Notes on the dissolved content of water in its effect upon fishes. *Bull. U. S. Bur. Fish.*, vol. 28, pp. 893–906.

Marza, V. D., Eugenie V. Marza, and Mary J. Guthrie. 1937. Histochemistry of the ovary of *Fundulus heteroclitus* with special reference to the differentiating oocytes. *Biol. Bull.*, vol. 73, pp. 67–92.

Matsubara, S. 1908. Goldfish and their culture in Japan. *Bull. U. S. Bur. Fish.*, vol. 28, pp. 381–398.

Matsui, Yoshiichi. 1925. On the wart growths of Japanese lionhead goldfish. *Annot. Zool. Japon.*, vol. 10, pp. 355–362.

——. 1934. Genetical studies on goldfish of Japan. *Jour. Imp. Fish. Inst.*, vol. 30, pp. 1–96.

Matsuura, Y. 1934. Influence of temperature upon the action of potassium chloride on the heart beats of *Oryzias*. *Jour. Fac. Science Tokyo Univ.*, vol. 3, pp. 509–516.

Matthews, Samuel Arthur. 1933. Changes in the retina of *Fundulus* after cutting the optic nerve and the blood vessels running to the eye. *Jour. Exp. Zool.*, vol. 66, pp. 175–191.

——. 1937. The development of the pituitary gland in *Fundulus*. *Biol. Bull.*, vol. 73, pp. 93–98.

——. 1938. The seasonal cycle in the gonads of *Fundulus*. *Biol. Bull.*, vol. 75, pp. 66–74.

——. 1939a. The effects of light and temperature on the male sexual cycle in *Fundulus*. *Biol. Bull.*, vol. 77, pp. 92–95.

——. 1939b. The relationship between the pituitary gland and the gonads in *Fundulus*. *Biol. Bull.*, vol. 76, pp. 241–250.

Meyer, Henry. 1938. Investigations concerning the reproductive behavior of *Mollienisia "formosa."* *Jour. Genetics*, vol. 36(3), pp. 329–366.

Miller, Laurence F. 1944. An experimental study of the effect of softened waters on fish. *Ecology*, vol. 25, pp. 249–253.

Mitsukuri, K. 1904. The cultivation of marine and freshwater animals in Japan. *Bull. U. S. Bur. Fish.*, vol. 24, pp. 266–273.

Mookerjee, H. K., and S. P. Basu. 1946. On the spawning habits and development of *Oryzias melastigma* (McClelland). *Calcutta Univ., Jour. Dept. Science*, new ser., vol. 2, pp. 22–50.

Mookerjee, H. K., G. N. Mitra, and S. R. Mazumdar. 1940. The development of the vertebral column of a viviparous teleost, *Lebistes reticulatus*. *Jour. Morph.*, vol. 67, pp. 241–270.

Morgan, T. H. 1927. *Experimental embryology*. Columbia University Press, New York. Pp. xi + 766.

Muenscher, W. C. 1944. *Aquatic plants of the United States*. Comstock Pub. Co., Ithaca. Pp. x + 374.

Nachmansohn, David. 1946. Chemical mechanism of nerve activity. *Annals N. Y. Acad. Science*, vol. 47(4), pp. 395–428.

Needham, J. G., and J. T. Lloyd. 1916. *The life of inland waters*. Comstock Pub. Co., Ithaca. Pp. 1–438.

Needham, J. G., et al. 1937. *Culture methods for invertebrate animals*. Comstock Pub. Co., Ithaca. Pp. xxxiii + 590.

Newman, H. H. 1907. Spawning behavior and sexual dimorphism in *Fundulus heteroclitus* and allied fish. *Biol. Bull.*, vol. 12, pp. 314–348.

Nichita, G. 1928. Contribution à l'étude de l'atresie folliculaire chez les Vertebres. Résultats obtenus sur *Girardinus guppyi*. *Arch. anat. micro. Paris,* vol. 24, pp. 33–72.

Nicholas, J. S. 1927. Application of experimental methods to the study of developing *Fundulus* embryos. *Proc. Nat. Acad. Science,* vol. 13, pp. 695–698.

Nicholas, J. S., and J. Oppenheimer. 1942. Regulation and reconstitution in *Fundulus. Jour. Exp. Zool.,* vol. 90(1), pp. 127–157.

Nigrelli, R. F. 1938. Fish parasites and fish diseases. I. Tumors. *Trans. N. Y. Acad. Science,* 2nd ser., vol. 1(1), pp. 4–7.

——. 1943. Causes of disease and death of fishes in captivity. *Zoologica,* vol. 28(22), pp. 203–216.

——. 1947. Spontaneous neoplasms in fishes. III. Lymphosarcoma in *Astyanax* and *Esox. Zoologica,* vol. 32, pp. 101–108.

Nigrelli, R. F., and C. M. Breder, Jr. 1935. Histological changes in the prolapsed intestine of a fish, *Mollienisia latip'nna* LeSueur. *Copeia,* (2), 68–72.

Noble, G. K. 1938. Sexual selection among fishes. *Biol. Rev.,* vol. 13, pp. 133–158.

——. 1939. The experimental animal from the naturalist's point of view. *Amer. Naturalist,* vol. 73(745), pp. 113–126.

Noble, G. K., and R. Borne. 1940. The effect of sex hormones on the social hierarchy of *Xiphophorus helleri*. Abst., *Anat. Rec.,* vol. 78, suppl. (4), p. 147.

Noble, G. K., and R. Borne. 1941. The effect of forebrain lesions on the sexual and fighting behavior of *Betta splendens* and other fishes. Abst., *Anat. Rec.,* vol. 79(3), suppl. 2, p. 49.

Noble, G. K., and R. Curtis. 1939. The social behavior of the jewel fish *Hemichromis bimaculatus* Gill. *Bull. Amer. Mus. Nat. Hist.,* vol. 76(1), pp. 1–46.

Noble, G. K., K. F. Kumpf, and V. N. Billings. 1938. The induction of brooding behavior in the jewel fish. *Endocrinology,* vol. 23(3), pp. 353–359.

Norman, J. R. 1948. *A history of fishes.* A. A. Wyn, New York. Pp. xv + 463.

Oguma, Kan, and S. Makino. 1932. A revised check-list of the chromosome number in vertebrata. *Jour. Genetics,* vol. 26, pp. 239–254.

Oka, T. B. 1931a. Effects of the triple allelomorphic genes in *Oryzias latipes. Jour. Fac. Science Imp. Univ. Tokyo,* ser. 4(2), pp. 171–178.

——. 1931b. On the processes of the fin-rays of the male of *Oryzias latipes* and other sex characters of this fish. *Jour. Fac. Science Tokyo Univ.,* vol. 2, pp. 209–218.

——. 1931c. On the accidental hermaphroditism in *Oryzias latipes. Jour. Fac. Science Imp. Tokyo Univ.,* ser. 4, vol. 2, pp. 219–224.

Ono, Yoshiaki. 1927. The behavior of cells in tissue culture of *Oryzias latipes* with special reference to the ectodermic epithelium. *Annot. Zool. Japon.,* vol. 11(2), pp. 145–148.

——. 1937a. Orienting behavior of *Oryzias latipes* and other fishes. *Jour. Fac. Science Tokyo Univ.,* vol. 4, pp. 393–400.

——. 1937b. Conditioned orientation of the fighting fish. *Jour. Fac. Science Tokyo Univ.,* vol. 4, pp. 401–412.

Oppenheimer, Jane M. 1936a. The development of isolated blastoderms of *Fundulus heteroclitus. Jour. Exp. Zool.,* vol. 72, pp. 247–269.

——. 1936b. Transplantation experiments on developing teleosts *Fundulus* and *Perca. Jour. Exp. Zool.,* vol. 72, pp. 409–437.

Oppenheimer, Jane M. 1936c. Processes of localization in developing *Fundulus*. *Jour. Exp. Zool.*, vol. 73(3), pp. 405–444.

——. 1937. The normal stages of *Fundulus heteroclitus*. *Anat. Rec.*, vol. 68, pp. 1–15.

——. 1947. Organization of the teleost blastoderm. *Quart. Rev. Biol.*, vol. 22(2), pp. 105–118.

Osorio-Tafall, B. F. 1943. Observaciones sobre la fauna acuatica de las cuevas de la region de Valles, San Luis Potosi (Mexico). *Rev. soc. mexicana hist. nat.*, vol. 4(12), pp. 43–71.

Parker, G. H. 1948. *Animal colour and their neurohumours*. Cambridge University Press. Pp. vii + 377.

Pelluet, D. 1944. Criteria for the recognition of developmental stages in the salmon, *Salmo salar*. *Jour. Morph.*, vol. 74(3), pp. 395–407.

Perry, Frances. 1938. *Water gardening*. Country Life, London. Pp. xvi + 353.

Philips, Fred S. 1940. Oxygen consumption and its inhibition in the development of *Fundulus* and various pelagic fish eggs. *Biol. Bull.*, vol. 78, pp. 256–274.

Phillipi, Erich. 1909. Fortpflanzungsgeschichte der viviparen Teleosteer *Glaridichthys januarius* und *G. decem-maculatus* in ihrem Einfluss auf Lebensweise, makroskopische und mikroskopische Anatomie. *Zool. Jahrb., Abteil. f. Anat. Ontogenie der Tiere*, vol. 27, pp. 1–94.

Phillips, A. M., Jr., and D. R. Brockway. 1948. Vitaminology. *Progr. Fish Cult.*, vol. 10(3), pp. 117–124.

Pickford, Grace E., and Ernest F. Thompson. 1948. The effects of purified mammalian growth hormone on the killifish [*Fundulus heteroclitus* (Linn.)]. *Jour. Exp. Zool.*, vol. 109(3), pp. 367–383.

Plehn, M. 1924. *Praktikum der Fischkrankheiten*. E. Schweizerbart'sche Verlagsbuchhandlung, Stuttgart. Pp. 1–179.

Pond, R. H. 1918. The larger aquatic vegetation. Pp. 178–209 in *Fresh water biology*. John Wiley & Sons, New York.

Potter, G. E., and A. B. Medlin. 1935. Organography of *Gambusia patruelis* (Baird and Girard). *Jour. Morph.*, vol. 57, pp. 303–316.

Price, J. W. 1934a. The embryology of the whitefish, *Coregonus clupeaformis* (Mitchill). I. *Ohio Jour. Science*, vol. 34, pp. 287–305.

——. 1934b. The embryology of the whitefish *Coregonus clupeaformis* (Mitchill). II. Organogenesis. *Ohio Jour. Science*, vol. 34, pp. 399–414.

——. 1935. The embryology of the whitefish *Coregonus clupeaformis* (Mitchill). III. The second half of the incubation period. *Ohio Jour. Science*, vol. 35, pp. 40–53.

——. 1939. The time-temperature relations in the incubation of the whitefish, *Coregonus clupeaformis* (Mitchill). *Jour. Gen. Phys.*, vol. 23(4), pp. 449–468.

Purser, G. L. 1937. Succession of broods of *Lebistes*. *Nature*, vol. 140, p. 155.

——. 1938. Reproduction in *Lebistes reticulatus*. *Quart. Jour. Micro. Science*, vol. 81(1), pp. 151–157.

——. 1941. Sex dimorphism in the fins of *Lebistes reticulatus*. *Quekett Micro. Club*, vol. 1, pp. 172–178.

Ralston, E. M. 1934. A study of the chromosomes of *Xiphophorus, Platypoecilus*, and of *Xiphophorus-Platypoecilus* hybrids during spermatogenesis. *Jour. Morph.*, vol. 56, pp. 423–433.

Rasmussen, E. 1948. Spawning and early development of *Epiplatys chaperi*. *Aquarium Jour.*, vol. 19(5), pp. 23–26.

Rasquin, P. 1946. On the reappearance of melanophores in blind goldfish. *Copeia*, (2), pp. 85–91.

——. 1947. Progressive pigmentary regression in fishes associated with cave environments. *Zoologica*, vol. 32(4), pp. 35–42.

Reed, H. D., and M. Gordon. 1931. The morphology of melanotic over-growths in hybrids of Mexican killifishes. *Amer. Jour. Cancer*, vol. 15(3), pp. 1524–1546.

Regnier, Marie-Theresa. 1938. Contribution à l'étude de la sexualité des cyprinodontes vivipares (*Xiphophorus helleri, Lebistes reticulatus*). *Bull. biol. France et Belgique*, vol. 72, pp. 385–493.

Reighard, J. E. 1890. The development of the wall-eyed pike *Stizostedion vitreum* Raf. *9 Biennial Rept. Mich. State Bd. Fish. Comm.*, Appendix, pp. 95–158.

——. 1906. The breeding habits, development, and propagation of the black bass. *10 Biennial Rept. Mich. State Bd. Fish. Comm.*, Appendix, pp. 1–63.

Rice, Victor A. 1942. *Breeding and improvement of farm animals*. 3rd ed. McGraw-Hill, New York. Pp. 1–750.

Richards, A. 1917. The history of the chromosomal vesicles in *Fundulus* and the theory of genetic continuity of chromosomes. *Biol. Bull.*, vol. 32, pp. 249–290.

——. 1935. Analysis of early development of fish embryos by means of the mitotic index. I. The use of the mitotic index. *Amer. Jour. Anat.*, vol. 56, pp. 355–363.

Richards, A., and R. P. Potter. 1935. Analysis of early development of fish embryos by means of the mitotic index. II. Mitotic index in the preneural tube stages of *Fundulus heteroclitus*. *Amer. Jour. Anat.*, vol. 56, pp. 365–393.

Roberts, Hervey. 1935. *Epiplatys (Panchax) chaperi*. *Aquarium*, vol. 4, pp. 10–12.

Robinson, E. J., and R. Rugh. 1943. The reproductive processes of the fish, *Oryzias latipes*. *Biol. Bull.*, vol. 84(1), pp. 115–125.

Roosen-Runge, E. C. 1938. On the early development—bipolar differentiation and cleavage—of the zebra fish, *Brachydanio rerio*. *Biol. Bull.*, vol. 75, pp. 119–133.

——. 1939. Karyokinesis during cleavage of the zebra fish, *Brachydanio rerio*. *Biol. Bull.*, vol. 77, pp. 79–91.

Rugh, Roberts. 1948. *Experimental embryology*. 2nd ed. Burgess Pub. Co., Minneapolis. Pp. 1–480.

Rust, W. 1941. Genetische Untersuchung über die Geschlechtsbestimmungstypen bei Zahnkarpfen unter besonderer Berücksichtigung von Artkreuzungen mit *Platypoecilus variatus*. *Zeit. f. Induk. Abstamm. Vererb.*, vol. 79, pp. 336–395.

St. Amant, L. S. 1941. The effect of castration and treatment with ethinyl testosterone on the development of the gonopodium of *Gambusia affinis*. Abst., *Anat. Rec.*, vol. 79, pp. 53–54.

Samokhvalova, G. V. 1933. Correlation in the development of secondary sexual characters and the sex gland in *Lebistes reticulatus* (in Russian with English summary). *Trans. Dynamics Devel.*, vol. 7, pp. 65–76.

Samokhvalova, G. V. 1935. The influence of the X-rays on the sex gonad and the secondary sexual characters in *Lebistes reticulatus* (in Russian with English summary and conclusions). *Trans. Dynamics Devel.*, vol. 10, pp. 213–228.

Sasaki, K. 1926. On the sex ratio in *Carassius auratus*. *Science Rept. Tohoku Imp. Univ.*, 4th ser., vol. 1, pp. 229–238.

Schäperclaus, Wilhelm. 1941. *Fischkrankheiten.* Gustav Wenzel & Sohn, Braunschweig. Pp. 1–296.

Schamberg, J. F., and B. Lucké. 1922. Fibrosarcoma of the skin in a goldfish (*Carassius auratus*). *Jour. Cancer Res.*, vol. 7(2), pp. 151–161.

Schlagel, S. R., and C. M. Breder, Jr. 1947. A study of the oxygen consumption of blind and eyed cave characins in light and in darkness. *Zoologica*, vol. 32(2), pp. 17–27.

Schlumberger, H. G., and B. Lucké. 1948. Tumors of fishes, amphibians and reptiles. *Cancer Res.*, vol. 8, pp. 657–754.

Schmey, Max. 1911. Über Neubildungen bei Fischen. *Frankfurter Zeit. f. Path.*, vol. 6, pp. 230–252.

Schmidt, Johs. 1919. Racial studies on fishes. III. Experiments with *Lebistes*. *Jour. Genetics*, vol. 8, pp. 147–153.

Schuett, F. 1934. Studies in mass physiology: the activity of goldfish under different conditions of aggregation. *Ecology*, vol. 15, pp. 258–262.

Schultz, L. P., and E. M. Stern. 1948. *The ways of fishes.* D. Van Nostrand, New York. Pp. xii + 264.

Schwier, Heinz. 1943. Vitalitätsuntersuchungen an normalen und albinotischen Makropoden. *Zool. Anz.*, vol. 143(1/2), pp. 33–44.

Scott, J. L. 1944. The effects of steroids on the skeleton of the poeciliid fish *Lebistes reticulatus*. *Zoologica*, vol. 29(7), pp. 49–52.

Scrimshaw, N. S. 1944a. Embryonic growth in the viviparous poeciliid, *Heterandria formosa*. *Biol. Bull.*, vol. 87(1), pp. 37–51.

——. 1944b. Superfetation in poeciliid fishes. *Copeia*, (3), pp. 180–183.

——. 1945. Embryonic development in poeciliid fishes. *Biol. Bull.*, vol. 88(3), pp. 233–246.

——. 1946. Egg size in poeciliid fishes. *Copeia*, (1), pp. 20–23.

Seal, William P. 1911. Breeding habits of viviparous fishes *Gambusia holbrookii* and *Heterandria formosa*. *Proc. Biol. Soc. Washington*, vol. 24, pp. 91–96.

Seeton, F. E. 1934. Another case of metal poisoning. *Aquarium*, vol. 3, pp. 18–19.

Seitz, A. 1940. Die paarbildung bei einigen cichliden. I. Die paarbildung bei *Astatotilapia strigigena* Pfeffer. *Zeit. Tierpsychol.*, vol. 4, pp. 40–84.

Self, J. T. 1940. Notes on the sexual cycle of *Gambusia affinis affinis*, and notes on its habits and relations to mosquito control. *Amer. Midland Naturalist*, vol. 23, pp. 393–398.

Shelford, V. E. 1918. Conditions of existence. Pp. 21–60 in *Fresh water biology.* John Wiley & Sons, New York.

Shlaifer, A. 1938. Studies in mass physiology: effect of numbers upon the oxygen consumption and locomotor activity of *Carassius auratus*. *Physiol. Zool.*, vol. 11(4), pp. 408–424.

——. 1939. An analysis of the effect of numbers upon the oxygen consumption of *Carassius auratus*. *Physiol. Zool.*, vol. 12(4), pp. 381–392.

——. 1940. The locomotor activity of the goldfish, *Carassius auratus* L., under various conditions of homotypic and heterotypic grouping. *Ecology,* vol. 21(4), pp. 488–500.

Shoemaker, H. H. 1944. A laboratory study of fish populations. *Trans. Amer. Fish. Soc.,* vol. 74, pp. 350–559.

Smith, D. C., and G. M. Everett. 1943. The effect of thyroid hormone on growth rate, time of sexual differentiation, and oxygen consumption in the fish, *Lebistes reticulatus. Jour. Exp. Zool.,* vol. 94, pp. 229–240.

Smith, George Milton. 1930. A mechanism of intake and expulsion of colored fluids by the lateral canals as seen experimentally in the goldfish. *Biol. Bull.,* vol. 58, pp. 313–321.

——. 1931. The occurrence of melanophores in certain experimental wounds of the goldfish (*Carassius auratus*). *Biol. Bull.,* vol. 61, pp. 73–84.

——. 1932a. Melanophores induced by X-ray compared with those existing in patterns seen in *Carassius auratus. Biol. Bull.,* vol. 63, pp. 484–491.

——. 1932b. Eruptions of corial melanophores and general cutaneous melanosis in the goldfish (*Carassius auratus*) following exposure to X-ray. *Amer. Jour. Cancer,* vol. 16, pp. 863–870.

Smith, H. M. 1909. *Japanese gold-fish.* W. F. Roberts Co., Washington. Pp. 1–112.

——. 1945. The fresh-water fishes of Siam, or Thailand. *Bull. U. S. Nat. Mus.,* vol. 188, pp. xi + 622.

Smith, H. W. 1930. Metabolism of the lungfish, *Protopterus aethiopicus. Jour. Biol. Chem.,* vol. 88, pp. 97–130.

Snow, J. W. 1918. The fresh-water algae. Pp. 115–177 in *Fresh water biology.* John Wiley & Sons, New York.

Solberg, A. N. 1938a. The susceptibility of the germ cells of *Oryzias latipes* to X-radiation and recovery after treatment. *Jour. Exp. Biol.,* vol. 78, pp. 417–439.

——. 1938b. The development of a bony fish. *Progr. Fish Cult.,* no. 40, pp. 1–19.

——. 1938c. The susceptibility of *Fundulus heteroclitus* embryos to X-radiation. *Jour. Exp. Zool.,* vol. 78, pp. 441–465.

Stanley, L. L., and G. L. Tescher. 1931. Activity of goldfish on testicular substance diet. *Endocrinology,* vol. 15, pp. 55–56.

Stanley, L. L., and G. L. Tescher. 1932. Weight of goldfish influenced by testicular substance diet. *Endocrinology,* vol. 16, pp. 153–154.

Stockard, C. R. 1907. The influence of external factors, chemical and physical, on the development of *Fundulus heteroclitus. Jour. Exp. Biol.,* vol. 4, pp. 165–201.

——. 1921. Developmental rate and structural expression; an experimental study of turns, double monsters, and single deformities, and the interaction among embryonic organs during their origin and development. *Amer. Jour. Anat.,* vol. 28, pp. 115–278.

Stone, L. S. 1940. Reimplantation and transplantation of eyes in anuran larvae and *Fundulus heteroclitus. Proc. Soc. Exp. Biol. Med.,* vol. 44, pp. 639–641.

Stoye, F. H. 1935. *Tropical fishes for the home.* Carl Mertens Pub., New York. Pp. 1–284.

——. 1947–1948. The fishes of the order Cyprinodontes. *Aquarium Jour.,* vol. 18(2), pp. 4–11; vol. 18(5), pp. 21–26; vol. 18(7), pp. 4–7, 34; vol. 18(9), pp. 9–14; vol. 18(10), pp. 13–19; vol. 18(11), pp. 14–19, 30; vol. 19(1), pp. 4–6, 26.

Sumner, F. B. 1943. A further report on the effects of the visual environment on the melanin content of fishes. *Biol. Bull.*, vol. 84, pp. 195–205.

Sumner, F. B., and N. A. Wells. 1935. Some relations between respiratory metabolism in fishes and susceptibility to certain anesthetics and lethal agents. *Biol. Bull.*, vol. 69, pp. 368–378.

Surber, E. W. 1948. Chemical control agents and their effects on fish. *Progr. Fish Cult.*, vol. 10, pp. 125–131.

Svärdson, Gunnar. 1943. Studien über den Zusammenhang zwischen Geschlechtsreife und Wachstum bei *Lebistes*. *Meddl. Kungl. Lantbruksstyrelsen,* vol. 21, pp. 1–48.

——. 1944. Polygenic inheritance in *Lebistes*. *Ark. f. Zool.,* vol. 36A(6), pp. 1–9.

——. 1945. Chromosome studies on Salmonidae. *Rept. Swedish State Inst. Fresh-Water Fish. Res. Drottingholm,* no. 23, pp. 1–151.

Svärdson, G., and T. Wickbom. 1942. The chromosomes of two species of Anabantidae (Teleostei), with a new case of sex reversal. *Hereditas,* vol. 28, pp. 212–216.

Swingle, H. S., and E. V. Smith. 1947. Management of farm fish ponds. *Bull. Alabama Polytech. Agr. Exp. Sta.* 254, pp. 1–30.

Takahashi, Keizo. 1934. Studies on tumors of fishes from Japanese waters. *Proc. Fifth Pacific Sci. Cong.,* vol. 5, pp. 4151–4155.

Tavolga, Margaret C. 1949. Differential effects of estradiol benzoate and pregneninolone on *Platypoecilus maculatus*. *Zoologica,* vol. 34, pp. 215–237.

Tavolga, William N. 1949. Embryonic development of the platyfish (*Platypoecilus*), the swordtail (*Xiphophorus*), and their hybrids. *Bull. Amer. Mus. Nat. Hist.,* vol. 94(4), pp. 161–230.

Tavolga, William N., and Ross F. Nigrelli. 1947. Studies on *Costia necatrix* (Henneguy). *Trans. Amer. Micro. Soc.,* vol. 66(4), pp. 366–378.

Tavolga, William N., and R. Rugh. 1947. Development of the platyfish, *Platypoecilus maculatus*. *Zoologica,* vol. 32, pp. 1–15.

Tchernavin, V. 1943–1944. The breeding characters of salmon in relation to their size. *Proc. Zool. Soc. London,* vol. 113(B)IV, pp. 206–232.

Thomas, L. 1931. Les tumeurs des poissons (étude anatomique et pathogénique) *Bull. assoc. franc. l'étude du cancer,* vol. 20, pp. 703–760.

Thorner, M. W. 1929. Recovery of the heart beat of *Fundulus* embryos after stoppage by potassium chloride. *Biol. Bull.,* vol. 56, pp. 157–163.

Timm, J. A. 1932. Oxygen in aquarium water in "Science is our hobby." *Aquarium,* vol. 1, pp. 131–132, 142.

Tinbergen, N., and J. J. A. Van Iersel. 1947. "Displacement reactions" in the three-spined stickleback. *Behavior,* vol. 1(1), pp. 56–63.

Tomita, G. 1936. Melanophore reactions to light during the early stages of the paradise fish *Macropodus opercularis*. *Jour. Shanghai Science Inst.,* sec. 4, vol. 2, pp. 237–264.

Toyama, K. 1916. On a few Mendelian characters. *Jour. Jap. Breeding Soc.,* vol. 1, pp. 1–9.

Tuchmann, H. 1936. Action de l'hypophyse sur la morphogenese et la differenciation sexuelle de *Girardinus guppii*. *C. r. soc. biol.,* vol. 122, pp. 162–164.

Tung, Ti-chow, Chin-ye Chang, and Yu-fung-yeh Tung. 1945. Experiments on the developmental potencies of blastoderms and fragments of teleostean eggs separated latitudinally. *Proc. Zool. Soc. London*, vol. 115(42), pp. 175–188.

Tunison, A. V. 1945. *Trout feeds and feeding*. U. S. Dept. Interior, Fish and Wildlife Service, Boston, Mass. Pp. 1–21.

Turner, C. L. 1937. Reproductive cycles and superfetation in poeciliid fishes. *Biol. Bull.*, vol. 72, pp. 145–164.

——. 1939. The pseudo-amnion, pseudo-chorion, pseudo-placenta and other foetal structures in viviparous cyprinodont fishes. *Science*, vol. 90, pp. 42–43.

——. 1940a. Pseudoamnion, pseudochorion and follicular pseudoplacenta in poeciliid fishes. *Jour. Morph.*, vol. 67(1), pp. 59–89.

——. 1940b. Superfetation in viviparous cyprinodont fishes. *Copeia*, (2), pp. 88–91.

——. 1941a. Morphogenesis of the gonopodium in *Gambusia affinis affinis*. *Jour. Morph.*, vol. 69(1), pp. 161–185.

——. 1941b. Regeneration of the gonopodium of *Gambusia* during morphogenesis. *Jour. Exp. Zool.*, vol. 87(2), pp. 181–209.

——. 1941c. Gonopodial characteristics produced in the anal fins of females of *Gambusia affinis affinis* by treatment with ethinyl testosterone. *Biol. Bull.*, vol. 80(3), pp. 371–383.

——. 1942a. A quantitative study of the effects of different concentrations of ethynyl testosterone in the production of gonopodia in females of *Gambusia affinis*. *Physiol. Zool.*, vol. 15, pp. 263–280

——. 1942b. Morphogenesis of the gonopodial suspensorium in *Gambusia affinis* and the induction of male suspensorial characters in the female by androgenic hormones. *Jour. Exp. Zool.*, vol. 91(2), pp. 167–193.

——. 1942c. Sexual dimorphism in the pectoral fin of *Gambusia* and the induction of the male character in the female by androgenic hormones. *Biol. Bull.*, vol. 83, pp. 389–400.

Umrath, K. 1939. Uber die Vererbung der Farbung und des Geschlechts beim Schleierkampffisch *Betta splendens*. *Zeit. f. Induk. Abstamm. Vererb.*, vol. 77, pp. 450–454.

Van Oordt, G. J. 1925. The relation between the development of the secondary sex characters and the structure of the testis in the teleost *Xiphophorus helleri* Heckel. *Brit. Jour. Exp. Biol.*, vol. 3, pp. 43–59.

——. 1928. The duration of the life of the spermatozoa in the fertilized female of *Xiphophorus helleri* Regan. *Tijdschr. Med. Dierh. Vereen*, vol. 3, pp. 1–4.

Vaupel, J. 1929. The spermatogenesis of *Lebistes reticulatus*. *Jour. Morph.*, vol. 47, pp. 555–587.

Walker, R., and G. C. Bennett. 1945. Size relations in the optic system of telescope-eyed goldfish. *Trans. Conn. Acad. Arts Science*, vol. 36, pp. 379.

Wallbrunn, H. M. 1948. Genetics of the *Betta splendens*. Abst., *Rec. Genetics Soc. Amer.*, vol. 17, pp. 62.

Walls, G. L. 1942. The vertebrate eye. *Bull. Cranbrook Inst. Science*, vol. (19), pp. xiv + 785.

Ward, H. B., and G. C. Whipple. 1918. *Fresh-water biology*. John Wiley & Sons, New York. Pp. ix + 1111.

Warmke, H. E. 1944. Use of the killifish, *Fundulus heteroclitus,* in the assay of marihuana. *Jour. Amer. Pharm. Assoc.,* vol. 33, pp. 122–125.

Waterman, A. J. 1939. Effects of 2.4-dinitrophenol on the early development of the teleost, *Oryzias latipes. Biol. Bull.,* vol. 76(2), pp. 162–170.

——. 1940a. Contractile activity of the extra-embryonic cells during early development of *Oryzias latipes.* Abst., *Anat. Rec.,* vol. 78, suppl., p. 135.

——. 1940b. Effects of colchicine on the development of the fish embryo, *Oryzias latipes. Biol. Bull.,* vol. 78, pp. 29–34.

Weisman, A. I., and C. M. Rothburd. 1938. The care and handling of the European bitterling as an experimental animal in endocrine research. *Endocrinology,* vol. 23(1), pp. 104–106.

Wells, N. A. 1935a. The influence of temperature upon the respiratory metabolism of the Pacific killifish *Fundulus parvipinnis. Physiol. Zool.,* vol. 8, pp. 196–227.

——. 1935b. Variations in the respiratory metabolism of the Pacific killifish *Fundulus parvipinnis* due to size, season, and continued constant temperature. *Physiol. Zool.,* vol. 8, pp. 318–336.

Wickbom, Torsten. 1941. The sex chromosomes of Cyprinodontidae and of teleosts in general with a list of new chromosome numbers in Cyprinodontidae. *Ark. f. Zool.,* Bd. 33, no. 10, pp. 1–6.

——. 1943. Cytological studies on the family Cyprinodontidae. *Hereditas,* vol. 29, pp. 1–24.

Wiebe, A. H. 1931. Notes on the exposure of several species of pond fishes to sudden changes in pH. *Trans. Amer. Micro. Soc.,* vol. 50, pp. 380–393.

Wilson, H. V. 1889 (1891). The embryology of the sea bass (*Serranus atrarius*). *Bull. U. S. Fish. Comm. 1889,* vol. 9, pp. 209–277.

Winge, O. 1922a. A peculiar mode of inheritance and its cytological explanation. *Jour. Genetics,* vol. 12, pp. 137–144.

——. 1922b. One-sided masculine and sex-linked inheritance in *Lebistes reticulatus. Jour. Genetics,* vol. 12, pp. 145–162.

——. 1923. Crossing-over between the X and the Y chromosome in *Lebistes. Jour. Genetics,* vol. 13, pp. 201–217.

——. 1927. The location of eighteen genes in *Lebistes reticulatus. Jour. Genetics,* vol. 18, pp. 1–42.

——. 1930. On the occurrence of XX males in *Lebistes* with some remarks on Aida's so-called "Non-disjunctional" males in *Aplocheilus. Jour. Genetics,* vol. 23, pp. 69–76.

——. 1932. The nature of the sex chromosomes. *Proc. Sixth Inter. Cong. Genetics,* vol. 1, pp. 343–355.

——. 1934. The experimental alteration of sex chromosomes into autosomes and vice versa, as illustrated by *Lebistes. C. r. Lab. Carlsberg,* ser. physiol., vol. 21, pp. 1–49.

——. 1937. Goldschmidt's theory of sex determination in *Lymantria. Jour. Genetics,* vol. 34, pp. 81–89.

Winge, O., and Eshen Ditlevsen. 1938. A lethal gene in the Y chromosome in *Lebistes. C. r. Lab. Carlsberg,* ser. physiol., vol. 22, pp. 203–210.

Winge, O., and Eshen Ditlevsen. 1947. Color inheritance and sex determination in *Lebistes. Heredity,* vol. 1, pp. 65–83.

Wolf, H. T. 1908. *Goldfish breeds and other aquarium fishes.* Innes & Sons, Philadelphia. Pp. 1–385.

Wolf, Louis E. 1931. The history of the germ cells in the viviparous teleost *Platypoecilus maculatus. Jour. Morph.,* vol. 52(1), pp. 115–153.

——. 1935. The use of potassium permanganate in the control of fish parasites. *Trans. Amer. Fish. Soc.,* vol. 65, pp. 88–100.

——. 1945. Dietary gill disease of trout. *New York State Cons. Dept. Fish. Res. Bull.,* vol. 7, pp. 1–30.

Wunder, W. 1930. Experimentelle Untersuchungen am dreistachligen Stichling (*Gasterosteus aculeatus* L.) wahrend der Laichtaeit (Kampfe, Nestbau, Laichen, Brutpflege). *Zeit. wiss. Biol. Abt. Zeit. Morph. u. Okel. Tiere,* vol. 16(¾), pp. 453–498.

Wylie, Philip. 1948. The Lerner Marine Laboratory. *Nat. Hist.,* vol. 57, pp. 312–319.

Yamamoto, T. 1931a. Studies on the rhythmical movements of the early embryos of *Oryzias latipes.* I. General description. *Jour. Fac. Science Imp. Univ. Tokyo,* vol. 2, pp. 147–152.

——. 1931b. Temperature constants for the rate of heart beat in *Oryzias latipes. Jour. Fac. Science Imp. Univ. Tokyo,* ser. 4, vol. 2, pp. 381–388.

—— 1934. On the rhythmic movements of the egg of goldfish (*Carassius auratus*). *Jour. Fac. Science Imp. Univ. Tokyo,* vol. 3(3), pp. 275–285.

——. 1936. Shrinkage and permeability of the chorion of *Oryzias* egg, with special reference to the reversal of selective permeability. *Jour. Fac. Science Imp. Univ. Tokyo,* vol. 4, pp. 240–261.

——. 1938. On the distribution of temperature constants in *Oryzias latipes. Proc. Imp. Acad. Tokyo,* vol. 14(10), pp. 393–395.

—— 1939a. Studies on the rhythmical movements of the early embryo of *Oryzias latipes.* IX. *Jour. Fac. Science Imp. Univ. Tokyo,* vol. 5(4,2), pp. 211–218.

——. 1939b. Changes in the cortical layer of the egg of *Oryzias latipes* at the time of fertilization. *Proc. Imp. Acad. Tokyo,* vol. 15, pp. 269–271.

——. 1939c. Mechanism of membrane elevation in the egg of *Oryzias latipes* at the time of fertilization. *Proc. Imp. Acad. Tokyo,* vol. 15, pp. 272–274.

——. 1940. The change in volume of the fish egg at fertilization. *Proc. Imp. Acad. Tokyo,* vol. 16, pp. 482–485.

Yudkin, Warren H. 1949. Thiaminase, the Chastek-paralysis factor. *Physiol. Revs.,* vol. 28(4), pp. 389–402.

Zahl, P. A. 1933. A successful Caesarean operation in fish. *Aquatic Life,* vol. 16, p. 444.

15· DROSOPHILA

WARREN P. SPENCER
College of Wooster
Wooster, Ohio

INTRODUCTION

The inclusion of a section on *Drosophila* in a book otherwise devoted exclusively to the care and breeding of vertebrates is a recognition of the actual and potential advantages of these flies as laboratory animals. They have been so widely and successfully used in genetics laboratories throughout the world, both in research and in laboratory courses, that a chapter on their care and breeding may seem unnecessary. However, it will be our purpose to present a summary of current methods of culture and to suggest certain improvements, which it is hoped may be of value both to the high-school and college teacher or student working for the first time with *Drosophila* and to workers in research laboratories.

The discussion will deal mainly with fundamental problems of culture. No attempt will be made to review all the techniques employed in such special fields as the cytogenetics of the salivary chromosomes, intra- and interspecific transplantation studies, radiation and other methods of artificial induction of mutations, and field and laboratory population studies.

While *Drosophila* have generally been used in genetic studies, they are valuable laboratory animals for experiments in insect physiology and animal behavior, in demonstrating life histories, mating reactions, host-parasitoid relationships, bio-assays of insecticides and other chemicals, and as food for many other laboratory-bred animals such as arachnids, fishes, and amphibians.

CULTURE ROOM

For rearing a few cultures of fruit-flies as demonstration material any room where the temperature range most of the day is from 20° C.

to 25° C. and the extreme fluctuations do not go below 10° C. or above 27° C. will serve quite well. Where cultures are reared outside of temperature cabinets they must never be exposed to direct sunlight, even for periods as short as one half hour. Such exposure is likely to result in killing or sterilizing the flies, owing to the rapid rise in temperature in the glass culture containers almost always used.

Where *Drosophila* are to be reared extensively for research or even for laboratory class use, the choice of a room both for rearing and for examining the flies is an important consideration. Too often in climates where high summer temperatures are encountered the *Drosophila* laboratory is found on an upper floor. Even with refrigerated cabinets for rearing the flies the worker finds it difficult to examine them where temperatures above 25° C. are encountered. At such high temperatures it is difficult to anesthetize the flies without killing them, and they come out of anesthesia so rapidly that they are difficult to handle. Fogging of microscope lenses and eyeglasses is also exasperating.

If possible the *Drosophila* laboratory, particularly in summer, should be located in a cool basement or first-floor room. If this is impractical an air-conditioned room large enough to accommodate all workers should be provided. In a basement room both diurnal and seasonal temperatures will generally fluctuate less. Incubators and refrigerating cabinets installed in such a room can be adjusted to the desired temperatures with a minimum use of thermostatically controlled heating and cooling units, which is in itself a safeguard against thermostat failure, and results in a more uniform temperature within the cabinet.

TEMPERATURE CABINETS

For studies in physiological genetics, developmental rates, life cycles, and even accurate linkage work rather exact temperature controls, with fluctuations of less than one centigrade degree, are needed. Even for keeping stock cultures and routine experimental work a series of temperature cabinets is a distinct advantage. They make possible the elimination of much work by keeping stocks at relatively low temperatures. In our laboratory stocks kept at 20° C. are now changed once in two months, and may be allowed to run even longer. At 25° C. they would need changing in three weeks to a month. On the other hand by placing experimental cultures in cabinets at optimum temperature much time or rather space can be saved. In a continuous research program on rapidly breeding forms the actual time that a generation requires makes little difference in the amount of work done. But for a longer time per generation more incubator space and more culture bottles are needed to accomplish the same amount of work per unit time. Where several workers are carrying on projects simultaneously this may be an important factor. Furthermore, in genetics courses which may run only a few months, the speeding up of experimental work is an advantage.

Bacteriological incubators may be used where relatively few cultures are being reared and these incubators are not otherwise in use. However, they have the following drawbacks. The heating elements have too much heat lag for operation at the temperatures required, and they are very expensive, considering their capacity. Where these incubators are used care should be taken to place the culture bottles as far as possible from the heating unit, preferably on an upper shelf. These incubators are usually not well ventilated and therefore should not be filled with *Drosophila* culture bottles in order to avoid dangerous accumulation of carbon dioxide and oxygen lack.

Where *Drosophila* work is carried on continuously it is much more satisfactory to have temperature cabinets designed for this work. The following considerations are important: (*a*) large size with plenty of air space giving adequate ventilation; (*b*) either double thermostatic control or thermostatically controlled heating unit of small enough capacity not to produce critically high temperature in case of thermostat failure; (*c*) good insulation and a heating unit with relatively little heat lag and as low capacity as possible to maintain the constant temperature desired.

A satisfactory temperature cabinet with ample room for five hundred half-pint culture bottles, including the features mentioned above, may be constructed for as little as $20.00. In figure 1 are shown two such temperature cabinets used for some years in our laboratory. The cabinet in the foreground is open to show interior construction; that in the background is closed and shows light lath holding the Cellotex insulation on the door. These cabinets were constructed from casket packs secured from a funeral director. Each box consists of a strong but light wooden framework to which is tacked three-ply ⅛-inch wood veneer paneling. These boxes vary in size. One of those shown has the following inside measurements: length 7 feet 4 inches, width 2 feet 6 inches, and depth 2 feet 1 inch. The lid is hinged to the box with three hinges to form a door, and three hooks are provided to hold the door shut. As the wood veneer is tacked on the inside of the framework, panels of Cellotex or other composition insulating material are cut of a size to fit snugly into the framework on the outside. These are held in place by light lath tacked over them and to the framework. The box is set on end and may be mounted on castors. The first shelf is placed at least 2 feet above the bottom. These shelves are made of 1-inch wire mesh tacked on wooden frames, which rest on supports nailed to the framework. Each shelf for an incubator of the above dimensions has a capacity of one hundred half-pint milk bottles, used as culture bottles. Five shelves can easily be used with ample room for removing culture bottles from the back of the shelf without disturbing those in front.

FIG. 1. Inexpensive incubators for *Drosophila*.

The heating units consist of electric light bulbs suspended near the bottom of the incubator. These have a small heat lag. A thermostat is placed on the back wall about the middle of the cabinet. One 60-watt light bulb holds the temperature of the cabinet, with its approximately 40 cubic feet of air space, about 3 centigrade degrees above the room temperature. We have used a 60-watt bulb in series with a bimetallic thermostat and other bulbs, which may be turned on, but are not in series with the thermostat. By carrying a light load on the thermostat danger of sparking and sticking is lessened. Then if sparking or sticking does occur the temperature will not go so high that serious damage is done. Where the outside room temperature fluctuates more than 3 centigrade degrees, a heavier load must be carried on the thermostat. Danger of overheating may then be minimized by the use of a relay in series with the thermostat. It is, of course, desirable to install these cabinets in a room where the diurnal temperature fluctuations are minimum, a basement room if possible. A Cellotex shelf is provided which may be fitted in at any level to cut down the size of the cabinet when full capacity is not needed. Such an incubator, with five hundred culture bottle capacity, gives surprisingly little fluctuation in temperature

from shelf to shelf. This is partly due to the long distance from the heating unit to the lowest shelf. Use of a small electric fan will further decrease this fluctuation. A container full of water may be set in the bottom of the incubator to maintain humidity. The cabinet, when empty, can easily be moved by one person. Fifteen years ago, when these cabinets were constructed, the materials, including the cabinet, cost less

FIG. 2. Front view of two temperature cabinets.

than $10.00. At present prices such an incubator can be made at a very nominal cost and will prove quite satisfactory for general work.

Through a grant from the Rockefeller Foundation a series of four large temperature cabinets were constructed and installed in our laboratory 11 years ago at a cost of about $1600.00. Their cost would be considerably more at the present time. A number of the unique features which have proven highly satisfactory could, however, be used in less expensive temperature equipment. Figure 2 shows a front view of two of the four cabinets. The four cabinets are built into one large unit, outside dimensions 15 feet long, 45 inches deep, and 76 inches high. Of heavy wood construction, this unit is installed in a cool basement room. Near by are placed a workbench

FIG. 3. Interior of temperature cabinet shown in figure 2.

and all equipment for examining flies. This arrangement makes for convenience and saves time, as all stock and experimental cultures are within easy reach.   The student genetics laboratory is housed in an adjacent room.

Figure 3 shows a full view of the interior of one of the cabinets with dimensions of 3 feet in width, 3 feet in depth, and 5 feet in height. Each cabinet is heavily insulated and lined with galvanized sheet metal.   Side walls and ceiling are baffled.   A battery of electric light bulbs in parallel along both sides above the ceiling baffle furnishes the heating unit.   In the center of the ceiling baffle is a circular opening in which is mounted a high-speed electric fan, which drives warm air down through the cabinet.   This air circulates from the floor back of the side wall baffles, the lower edges of which are 1 inches above the floor.   The air circulates up behind the side wall baffles and over the battery of light bulbs. The cabinets shown in the figure have only heating units.   A third cabinet is provided with both heating and refrigerating units, and a fourth cabinet with only refrigerating unit.   The cooling system for each of the two cabi-

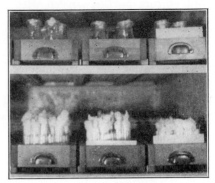

Fig. 4.   Detail of two shelves and culture drawers in cabinet.

nets consists of a ¼-horsepower Reco-Mills condensing unit for use with "Freon," mounted above the cabinet, and a set of copper and aluminum finned coils above the ceiling baffle.

Removable shelves of a skeleton framework of heavy angle iron, with four tracks in each shelf for culture trays, rest on metal clips which can be set at any height.   Each culture tray or drawer consists of a pine framework with bottom of ½-inch mesh metal screening, hardware cloth.   On the underside of each tray two hardwood strips are nailed, and on these the trays slide in the shelf tracks.   Figure 4 is a detail of two shelves filled with culture drawers.   On the side wall to the left, at *C* in the figure, are the two thermostats in series.   This double thermostat operates a relay, and there is practically no chance of thermostat failure.   In the cabinet provided with heating and cooling units both are under simultaneous thermostatic control, and never operate against one another.

Very sensitive thermostatic control with toluene-filled glass grids have been used in some *Drosophila* laboratories.   However, such sensitive control within the temperature cabinet does not insure an equally sensitive control within the culture bottles. As Sturtevant (1937) has suggested, temperatures within the culture bottle may be higher than in the cabinet, because of fermentation of culture medium which contains yeast, and the activity of fly larvae and adults. These factors no doubt also produce minor temperature fluctuations, and a control within ½ centigrade degree seems adequate for any *Drosophila* work. Further refinements are useless.

Each culture drawer carries thirty-three half-pint or forty quarter-pint bottles. Each cabinet can easily hold over six hundred half-pint culture bottles. Even with the cabinet filled to capacity with trays there is good air circulation. Two ventilators on the door may be opened when large numbers of cultures are being reared. Owing to the size of the cabinets lack of proper ventilation has never caused trouble. Thermostats on each cabinet may be set over a range of about 20 centigrade degrees and the total range is from 5 degrees to 50 centigrade degrees for the four cabinets. This equipment has been in almost constant use for 11 years with no replacements needed and no thermostat failure.

It would be impossible in the scope of this chapter to describe temperature cabinet designs in use in other laboratories. These range from constant temperature rooms with series of smaller units installed in them at a cost of several thousand dollars to small incubators for a few dozen cultures. Equipment will vary with the capacity needed, the range of temperatures desired, and the sensitivity necessary for special investigations. For general work, where a basement room is available, satisfactory temperature controls can be set up, as indicated above, at small expense. Whatever the design adopted attention should be given to avoiding overheating through thermostat failure and to providing ample air space for ventilation.

CULTURE CONTAINERS

Glass containers are almost always used for culturing the flies. The cheapest and for many purposes the most satisfactory containers are half- and quarter-pint milk bottles, which, when purchased in quantity, cost only a few cents each, are not easily broken, give good visibility, and are of convenient size. In our laboratory and in most other places stock cultures are carried in half-pint milk bottles. These are sufficiently large so that many parent flies, fifty or more, can be used, thus avoiding failure of the culture through accidents to individual flies or sterility of some individuals. However, for some small species which are less prolific than *Drosophila melanogaster* smaller culture containers such as glass vials are superior, unless very large numbers of parents are used. For experimental work on *D. melanogaster,* when pair matings are used, 1-ounce creamers are very satisfactory, providing yeast-enriched food (see below) is used. Creamers are much cheaper than glass vials, are not so easily broken, and have the same advantage as vials in taking up little space in the incubator. Creamers are less satisfactory than vials for slow-breeding species even when these are small flies, as the medium in them tends to dry out before a generation comes through. We have found glass vials 4 inches long by 1 inch in diameter to be most satisfactory, though some laboratories prefer smaller sizes. Cotton plugs for vials or bottles may be sterilized and used repeatedly if covered with a layer of coarse cheesecloth. The plugs should fit snugly to prevent escape or entrance of flies. Pasteboard caps for milk bottles and creamers are cheap and convenient covers, but should be perforated for ventilation. In our laboratory milk-bottle caps have a hole $\frac{1}{3}$ inch in diameter punched out with a large desk punch and the hole plugged with cotton. This insures adequate ventilation. In some laboratories caps are perforated with multiple needle holes. A dental scalpel may be used for cutting two or three narrow slits in creamer caps. This method gives better ventilation than needle holes, with less time involved in cutting the slits. Some laboratories prefer cotton plugs for bottles. These, however, are more expensive. The use of tight cotton plugs to avoid mite infestation is recommended by some workers, but this in itself is not adequate protection against mites.

Creamers and creamer caps may be secured from any wholesale restaurant supply company. Bottle and creamer caps may also be ordered directly from manufacturers. One of these is the Ohio Bottle Cap Company, Piqua, Ohio.

For shipping live flies to other laboratories glass vials containing culture medium have generally been used. These have the disadvantage of breakage en route unless carefully packed; the weight of vials and adequate packing makes their shipment by air mail quite expensive. The following method for mailing live *Drosophila* is very satisfactory. Aluminum tubing, $\frac{1}{2}$ inch outside diameter, 0.022 inch wall, is cut into $3\frac{1}{2}$ inch lengths. The usual culture medium (see below) is poured to a depth of $\frac{3}{4}$ inch into a beaker or other container, and as many aluminum tubes as will fit into the container are set on end in the hot medium. If a container with a tight cover is used, such a supply of tubes may be set away and kept for weeks in the refrigerator.

To mail *Drosophila*, a tube is broken out of the agar medium, a small cork inserted in the end of the tube containing the medium, flies shaken into the tube by use of a funnel, or etherized flies dropped into the tube. A cotton stopper is then inserted. To make sure that flies do not stick to the aluminum wall the tube may be lined with a cylinder of paper. These tubes may be labeled with India ink. Two such tubes may be sent in an air-mail letter without further packing between points in the United States for the usual 6¢ postage. Flies were sent through by air from Rio de Janeiro, Brazil, to our laboratory in this way, a distance of some 6000 miles. This method of shipping makes possible the use of air-mail service at small expense, and should prove particularly valuable to field collectors. Aluminum tubing of this or other sizes may be purchased from the Aluminum Company of America, 2210 Harvard Ave., Cleveland, Ohio. The cost for each of the tubes described above is only a few cents, and they can be washed and used many times.

CULTURE MEDIUM

In the early work on *Drosophila* the culture medium consisted of ripe banana allowed to stand for a couple of days in a covered jar after being inoculated with yeast. Improvements have been introduced as follows: (*a*) the addition of agar or other material to stiffen the food and make it easier to shake flies from the culture bottles; (*b*) substitution of molasses or syrup and corn meal or other ingredients which are more uniform in quality and less expensive than bananas; (*c*) addition of some antiseptic preservative to check mold and harmful bacterial growth, which is at the same time non-toxic to fly larvae and adults; (*d*) and along with (*c*) the addition of killed brewer's yeast to the medium, as antiseptics also hinder the growth of yeasts on which fly larvae feed. Attempts have been made to develop synthetic and highly standardized media but thus far with indifferent success. The food requirements and texture of the food for adults and larvae differ. For keeping stocks general-purpose media have been developed, which are relatively satisfactory for both larvae and adults.

In figure 5 on the bottom shelf are shown the ingredients which we use for fly media. From left to right they are a large container of brewer's yeast; a cake of baker's yeast; a pound box of Tegosept, antiseptic preservative; glass jars full of corn meal; powdered agar; ½-gallon jars of Karo syrup. Baker's yeast in pound cakes is kept in the refrigerator. All other ingredients are in closed containers and will keep indefinitely without spoiling.

The culture medium is prepared as follows: 4 liters of water are put in an aluminum kettle, and to this are added 25 grams of powdered agar. This is placed over a gas fire. Figure 6 shows the gas stove used for preparing medium and for oven-sterilizing old culture bottles. When the water boils 60 cc. of brewer's yeast are stirred in and boiled for about 15 minutes. This is an important step, and unless the yeast is

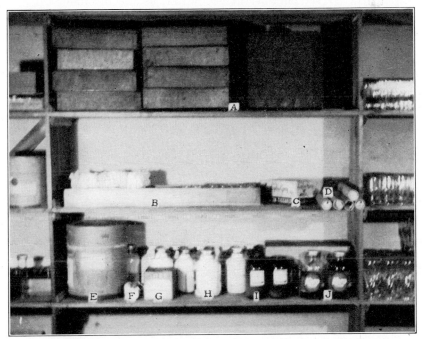

Fig. 5. Ingredients used in culture media (see text).

all killed by boiling the medium will be unsatisfactory. Five hundred cubic centimeters of Karo syrup are now added. The amount may vary over rather wide limits, and there is less waste if this is poured directly from the container into the boiling medium. Five hundred cubic centimeters of corn meal are placed in a pan and thoroughly mixed with about 500 cc. of water until all the meal is moistened. This is then stirred into the boiling medium, and 20 cc. of dry Tegosept-M, the methyl ester of p-hydroxybenzoic acid, are immediately added. The corn meal and the preservative are stirred into the medium, and the mixture boiled for another 5 or 10 minutes.

The hot mixture is immediately poured into culture bottles, creamers, or vials. The recipe given above will prepare about seventy-five half-pint bottles with medium poured to a depth of 1 inch. In some laboratories medium is wasted by pouring too much into the culture bottles. Neither larvae nor adults use the medium at the bottom of the container. Vials are poured to a depth of about ¾ inch and creamers to ½ inch or less. Twice the above formula may be prepared, but for larger quantities either two lots should be made up or the kettle placed in a larger container in hot water, as the medium begins to stiffen as it cools and is difficult to pour.

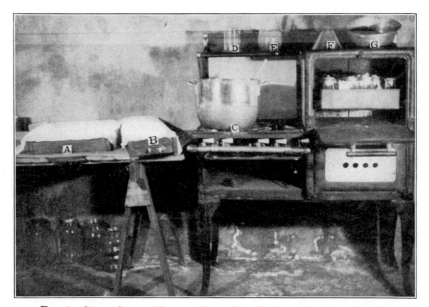

Fig. 6. Stove for cooking media and sterilizing old culture bottles.

For pouring into bottles a container with a narrow spout or lip, such as a small tea kettle, will be found useful. We have devised the apparatus shown in figure 7 for pouring medium into vials and creamers. A rectangular base metal support is fastened rigidly near one end of a board about 24 by 8 inches (board not shown in diagram). Cleats are nailed to the underside of this board at the two ends, raising it off the table so that a large Petri dish may be slipped under it. An extension clamp is fastened by a clamp holder to the metal support. In the extension clamp is placed a plastic funnel (B), one-pint size or larger, with a short stem of inside diameter about 1 centimeter. A hole slightly smaller than the creamer or culture vial is bored through the board directly under the stem of the funnel. A wooden dowel pin (A) is beveled down so that it fits snugly into the stem of the funnel. Two small strips of wood are tacked to the board on either side of the hole, forming a small V-shaped guide. With the stick or dowel pin, which should be about 8 inches long, in place in the funnel stem, the funnel is filled with hot medium. Then a vial or creamer is placed under the funnel stem, using the V-shaped guide for placing it quickly and accurately. The stick is then raised and pushed back into the funnel stem, delivering the desired amount of medium. With one hand the worker raises and lowers the stick, and with the other handles the vials or creamers. Any drip from the funnel stem drops through the hole into the Petri dish placed beneath the board. Even when medium becomes too stiff to pour, it can be pushed through the funnel stem with the stick. With a little practice it is possible to pour one hundred creamers in 5 minutes with this device. The only parts to be cleaned are the plastic funnel and the dowel pin.

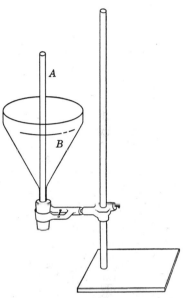

Fig. 7. Device for pouring medium into vials (see text).

After cultures have been poured they should be covered with cheese-cloth to keep out any stray flies, and allowed to cool before capping or stoppering. This step is important, as it prevents condensation of moisture on the inside of the culture vessel and the surface of the medium as the warm, humid air cools. In this condensed moisture parent flies are likely to become mired. After cooling, the culture bottles may be fitted with covers and placed in a cold cabinet where they will keep for several days. Shortly before use a drop of thick live baker's yeast suspension in water should be pipetted onto the surface of the medium. This suspension should be just thin enough to be sucked up into a pipette. A third of a double sheet of cleansing tissue or a piece of paper toweling should be pushed down into the medium at one side of half-pint or quarter-pint culture bottles. This paper furnishes parent flies a place to rest when not feeding or ovipositing and cuts down loss from flies becoming mired in the culture medium. Etherized flies may be dropped on this paper in making up cultures.

There is no need to cut out a chunk of food medium, as formerly practiced in many laboratories to allow escape of carbon dioxide from under the food cake. With the use of thoroughly killed yeast and of antiseptic preservative food cakes will not ferment and push up against the stopper. In vials and creamers a small amount of paper may be added, though this is generally not necessary if parent flies have been aged before starting an experimental culture. However, with weak mutant stocks paper in vials is an advantage. Paper also furnishes a place for larvae to pupate.

The amount of agar used in culture media may vary somewhat. In very dry climates somewhat less may be used; when humidity is high more will be required. Enough agar should be used to keep the food cake from shaking loose when culture bottles are jarred to shake flies into another bottle or etherizer. A fairly satisfactory and much cheaper substitute for agar may be purchased from the Krim Ko Corporation, New Bedford, Mass., under the trade name of Carragar, a preparation from the seaweed, *Chondrus crispus*. Medium prepared from this substance, however, tends to liquefy as larvae develop. Addition of extra cleansing tissue is recommended when Carragar is used.

Until recently most *Drosophila* laboratories in the United States have used Moldex-A, the sodium salt of p-hydroxybenzoic acid, as a mold preventative. As this product is no longer on the market, substitute methods of cutting down mold growth have been adopted. In several of the larger laboratories propionic acid is added to the medium and a special strain of yeast, tolerant to this acid, is grown and used to inoculate cultures. This procedure is laborious as it requires constant culturing of the yeast strain. We have found the methyl ester of p-hydroxybenzoic acid fully as satisfactory as Moldex. It is nontoxic to flies and an excellent mold and bacteria preventative. This may be purchased from the Goldschmidt Chemical Corporation, 153 Waverly Place, New York 14, N. Y., under the trade name of Tegosept-M at the current price of $2.20 per pound. Other esters of this same acid may be purchased from the same company. They have stronger antiseptic action but the disadvantage of being less soluble in aqueous media. Brewer's yeast may be purchased from the Standard Brands Incorporated, 595 Madison Ave., New York 22, N. Y., at about 25¢ per pound. Baker's yeast may be secured in pound cakes at any bakery; it must be kept refrigerated and not allowed to dry out. Where large quantities of media are used a cheap grade of molasses, purchased in kegs, will be much less expensive than Karo syrup. The amount of brewer's yeast may be varied according to the use of the culture medium. For stock cultures the concentration mentioned above is recommended. With much higher concentrations too many flies come through; cultures become overcrowded with adults and need to be changed more frequently. With experimental cultures addition of more yeast may give better results. A special technique giving very high yields will be described below.

HANDLING FLIES

Figure 8 shows the standard equipment for handling *Drosophila*. The following items of equipment are shown: (*A*) a good wide-field

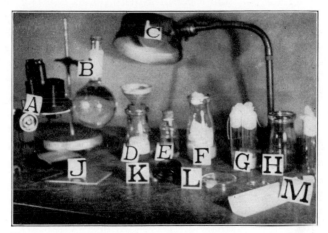

Fig. 8. Standard equipment for examining flies (see text).

binocular microscope; (*B*) a flask filled with water or glycerin for focusing light and absorbing heat; (*C*) light source; (*D*) etherizer; (*E*) bottle of ether; (*F*) culture bottle; (*G*) culture vials; (*H*) fly morgue; (*I*) empty vial and creamer; (*J*) counting plate; (*K*) rubber castor; (*L*) re-etherizer; (*M*) folded card for transferring flies.

While *Drosophila* may be observed and some mutant characters recognized with the naked eye or a hand lens, a good wide-field binocular microscope is indispensable for any extensive work. The magnification chosen will vary with the characters being studied; a magnification of about fifteen diameters is recommended for most work. Be sure binocular prisms are properly aligned to avoid eyestrain. Unless the lenses of the eyepieces are countersunk a small rubber collar should be placed around the eyepiece if the worker wears spectacles.

Otherwise both eyepieces and spectacles will eventually become scored if much work is done.

As a light source a desk lamp with 100-watt bulb is satisfactory. Individual workers may prefer other types of lamps. In any case a strong light is needed. The light rays are focused by placing a spherical flask filled with water or glycerin between the light and the microscope field. The flask should be adjusted to give a spot of intense light about 2 inches in diameter on the counting plate, with the center of this spot at the center of the microscope field. Too frequently beginners neglect the careful adjustment of light for maximum efficiency. Many mutant characters, particularly eye and body colors, show up very differently under different lighting and magnification. Every worker, whether beginning student or veteran researcher, should do his work in so far as possible with the same microscope and illuminating equipment at all times. Individuals will differ in the intensity of the light and the angle of the rays at which they can observe most accurately and with the least eyestrain. Difficult eye and body colors can often be classified much more accurately by viewing surfaces obliquely rather than directly. Bridges (1936, *D. I. S.* 6), probably the keenest of all drosophilists in classifying difficult mutant types, has recommended that the light rays strike the microscope field at a 45-degree angle from the vertical.

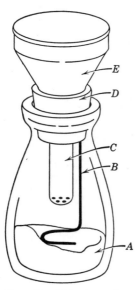

Fig. 9. Etherizer (see text).

Flies may be anesthetized for study by shaking them from the culture bottle into a half-pint milk bottle or other glass container of convenient size and quickly covering the bottle with a cotton stopper to which a little ether has been added. This method is satisfactory, and by varying the amount of ether the anesthesia may be applied very gradually. However, a fresh supply of ether must be added to the stopper quite frequently. To avoid this waste of ether and the discomfort to the worker from excess ether fumes several types of etherizers have been devised. The one shown in figure 9 has been adopted in our laboratory and is a simplified model of a design first described by Muller (*D. I. S.* 2). A quarter-pint milk bottle or wide-mouthed specimen bottle is used as the ether chamber. Cotton (*A*) is packed tightly in the bottom of the bottle. A paper clip is bent into the shape shown at (B), and the straight end pushed into the cork (D). This keeps the cotton from shaking loose when flies are shaken out of the etherizer. (C) is the long end of a large gelatin capsule of ¾ inch diameter pushed into a hole bored in the cork. With a hot needle several small holes have been punched in the end of the capsule. A small aluminum funnel (E) fits into a larger hole bored through the upper end of the cork to meet the smaller hole below. The cork is removed, a pipette full of ether dropped on the cotton, and the cork replaced. The funnel is placed in the hole in the cork and flies shaken down into the gelatin capsule by removing the cover of a culture bottle, inverting the bottle quickly over the funnel, and tapping the etherizer repeatedly on a rubber castor such as shown at (K) in figure 8. Ether fumes from the bottle diffuse through the holes in the capsule, and flies are quickly etherized. The aluminum funnel is lifted out, and the anesthetized flies poured out on the counting plate for examination. With this etherizer flies are clearly visible in the capsule, and overetherization may be avoided with reasonable care. The small aluminum funnels are on the market at about 15¢ each, and large gelatin capsules may be purchased at a pharmacy for about 5¢ each. Etherizers of this design can be constructed in a few minutes. The gelatin capsule, which is fitted into the cork, can easily be removed and cleaned and will last for months. One pipette full of ether is sufficient for 2 hours or more. More elaborate etherizers of somewhat similar design have been used in other laboratories. All these etherizers conserve ether and anesthetize the flies rapidly. Care must be taken in using them not to overetherize flies. The first sign of overetherization is a straightening out of the legs, followed by folding of the wings into a vertical position over the thorax. Flies in this condition often do not survive. Whether flies are to be saved or not, overetherizing is poor technique as the flies are much more difficult to manipulate on the counting plate when in this condition.

Most flies, after examination and recording, are discarded. A discard bottle or morgue, (H) in figure 8, partly filled with no. 10 motor oil, is kept on the desk. A more convenient arrangement is to have a large hole bored in the table top; under this a can or bottle of motor oil is fixed. Over this hole a guard of coarse wire may be placed to catch objects accidentally dropped.

The counting plate should be smooth, rigid, and of convenient size and suitable color. White or a light color is generally preferred. Porcelain is often used. My own choice is a light yellow plastic plate about 5 inches by $3\frac{1}{2}$ inches. A strip of cardboard, (M) in figure 8, somewhat longer than the plate and folded lengthwise, is very convenient for pushing flies into rows on the counting plate for examination. Flies may be pushed off the plate onto this card for transfer to the morgue or to a culture bottle.

Some sort of "fly-pusher" or manipulator is needed for turning, separating, and counting flies on the plate. Some workers prefer a fine camel's-hair brush; others use a flat strip of copper, cut so that when held in the hand at an oblique angle to the counting plate a straight edge about $\frac{1}{2}$ inch long at the lower end fits against the plate. Flies can easily be pushed into desired positions. Light dental forceps or insect forceps may also be used.

The beginner must first learn to sex the flies. While *Drosophila melanogaster* shows marked sexual dimorphism, every student should learn to determine sex by the external genitalia. When flies are turned on their backs the posterior end of the abdomen of the female is seen to come to a point in the ovipositor plates. In the male two toothed "claspers," black and heavily chitinized, lie parallel and near the tip of the abdomen. The posterior end of the male abdomen is rounded. In mature flies testis color shows through the ventral surface of the abdomen, but this character cannot be seen in young flies. Four prominent sternites appear on the undersurface of the male abdomen, and six sternites on the female abdomen. As students may have occasion to examine species other than *D. melanogaster*, they should not rely on such secondary sexual characters as the sex-combs on the fore-tarsi of males and the black pigmentation of the posterior tergites of the male abdomen, though these characters are useful in recognizing the sex in this species when unetherized flies are examined through the glass wall of the culture vessel. In teaching students to recognize the sexes teachers will find it convenient to use the F_1 flies of a sex-linked cross in which some easily observed mutant character marks all males and none of the females.

In making cross-matings between two stocks care must be taken that all females used are virgin. For many of the slower-breeding species this is no problem, as females are not sexually mature for two days in some species and even longer in others, after emergence from the pupa case. Even with these species it is well to shake all old flies out of the stock bottles and use females which emerge within two days. For *Drosophila melanogaster* matings, all flies should be shaken from the stock bottle and only females emerging within a six-hour period should be used, as both males and females may become sexually mature within a half day after emergence. In *Drosophila immigrans* old males will mate with young females within an hour of their emergence, and the males become sexually mature about two or three hours after emergence. The laborious task of isolating pupa cases into empty vials in order to be certain that emerging females are virgin is not recommended, although practiced by some workers. If one or more double sheets of cleansing tissue paper are soaked in a suspension of baker's yeast in water, one part of yeast to about five of water, and dropped into a stock bottle containing many larvae, the later emergence of many adults in a short time interval of a few hours will be assured. With this technique many healthy virgin females may be easily collected.

Either before or after males and females are placed together in making a cross-mating, the flies should be aged in a container with fresh food for two days before being placed in the culture bottle. This procedure insures a good output of eggs shortly after parent flies are placed in the culture bottle. Care in this respect cuts down mold and bacterial contamination. For many slower-breeding species aging for about a week and then transfer to a fresh culture bottle are recommended. By shaking parent flies into fresh culture bottles every other day a series of cultures may be started. Flies treated in this way will give astounding yields, as the transfer to fresh food insures optimum nutrition and heavy egg production. In certain experiments many flies may be aged together, and then pair or small mass matings may be made.

To insure maximum yields of large flies and to reduce the length of the emergence period, after sufficient time has elapsed to allow all eggs to hatch, cleansing tissue soaked in baker's yeast suspension in water should be added to the culture bottle. The paper should be dipped into a beaker of yeast suspension, mixed as described above, excess fluid allowed to drain off, and then the paper added to the culture bottle, but not pressed down into the medium. Larvae tunnel through the paper and feed on the yeast. If the paper is placed in the bottle properly it will form a mat on top of the culture medium, and in this mat the larvae pupate. There is relatively little danger of the paper shaking loose from the medium when newly emerged flies are shaken from the culture bottle. This method is equally applicable to small culture vials and creamers (Spencer, 1943). In much of the early work on *Drosophila* great care was taken that larvae were not crowded, so that only a few flies were secured from a culture. However, by using a concentrated food supply large numbers of healthy flies may be reared from one culture without larvae being starved, even though they are physically crowded. Table 1 gives the data on a

TABLE 1

Drosophila hydei Flies Emerging in Five Culture Vials in the Second Generation of a Cross of the Multiple Mutant Stock—Scarlet (Eye-color in Chromosome II); Scabrous (Eye-texture in III); Pearly (Body-color in IV); Javelin (Bristle Mutant in V) × Bronze (Eye-color in III)

All mutants are recessive and because of absence of crossing-over in male, scabrous and bronze give a ratio of 1 scabrous:2 wild:1 bronze. The other three mutants give the familiar tri-hybrid ratio, which, when combined with 1:2:1 gives ratio shown in table. The results of the cross are presented to show that the yeast suspension-cleansing paper technique allows for rearing large numbers of flies in small space without distortion of Mendelian ratios by larval competition. 654 of the flies emerged in the first 48 hours of the emergence period.

Ratio	Classes	Observed Frequency	Expected Frequency
54	Wild type	166	162
27	Scabrous	70	81
27	Bronze	83	81
18	Scarlet	59	54
9	Scabrous-scarlet	25	27
9	Bronze-scarlet	32	27
18	Pearly	54	54
9	Scabrous-pearly	19	27
9	Bronze-pearly	24	27
18	Javelin	62	54
9	Scabrous-javelin	37	27
9	Bronze-javelin	25	27
6	Scarlet-pearly	23	18
3	Scabrous-scarlet-pearly	10	9
3	Bronze-scarlet-pearly	9	9
6	Scarlet-javelin	14	18
3	Scabrous-scarlet-javelin	5	9
3	Bronze-scarlet-javelin	7	9
6	Pearly-javelin	20	18
3	Scabrous-pearly-javelin	7	9
3	Bronze-pearly-javelin	5	9
2	Scarlet-pearly-javelin	8	6
1	Scabrous-scarlet-pearly-javelin	3	3
1	Bronze-scarlet-pearly-javelin	3	3
256	Total	770	768

cross in *Drosophila hydei* in which the parent flies were heterozygous for five recessive mutant genes lying in four chromosome pairs, bronze and scabrous being in the two third chromosomes respectively. The flies in the table were reared in five culture vials of 75 cc. capacity each, using the paper-yeast suspension method described. *Drosophila hydei* is much larger than *D. melanogaster*. With an average output per vial of over 150 flies all five mutants approached closely to Mendelian expectations; even triple and quadruple recessives came through in expected frequencies and a chi-square test indicates that fluctuation from expected Mendelian ratio was due to chance. Emergence of about 650 of the flies occurred in the first 48 hours after the first fly emerged. It is clear that the culture method described conserves space, saves time, and gives excellent results. When this method is used adults should be removed and examined within 2 days after emergence, as the food supply and space for them are limited in the vials.

The method of culture described above is also excellent for securing larvae for salivary chromosome studies. When the first larvae begin to pupate the paper may be removed from the vial and placed in a shallow dish of normal saline solution. An adequate supply of well-fed mature larvae can be shaken out of the paper and collected.

COLLECTION OF EGGS

In many experiments it is important to collect large numbers of eggs over a short time period. First a good supply of young flies should be secured. These should be aged for the appropriate time, several days to over a week, depending on the species. During the aging period the flies should be shaken into fresh culture bottles at two- or three-day intervals. In the aging bottles several drops of heavy baker's yeast suspension should be placed on the food surface. The amount of food surface and not the volume of food in the aging bottle is important. To increase food surface a wad of stiff paper or of cotton may be placed in the empty culture bottle and the hot medium poured in on top of this. In a culture bottle prepared in this way the food surface can be doubled. After aging and heavy pre-feeding, the eggs may be collected by the use of several devices. One simple method consists of pouring drops of hot food medium on the undersurface of bottle caps, and placing a drop of heavy yeast suspension on this food after it has cooled. A large supply of these caps can be prepared and stored for several days in a covered container in the refrigerator. Flies are shaken into an empty culture bottle, one of these caps placed on the bottle, and the bottle set in an inverted position. The egg yield will be increased by scoring the food surface with a fork or other instrument. This makes many tiny depressions or valleys in the food surface; at these points the humidity is high and this helps to elicit the ovipositing reaction. Flies will not deposit eggs unless the humidity next the food surface is high. An even better method of securing eggs is to use glass cylinders open at both ends. A cylinder about 2 inches in inside diameter and 3½ inches long is recommended. These can be secured from companies selling glassware by special order. Small Petri dishes or Syracuse watch glasses are filled with hot medium, yeasted, and stored in closed containers. A cotton plug is placed in one end of the cylinder, flies are shaken into the cylinder, a culture dish is quickly placed over the open end, and the cylinder inverted over the dish. The cotton plug can then be pushed down relatively close to the food surface, thus increasing the humidity in the egg-laying chamber. When a new culture dish is added the cotton plug should first be drawn out to the end of the cylinder, and flies shaken down in this end while the first dish is removed and a second one added. If dishes are changed once a day the flies will continue to produce large numbers of eggs daily for a period of two weeks or more. The addition of a little lampblack to the food medium while it is being prepared darkens the medium and makes it easier to count eggs if this is desired. In Pearl's early work on fecundity he stressed the importance of not crowding the flies if large numbers of eggs were to be secured. Actually, the space per fly in the container is of minor importance; the food supply per fly of major importance. If fresh food is regularly supplied enormous numbers of eggs may be secured. Fifty pairs of *D. melanogaster* may well be used in a cylinder of the size suggested. It is always important to simplify technique as much as possible, and the methods described above give maximum egg yield for the time and work involved.

PESTS

In *Drosophila* laboratories there are a few flies let loose accidentally from time to time in the course of experimental work. These actually constitute a pest, as they may be a source of contaminating cultures. Organic débris and other materials where such flies can feed and particularly where they can breed should be removed promptly from the laboratory or kept in fly-proof containers. Particular care should be taken that all culture bottles are securely stoppered with no crevices open through which such flies can gain access to the culture. In a humid climate in the summer different species of Muscidae may deposit eggs around cotton plugs or bottle caps. The eggs may hatch here, and the young larvae crawl down into the culture medium, where they quickly take over and are likely to kill off all *Drosophila* larvae, possibly by utilizing available food. *Drosophila busckii* will also contaminate fly cultures in this way. Where flies cannot be screened out of the room and thus constitute a pest all culture bottles should be covered with cheesecloth or otherwise screened. Contamination of this sort is relatively rare, but may prove troublesome in basement rooms in a humid summer and should be guarded against.

More frequently bacterial infections of stocks may occur. These infections may produce a slimy film over the food surface; adult flies become mired, eggs never hatch, and larvae are killed off. When such an epidemic occurs, old culture bottles should be thoroughly heat sterilized and the medium from such bottles should be disposed of immediately. Empty culture containers should be heat sterilized before adding medium. Fly stocks, normally kept at low temperature, should be brought through as rapidly as possible by supplying extra yeast suspension and rearing them at optimum temperature. Repeated transfer of parent stock flies to fresh culture bottles should be made. If this routine is followed, the bacterial contamination will soon be cleared up.

Molds are a still greater source of trouble. Especially for slow-breeding species the use of an antiseptic preservative in the medium is of paramount importance. During a period of mold contamination stock cultures may well be reared at higher temperatures, as mold growth is not retarded as much as larval growth by low temperature. The flies, if properly handled, will get the better of mold infections. They should be transferred one or more times to fresh culture bottles and the old bottles sterilized. Once a good stand of larvae is present in a culture bottle, they feed on the mold as well as yeast, and the mold has little chance to spread and to form spores. Molds are likely to kill off a fly culture when only a few parent flies, not properly aged, are added to a culture bottle accidentally inoculated with mold spores from the flies themselves, the air or contaminated cotton plugs, bottle caps, or paper. Keeping such a culture at low temperature is inviting disaster. Many species of *Drosophila* can be cultured more successfully either by rearing in small culture vials or by using large numbers of parent flies if half-pint bottles are used. Where there is a plentiful supply of adults and larvae, mold has little chance of killing off the culture. Even adult flies, which are so covered with mold spores that they would soon die, may be used to recover a moldy stock if they are repeatedly transferred daily to fresh culture bottles.

By far the most serious menace to fly culture is contamination with Tyroglyphid mites of the genus *Histiostoma*. If more than one species is involved, they are certainly closely related. The complex life cycle of this mite includes a parthenogenetic stage in which both males and females may be produced, and a non-feeding hypopus stage which is migratory and attaches itself to adult *Drosophila* and other insects. In spite of the fact that these mites are endemic in some *Drosophila* laboratories where much of the best experimental work has been done, they constitute an ever-present source of serious trouble and many valuable stocks of mutants have been lost through their depredations. When a stock bottle becomes heavily infested, the adult mites undoubtedly compete with fly larvae for available food and the hypopi may attach themselves in such numbers to adult flies as to kill, weaken, or sterilize them. The external genitalia of the fly may be so covered with mites as to make copulation mechanically impossible; wings and legs may be loaded down so that the fly cannot move about normally; mouth parts may be covered with mites, thus interfering with normal feeding. Many valuable multiple mutant stocks, particularly where the flies under optimum conditions are less hardy than wild stocks, have been lost by allowing such conditions to develop.

Many of the measures proposed by various workers for the control of mites are practically useless; others have some merit but are ineffective in themselves, and some which may have resulted in the elimination of mites in a given epidemic are nevertheless not reliable. A sound approach to this problem is based on the recognition of certain facts apparently often overlooked. One hypopus, not two, is all that is necessary to start a mite epidemic, because of the parthenogenetic development of both males and females during the life cycle. The mites which start an epidemic will seldom be seen, and those which continue the endemic condition after a heavy infestation has been partly cleaned up will not be seen. One hypopus, attached to some concealed part of an adult fly, can easily be overlooked even though the fly is examined carefully under the binocular microscope. Wild flies collected from their natural habitats may carry the hypopus stage, although actually they seldom do. Stocks received from other laboratories may be mite infested, and if the infestation is light it may go unnoticed even though the stocks are examined for mites. The stocks of one careless or indifferent worker in the laboratory may serve as a constant source of reinfestation of stocks of other workers.

Where a mite infestation, either endemic or epidemic, is to be cleaned up, the following procedure should be followed by every worker in the laboratory. All old cultures not required for breeding stock should be collected and thoroughly sterilized by heat. The practice of dipping old and infested cultures in carbon tetrachloride or other chemicals is both expensive and unreliable. Many laboratories keep three sets of stock cultures of different ages to guard against loss of stocks. This practice must be discontinued, as the third set of cultures always furnishes a source of reinfestation in a laboratory harboring mites. The second set of cultures should also be discarded and heat sterilized as soon as larvae appear in the last cultures made up. Loss of valuable stocks should be guarded against by replication and not by keeping old stocks. All shelves where stocks are kept, utensils used in handling flies, and garbage pails should be wiped off with a solution of phenol, 1:500, the most effective killing agent for hypopi. Parent flies should be repeatedly transferred about every week for at least three transfers to a new set of culture bottles, and the old bottles sterilized by heat as soon as larvae appear in the new bottles. Under no condition should any stock bottle be allowed to stand around until the hypopus or migratory stage of the mite appears. This stage, barely visible to the naked eye, can readily crawl out of a culture bottle capped with a paper cap or covered with a cotton plug wrapped in cheesecloth. The use of tight cotton plugs to keep mites from spreading is an admission that the hypopi may be there, and certainly these plugs will be removed sooner or later, giving hypopi a chance to escape.

The success of a drive against mites depends upon breaking up the life cycle and not allowing new hypopi to develop. As hypopi leave an adult fly within two weeks of the time they first attach themselves, and then crawl down into the medium to complete the life cycle, and as the adults of all species of *Drosophila* live and breed much longer than two weeks, any stock of *Drosophila* can be cleared of mites by the simple expedient of transferring the adults three times at weekly intervals to fresh culture bottles. If this procedure, along with thorough heat sterilization of all discarded culture bottles, is followed simultaneously for all stocks in the laboratory, including experimental cultures, the mite infestation will be not only brought under control but totally eliminated. This result will not follow, however, if somewhere in the laboratory wild flies are allowed to breed on accumulated waste, barrels of formalin specimens, or other organic débris. The procedure outlined will require a lot of work, including the preparation of several times as much culture medium as is normally prepared in the laboratory over the period of a month during which the infestation is being eliminated. No measures short of elimination are adequate. During this time not a single fly need be examined under the microscope for the presence of mites. At the end of a month, if every worker has cooperated, there will be no mites in the laboratory. A large sample of stock bottles may then be set aside in an isolated room and allowed to stand for several weeks. Inspection of these bottles will show that there are no mites present.

In many laboratories mite infestations persist because halfway measures for cleaning them up are adopted. Some workers may be constantly battling the mite infestation. An indifferent worker allows the hypopi to accumulate in old cultures and infest tables, chairs, and shelves. The janitor goes through this room to others, spreading the hypopi with his duster or cleaning rag. Selection of adults, pupae, or larvae which do not carry hypopi or mite eggs; washing larvae or pupae to remove mite eggs or hypopi; controlling mites by placing culture bottles in soap solution or other chemicals; design of culture covers to prevent the spread of hypopi may help to reduce the mites, but these methods are fundamentally wrong in their approach to the solution of the problem. The work involved in eliminating mites from a laboratory is actually much less than that expended from time to time or constantly in the use of these ineffectual halfway measures.

After mites have been eliminated, sterilization of all discarded culture bottles with heat, isolation of stocks from other laboratories, and the repeated transfer of adults from these stocks to fresh culture bottles until these are free from any mites which may have been present, and general cleanliness, including the prompt disposal of garbage and other organic débris, should be practiced.

Several other species of mites without an hypopus stage also infest *Drosophila* cultures. These are less serious pests but should be eliminated. This can be done by rapid transfer of stocks, general cleanliness, and elimination of potential food supplies, particularly the accumulation of moist organic material.

MAINTENANCE OF STOCKS

As suggested above, stocks are generally kept in half-pint milk bottles. Many more adult flies should be added than are usually necessary to start a culture. This procedure insures against the accidental death of some adults and the sterility of others. Furthermore the stock-keeper should learn that certain stocks have a low fecundity and use more parent flies for these stocks. Each stock bottle should be given a number corresponding to a number in the stock list rather than marking the stock with the complete genetic formula. The practice of keeping three sets of stocks of different ages should be discontinued in favor of keeping two sets of stocks. To insure against loss of valuable stocks they should be replicated, as many as four bottles of the same age being made up for some stocks. In our laboratory the amount of replication depends upon the value of the stock and its vigor. Where only one species is maintained the temperature chosen should be as low as possible without danger of serious mold contamination. As we rear a number of species the stock cabinet is kept at about 20° C., a temperature at which all species seem to thrive. If *D. melanogaster* alone were kept this temperature would be reduced to about 17° C. At the temperature maintained we make up fresh stocks of *D. hydei, robusta, funebris,* and other slow-breeding species once every two months. *Drosophila melanogaster* needs to be cultured at somewhat more frequent intervals, as it develops more rapidly.

Details on the use of special methods, such as balanced stocks for maintaining mutants which are lethal in homozygous form, are beyond the scope of this discussion. This caution, however, seems in order. Most balanced stocks should be examined at least twice a year by etherizing flies and making sure that the stock has not changed in genetic structure. Rare cross-overs do occur and often the new combination has a marked selective advantage over the balanced stock, leading to the eventual loss of mutants if flies of the right constitution are not selected. Generally homozygous stock flies may be shaken without etherization into fresh culture bottles, but special care should be taken that contamination does not occur at the time stock transfers are made. In some laboratories stocks are etherized and examined every time new cultures are made up. This is time consuming, but should probably be done at least once a year to catch any contamination which might have occurred. Every effort should be made to eliminate work in connection with stock-keeping without running undue risk of losing valuable stocks.

There are no doubt improvements still to be introduced which will further reduce the time required in keeping stocks which are seldom used but too valuable to discard. We have at present a stock of *D. melanogaster* which has been cultured without transfer for 16 months in a vial 1 inch in diameter by 4 inches in height, simply by adding a little fresh food about once a month. If long vials or large test tubes were used, it might be possible to maintain such stocks for as long as 2 years in this way without transfer.

CONCLUDING STATEMENT

In the brief scope of this chapter no attempt has been made to review the literature on *Drosophila* culture or to give an historic account of the development of improved techniques. It has rather been our purpose to present a helpful résumé of present methods. Material from the volumes of *Drosophila Information Service,* issued by the Department of Genetics, Carnegie Institution of Washington, Cold Spring Harbor, under the editorship of the late Dr. C. B. Bridges and of Dr. M. Demerec, has been drawn upon with mention of the author. We have not made reference to volume and page, as *D. I. S.* is not generally available. An excellent pamphlet by Dr. M. Demerec and Dr. B. P. Kaufmann (see bibliography) may be secured from the Carnegie Institution, 1530 P St., N.W., Washington 5, D. C., containing, in addition to some of the material covered in this chapter, an account of the technique of making salivary chromosome preparations and permanent slides and a series of laboratory experiments or exercises on the genetics of *D. melanogaster.* All students of *Drosophila* owe much to the late Dr. Calvin Bridges, who was ever seeking to improve techniques of culture as well as to develop mutant stocks useful in genetic studies. His contributions in these fields have made possible much of the modern work on theoretical genetics. My understanding of the biology of *Drosophila* has come largely through many helpful discussions on the subject with Dr. A. H. Sturtevant and Dr. Harrison Stalker. The short bibliography appended may prove helpful to those who are interested in a further study of *Drosophila* culture. Thanks are due to Miss Vivian Pigossi for preparing the drawings of apparatus and to Mr. Albert Spritzer for taking the photographs.

In conclusion it may be said that, except for special investigations where elaborate methods may serve to give more exact data, techniques of culture should be kept as simple as possible in the interest of saving time, labor, and expense. It is well, however, for the worker whose primary interest is the use of *Drosophila* as a tool to pay some attention to the biology of the animal.

REFERENCES

Bridges, C. B. 1932. Apparatus and methods for *Drosophila* culture. *Amer. Naturalist,* vol. 66, pp. 250–273.

Bridges, C. B., and M. Demerec. 1934–1947. *Drosophila Information Service,* vols. 1–21.

Demerec, M., ed. 1950. *The biology of Drosophila.* John Wiley & Sons, New York.

Demerec, M., and B. P. Kaufmann. 1943. *Drosophila guide.* 3rd ed. Carnegie Inst. of Wash., Washington, D. C. 51 pp.

Lebedeff, G. A. 1937. Methoden zur Züchtung von *Drosophila. Handb. biol. Arbeits.,* Abt. 9, Tl. 3, H. 7, pp. 1115–1182.

Spencer, Warren P. 1943. *Drosophila* culture with a minimum of agar. *Ohio Jour. Science,* vol. 43, pp. 174–175.

Sturtevant, A. H. 1937. Culture methods for *Drosophila.* Pp. 437–446, in *Culture methods for invertebrate animals.* Comstock Pub. Co., Ithaca, N. Y.

16· THE CONTROL OF LABORATORY PESTS AND PARASITES OF LABORATORY ANIMALS

W. E. DOVE
U. S. Industrial Chemicals, Inc.
Baltimore, Maryland

INTRODUCTION

The production and use of healthy animals that are free of parasites, and organisms that are transmitted by parasites, are matters of first importance to every laboratory. Even though all the best principles of breeding, feeding, sanitation, and care of animals are in constant practice, an infection of a louse-transmitted *Bartonella* can seriously affect experiments during the critical period. The careful experimentalist desires to eliminate parasites because they live at the expense of the animals, and because they are under suspicion as vectors of disease-producing organisms. He also wishes to eliminate any pests that can contaminate the food of the laboratory animals.

Of equal importance is the question of the safety of the materials that are used for control of the parasites and other pests. If it is necessary for the user of an insecticide or rodenticide to be cautioned against eating, breathing, drinking, or absorbing the material because of its ill effects on the health of man or animals, the statement is definitely an admission that the material is not satisfactory for use about laboratory animals. The experimentalist is interested in rodenticides and insecticides that can be handled effectively and economically without any hazards. There is no excuse for a laboratory to choose an insecticide or rodenticide that offers a hazard, if a safer material is available for accomplishing the same purpose.

478

EXTERNAL PARASITES THAT LAY EGGS ON THE ANIMALS

Four kinds of parasites remain on animals throughout their existence and lay eggs upon the animals.

1. The sucking lice (Anoplura), which are blue in color, attach to the skin and feed upon the blood of the animal. The longest period of incubation known for any species of blue lice is 25 days, the average being about 10 days.

2. The biting or chewing lice (Mallophaga) are brownish or yellowish in color and crawl freely over the animals. They feed primarily upon the epidermis, though some species eat some delicate portions of feathers. When they feed in clusters on the skin they are capable of leaving areas of the true skin exposed so that these become very sensitive and dry. Such areas are not only uncomfortable to the animal but also can serve as places of entry for bacteria or other organisms. The injuries caused by chewing lice sometimes justify the use of a light application of a soothing ointment or a vegetable oil to aid in healing. The incubation period usually varies from about 5 to 10 days.

3. Keds or "sheep ticks" (*Mallophagus ovinus*) feed upon the wool, skin, and blood of sheep. They are capable of living off the animal for a few days but can reproduce only when permitted to remain on the sheep.

4. Three kinds of mangy skin are caused by four species of mange mites that live and reproduce in the skin. Sarcoptic mange is the most common, and is recognized by the presence of *Sarcoptes* mites in scrapings made from the skin. In making the scrapings for this species a slight amount of bleeding usually occurs. Demodetic mange is characterized by the presence of nodules that contain all stages of these elongated mites. The canine *Demodex* are about five times as long as *Demodex bovis* in cattle. Infested dogs rupture the nodules with their teeth, and numbers of *Demodex* eggs may be found in fecal specimens. Both psoroptic and chorioptic mites produce thick scab lesions that are usually restricted to the legs, base of the tail, and the escutcheon of cattle. The mites are found quite easily beneath one of the mange scales. If the mites possess pointed mouth parts they are readily recognized as *Psoroptes*, and if they have broad mouth parts they belong to the genus *Chorioptes*. One species of *Psoroptes* often occurs in the middle ear of rabbits and sheep, where it is protected by the wax of the ear.

CONTROL OF LICE AND MITES

All species of lice are readily controlled by the use of dusts or emulsions made of piperonyl butoxide and pyrethrins. For Mallophaga and for migrating Anoplura spot treatments made on the animals or birds gradually bring the infestations under control. When complete control is desired from spot treatments, a second application is desirable, about 10 days after the first one is made. Complete coverage of the animal with a dust containing 0.025 per cent pyrethrins and 0.5 per cent piperonyl butoxide has given practical control of the short-nosed blue louse of cattle for 125 days, and has eliminated two species of *Mallophagus* on goats. Dips and sprays lend themselves to rapid and complete coverage of the animals. When made with dilutions of 1 to 100 from a concentrate containing 10 per cent piperonyl butoxide and 0.5 per cent pyrethrins, single treatments of animals have consistently given practical control of lice for 120 days or longer in tests involving about 8000 range cattle (Snipes, 1948). For control of lice on hogs in pastures (Moore, 1947) oil solutions containing 0.5 per cent piperonyl butoxide and 0.05 per cent pyrethrins were used as spot treatments for complete control. Two treatments were applied to burlap on a rubbing post in the pasture, with an interval of 10 days between treatments of the burlap. The hogs rubbed the itching portions. As the lice moved about on the animals they came into contact with the treated spots and were killed. On human lice more than 50 per cent mortality was obtained with a dust containing 0.025 per cent piperonyl butoxide and 0.0025 per cent pyrethrins, whereas a powder containing the same concentrations of pyrethrins and N-isobutyl undecylenamide, the best synergist developed in World War II research and used with success in the Army pyrethrum louse powder, killed only 18 per cent of the lice. In view of the high specificity of different combinations and concentrations of piperonyl butoxide and pyrethrins to different species of lice, including the use of dusts for lice on rats, there is no hesitancy in recommending such combinations for all lice of animals.

Perhaps the most promising treatment for all mange mites thus far tested consists of the following: piperonyl butoxide, 0.8 per cent; pyrethrins, 0.08 per cent; Ultrasene, 2.8 per cent; Tween 80, 2.0 per cent made to 20 ml. with benzyl benzoate. This solution is diluted with 60 ml. of isopropyl alcohol and 20 ml. of water to form an emulsion. After the mite-infested portions of the body are washed in warm water to soften the scabs, the mange treatment is used to wet thoroughly the infested areas of the skin. Two or three applications at intervals of one week have completely eliminated all *Demodex* from dogs and cattle. Sarcoptic mange has been controlled by the use of the same material with a single treatment. *Psoroptes* in the ears of laboratory rabbits have responded to treatments made at the same strength as those employed for *Sarcoptes,* and a single treatment has been sufficient to eliminate this parasite in all rabbits treated.

PARASITES THAT LAY EGGS ON THE SOIL

Two species of ticks sometimes become serious pests of laboratory dogs; the brown dog tick, *Rhipacephalus sangineous* Koch, and the American dog tick, *Dermacentor andersoni* Stiles. The latter serves as a vector of spotted fever of man in the eastern states. It develops from the larval to the nymphal stage on field mice, where it normally drops from the animal and molts. Later it normally attaches to dogs but will feed upon man. Before the development of piperonyl butoxide, dogs infested with these ticks were dusted with ground derris containing 5 per cent rotenone. It was necessary to repeat the treatments in order to keep enough dust on the animal. Both species may now be controlled by dipping or thoroughly spraying dogs with a diluted emulsion containing piperonyl butoxide 1 per cent and 0.1 per cent of pyrethrins. Another formula has been used on a limited number of animals, and it promises to be equally as effective at a reduced cost. It consists of piperonyl butoxide 4 per cent and rotenone 1 per cent, with a suitable emulsifier. It is diluted at the rate of one volume of the concentrate with 19 volumes of water, and is also applied either as a dip or spray for complete coverage of the animal. In the use of either formula the same diluted material is also sprayed on the floors and walls of the animal room in order to kill any living ticks that may be present. In applying the spray it is important to cover the baseboards or the floors near the baseboards thoroughly, so that molted nymphs of the brown dog tick migrating down the walls will go across the treated surface and will be killed.

Fleas lay eggs while they are on animals, but the eggs drop to the soil or bedding and hatch into flea maggots. In dust or dry bedding the maggots find ideal conditions for their development. For a rapid elimination of all fleas one should treat the bedding of the animal at the same time the animal is treated. One may use either the dip or spray suggested above for control of lice, or he may apply a dust consisting of 0.5 per cent piperonyl butoxide and 0.05 per cent pyrethrins. The same dust may be blown into rat burrows and on floors for control of fleas on rats and other animals.

If the animal infested with fleas has been permitted to run in a yard or runway, one can successfully treat the animal with a liquid soap containing 1 per cent piperonyl butoxide and 0.1 per cent pyrethrins. About 30 ml. is worked into a lather on an average-size dog and is allowed to become dry on the hair. The animal is allowed to visit different places where fleas are emerging so that the fleas will be picked up and promptly killed by the treatment. One treatment of this kind remains effective upon the animal for about five days, after which it should be repeated for elimination of other fleas that are constantly emerging from the soil. If the dry soap presents an objectionable appearance, one can apply it every day or two and rinse the animals in the normal manner. This will kill the fleas present but will not afford a lasting residue on the hair.

INTERNAL PARASITES OF ANIMALS

In a survey of 2636 rats from a refuse dump near Baltimore, Luttermoser (1936) found different groups of helminths in the proportions shown in table 1 for adult and juvenile rats.

TABLE 1

INCIDENCE OF INFESTATIONS OF HELMINTHS IN 2636 BALTIMORE RATS, *Rattus norvegicus*

Taxonomic Group	Name of Parasite	2500 Adult Rats Positives	Percentage of Incidence	136 Juvenile Rats Positives	Percentage of Incidence
Cestoda	*Cysticercus fasciolaris*	480	19.2	1	0.7
	Hymenolepis diminuta	414	16.6	9	6.6
	Hymenolepis nana var. *fraterna*	286	11.4	5	3.7
Nematoda	*Trichosomoides crassicauda*	2259	90.4	44	32.35
	Capillaria hepatica	2140	85.6	29	21.5
	Strongyloides ratti	508	20.2	2	1.5
	Nippostrongylus muris	421	16.8	0	0
	Heterakis spumosa	41	1.6	1	0.7
	Trichinella spiralis	30	1.2	0	0
	Syphacia obvelata	12	0.48	0	0
Acanthocephala	*Moniliformis moniliformis*	16	0.64	0	0

These findings suggest that the presence of wild rats in animal rooms might introduce these parasites into the colony of experimental rats. They also suggest the importance of rat proofing the animal building and some additional methods for preventing rats from obtaining infestations of helminths. Complete drying of the floors readily results in destruction of the rhabdidiform and infective larvae of nematodes. The use of hot water on the floors kills infective larvae and coagulates the albumen of the tapeworm eggs. Regular and frequent scrubbing of the floors with hot water and soap removes *Ascaris* eggs of dogs or cats. When dogs and cats are kept away from the rats and rabbits, there is less opportunity for *Cysticercus* in the smaller animals, also little chance that adult tapeworms will develop in the dogs and cats. Only well-cooked meat should be supplied to experimental animals. This avoids development of tapeworm, a cysticercus or a *Trichinella* infestation.

ARTHROPOD PESTS

Different species of insects occur in the laboratory and about the feed rooms, where they have an opportunity to contaminate feeds and annoy the animals. It is possible for mosquitoes and biting midges to transmit bird malaria, and heartworm of dogs; for some grain weevils to serve as an intermediate host for a tapeworm of rats; or for one species of cockroach to serve as an intermediate host of Manson's eyeworm of chickens. If one prevents the development of these pests and applies the good principles of sanitation and animal management, the full advantage of a safe and efficient insecticide can be obtained in animal buildings.

PREVENTION OF DEVELOPMENT OF PESTS

Despite claims to the contrary that may have been misleading, prevention remains the keynote of control. If pests that normally develop about the premises of the animal building are not prevented from developing, the continuing emergence from breeding places on the premises can offset the advantages obtained from proper application of a good insecticide. It is important to remove droppings of animals promptly from the premises, and to dispose of them at regular intervals to insure drying of manure before any flies have a chance to develop in it. When houseflies, stableflies, and eye gnats are not allowed to build up in accumulations of manure and straw, fly control becomes a matter of killing flies that "hitch hike" on trucks or migrate from outside breeding places. If wild rats and mice are eliminated by rat proofing the buildings rather than by permitting visits from dogs and cats, it is possible to avoid introduction of fleas, mites, and lice of rats.

INSECTICIDES AND RODENTICIDES

Any insecticide or rodenticide used about feeds or the animals should be free from hazards of toxicity to warm-blood animals. This not only is good insurance against errors in applications by untrained or inexperienced persons, but also eliminates any possibility of poisoning valuable laboratory animals.

Before World War II, pyrethrum sprays were used on surfaces of huts for control of malaria mosquitoes in India by Russell and Knipe of the Rockefeller Foundation. Because there was a short supply of this safe insecticide during the war it did not find general use in surface treatments for control of flies and other insects. Now that it is being imported in quantities ample to meet all reasonable needs, and safe insecticides which are activated and extended by the addition of small amounts of pyrethrum extracts are in production, it is reasonable to expect that laboratories with safety uppermost in mind will use increased amounts of such combinations.

Ground pyrethrum dust has long been used safely on dry floors and in cupboards for control of ants, fleas, mites, silver fish, and scorpions, but it has been more universally used for control of cockroaches. For all these purposes, the dusts impregnated with pyrethrum extracts are more effective than are the ground pyrethrum flowers. The quick killing action on roaches and other insects afforded by pyrethrum dusts, and the complete safety of the materials about food products, have had much to do with their wide and successful use in food-processing plants and in food-handling establishments.

The development of piperonyl butoxide for use with small amounts of pyrethrins (Wachs, 1947; Dove, 1947; Donohoe, 1948) has a very special appeal to laboratories that maintain animals for experimental purposes. Studies on the acute toxicity of this compound show that it is necessary to use 7½ cc. of piperonyl butoxide per kilogram of body weight in order to obtain an MLD 50 for rats, and that the same order of toxicity also applies to other warm-blooded animals. This means that one may use 37½ times as much of this chemical and still be in the same range of safety as for many insecticides. Actually, only about ⅕ the quantity is needed in insecticides, so that on the basis of use the toxicity to warm-blooded animals is about $\frac{1}{200}$ of that of some other insecticides.

Chronic toxicity tests are being made on rats at Baltimore. Groups of twelve pairs of Wistar strain rats have tolerated 10,000 parts per million of piperonyl butoxide in all food eaten by them. The tests were started at weaning time. At the end of 96 weeks three successive generations have been reared without any mortality due to the chemical. Similar rats have already tolerated for 96 weeks 1000 parts per million of piperonyl butoxide and 200 parts per million of pyrethrins, with no mortality and no effect upon reproduction. Next to safety an insecticide must be efficient in killing insects.

AEROSOLS AND SPACE SPRAYS

Aerosols or finely atomized sprays are especially effective against free flying insects and are dependable for quickly killing such insects in buildings. Aerosol bombs containing only piperonyl butoxide and pyrethrins as active ingredients are available and can be effectively used at 3 grams of the contents to 1000 cubic feet of space. Flies that come into the buildings from distant places as well as any emerging in the buildings are killed within 30 minutes. Both aerosols and finely atomized sprays will also kill grain moths, mosquitoes, midges, gnats, leaf hoppers, and different species of insects that are attracted to lights in buildings. Electric sprayers and fog generators are finding use for similar purposes. These treatments also can be used to kill cockroaches, but require exposure periods of 4 to 6 hours. One uses about 5 times the dosage that is required for flying insects.

Good space sprays, finely atomized sprays, and aerosols consist of particles that usually range from about 10 to 30 micra in diameter. Droplets larger than these fall rapidly to the floor. The small droplets in this range float in the air and remain effective for one-half hour or longer against different flying insects. All of these methods are most advantageously used when it is desirable to kill the flying insects quickly. Their use makes it possible for one to promptly remove the dead insects so that animals cannot eat them.

SURFACE SPRAYS

In contrast to space sprays, surface sprays are applied with large droplets which do not float in the air. They are used on surfaces in buildings and about entrances for producing residues that will kill insects that rest upon them. In applying a surface spray any drifting of the spray particles to the floor is considered an objectionable feature. Satisfactory applications are made from a sprayer nozzle having an opening of about 0.060 inch, and with a pressure of 25 to 40 pounds per square inch. Experienced pest-control operators often show a preference for a fan-shaped spray rather than a cone-shaped one. Some of them prefer a tubular tank sprayer with a capacity of 3 gallons, equipped with Spraying Systems Company 8001 or 8002 fan-type nozzles. About 5 ml. per square foot or about 1 gallon to 750 square feet is the usual rate of application that will give no run-off but a good coverage of the surface. When such sprays deposit 5 mg. of pyrethrins and 50 mg. of piperonyl butoxide per square foot of surface they can be used effectively against flying insects about the entrances of buildings, and on the inside walls. They should not be applied directly above the animal cages or feeds so that dying insects can be ingested by the animals. An emulsion concentrate containing 10 per cent piperonyl butoxide and 1 per cent pyrethrins, diluted with 9 equal volumes and applied at the rate of 1 gallon to 750 square feet, will give a deposit of approximately 5 mg. of pyrethrins and 50 mg. of piperonyl butoxide per square foot. Such a treatment is applied only to the favorite resting places of flies, as indicated by fly specks on the sharp corners of posts, beams, and window glass. A dilution of the 10:1 concentrate with 4 volumes of water, applied at the same rate, will give protection against flies for about 6 weeks, instead of for 3 weeks, and will be more rapidly effective against roaches.

Both these applications are designed for obtaining residues on surfaces that will kill flies within 30 minutes. At the end of 4 months each of these spray treatments in buildings has continued to kill flies. From a large number of practical tests with these applications it is conservative to say that either of them will kill flies inside buildings for an entire season in the northern portion of the United States and that about three treatments will give practical control of flies for an entire season in the southern part of the United States.

Fruit flies (*Drosophila* sp.) can develop so rapidly in an old neglected orange or cantaloupe that they are not killed as fast as they emerge. It is advisable to eliminate such breeding places each day.

A specially prepared roach concentrate may be diluted with an odorless base oil. It has a ratio of 1 part of pyrethrins to 5 parts of piperonyl butoxide, and may be diluted with 9 equal volumes of oil for use as a direct spray. It gives a knockdown of most of the roaches within 30 minutes. Even the most resistant last-stage nymphs of the

American roach show a mortality of more than 60 per cent during the first 24 hours, in tests made by the Beltsville proposed method. The progressive kill which follows a single treatment gives about 100 per cent of the last-stage nymphs of the American roach in 96 hours. The Pyrenone roach concentrate is favored by persons who insist upon a high degree of efficiency, complete safety, and rapidity of kill. Repeated use of this spray tends to build up some very effective and long-lasting residues, which can be used to distinct advantage in the special places that are most favorable for cockroaches.

REFERENCES

Snipes, B. Thomas. 1948. Field experiments with Pyrenones for one-treatment control of lice on cattle. *Agr. Chemicals,* September.

Moore, Donald H. 1947. Effectiveness of piperonyl butoxide and pyrethrins as a practical treatment for hog lice. *Jour. Parasitol.,* vol. 33, no. 5, pp. 439–443.

Luttermoser, George W. A. 1936. Helminthological survey of Baltimore house rats (*Rattus Nowegicus*). *Amer. Jour. Hyg.,* vol. 24, no. 2, pp. 350–360.

Wachs, H. 1947. Synergistic insecticides. *Science,* vol. 105, no. 2733, pp. 530–531.

Dove, Walter F. 1947. Piperonyl butoxide, a new and safe insecticide for the household and field. *Amer. Jour. Trop. Med.,* vol. 27, no. 3, pp. 339–345.

Donohoe, Eileen Lindenberg. 1948. Pyrenones in roach control practice. *Pests and Their Control,* vol. 16, no. 7, pp. 9–11 and 48.

INDEX

489

MIDDLEBURY COLLEGE
LIBRARY

MIDDLEBURY COLLEGE
LIBRARY